**This book is to be returned on or before
the last date stamped below.**

3 NOV 1988

27 NOV 1990

-9 NOV 1992

17 MAY 1993

16. JUN 1995

5 OCT

- 3 APR 2006

0 2 NOV 2009

HAY & REID

# The ANATOMICAL and MECHANICAL BASES of HUMAN MOTION

James G. Hay
University of Iowa

J. Gavin Reid
Queen's University, Canada

# The ANATOMICAL and MECHANICAL BASES of HUMAN MOTION

Prentice-Hall, Inc., Englewood Cliffs, New Jersey 07632

*Library of Congress Cataloging in Publication Data*

HAY, JAMES G.
    The anatomical and mechanical bases of human motion.

    Includes bibliographies and index.
    1. Human mechanics.   2. Kinesiology.   I. Reid,
J. Gavin.   II. Title.
QP303.H389        612'.76        81-8601
ISBN   0-13-035139-3          AACR 2

Interior design and editorial/
    production supervision by Chrys Chrzanowski
Art layout and page makeup
    by Amy Rosen
Cover design by 20/20 Services, Inc.
Cover logo courtesy Organization Committee
    for gymnastics meeting of Switzerland,
    Geneva 1978
Manufacturing buyer: Harry P. Baisley

Printed in the United States of America

10   9   8   7   6   5   4   3   2   1

ISBN   0-13-035139-3

PRENTICE-HALL INTERNATIONAL, INC., *London*
PRENTICE-HALL OF AUSTRALIA PTY. LIMITED, *Sydney*
PRENTICE-HALL OF CANADA, LTD., *Toronto*
PRENTICE-HALL OF INDIA PRIVATE LIMITED, *New Delhi*
PRENTICE-HALL OF JAPAN, INC., *Tokyo*
PRENTICE-HALL OF SOUTHEAST ASIA PTE. LTD., *Singapore*
WHITEHALL BOOKS LIMITED, WELLINGTON, *New Zealand*

**To our wives,**

Hilary and Patricia

# Contents

**Tables** xii

**Preface** xiii

**Chapter 1: Why Study Kinesiology?** 1

## Part I:   Basic Anatomical Concepts

**Chapter 2: Introduction** 5

Surface Anatomy   5
Spatial Terms   6
Exercises   14
Recommended Reading   15

**Chapter 3: The Skeletal System** 16

Bones   16
Joints   24
Exercises   31
Recommended Readings   32

**Chapter 4: The Muscular System** 33

Characteristics   33
Types of Muscular Tissue   33
Muscle Structure and Attachments   34
Muscular Contraction   41

*Cooperative Action of Muscles  46*
*The Study of the Action of Muscle  47*
*Exercises  48*
*Recommended Readings  48*

## Chapter 5:  The Neck and Trunk  50

*The Vertebral Column  50*
*The Thorax  62*
*The Pelvic Girdle  64*
*The Shoulder Girdle  67*
*Exercises  71*
*Recommended Readings  71*

## Chapter 6:  The Upper Extremity  73

*The Shoulder Joint  77*
*The Elbow & Radioulnar Joints  82*
*The Wrist Joint  87*
*The Carpometacarpal Joints  89*
*The Metacarpophalangeal Joints  90*
*The Interphalangeal Joints  90*
*Exercises  90*
*Recommended Readings  91*

## Chapter 7:  The Lower Extremity  92

*The Hip Joint  95*
*The Knee Joint  100*
*The Ankle Joint  103*
*Exercises  109*
*Recommended Readings  109*

# Part II:   Basic Mechanical Concepts

## Chapter 8:  Human Motion  113

*Linear Motion  113*
*Angular Motion  115*
*General Motion  116*
*Exercises  117*
*Recommended Reading  117*

## Chapter 9:  Describing Linear Motion (Linear Kinematics)  *118*

*Distance and Displacement    118*
*Speed and Velocity    119*
*Acceleration    121*
*Vectors and Scalars    124*
*Projectile Motion    129*
*Exercises    139*
*Recommended Readings    144*

## Chapter 10:  Describing Angular Motion (Angular Kinematics)  *145*

*Angular Distance and Angular Displacement    145*
*Angular Speed and Angular Velocity    146*
*Angular Acceleration    147*
*Angular Motion Vectors    147*
*Velocity and Angular Velocity    149*
*Exercises    150*
*Recommended Reading    151*

## Chapter 11:  Explaining Linear Motion (Linear Kinetics)  *152*

*Inertia    152*
*Mass    153*
*Force    153*
*Momentum    153*
*Newton's Laws of Motion    154*
*Newton's Law of Gravitation    158*
*Weight    159*
*Friction    160*
*Impulse    164*
*Impact    166*
*Pressure    179*
*Work    182*
*Power    183*
*Energy    181*
*Exercises    187*
*Recommended Readings    192*

## Chapter 12: Explaining Angular Motion (Angular Kinetics)   *194*

*Eccentric Force   194*
*Couple   195*
*Moment   196*
*Equilibrium   198*
*Center of Gravity   202*
*Stability   216*
*Moment of Inertia   219*
*Principal Axes   221*
*Angular Momentum   222*
*Newton's Laws of Angular Motion   223*
*Transfer of Angular Momentum   231*
*Exercises   231*
*Recommended Readings   237*

## Chapter 13: Fluid Mechanics   238

*Flotation   238*
*Relative Motion   241*
*Fluid Resistance   242*
*Exercises   252*
*Recommended Readings   253*

## Chapter 14: Free Body Diagrams   255

*Exercises   258*

# Part III:   Practical Application

## Chapter 15: Qualitative Analysis   *261*

*Definition of Terms   261*
*Methods of Analysis   262*
*Qualitative Analysis   263*
*Prerequisite Knowledge   263*
*Basic Steps   265*
*Development of Model   266*
*Observation of Performance   279*
*Identification of Faults   286*

*Evaluation of Faults   294*
*Instruction of Performer   298*
*Exercises   301*
*Recommended Readings   301*

**Chapter 16: Examples in the Use of Qualitative Analysis   303**

*Running   303*
*Diving   318*
*Serving in Tennis   334*
*Conclusion   350*
*Exercises   350*

## *Appendices*

**Appendix A: Elementary Anatomy   385**
**Appendix B: Elementary Mathematics   408**
**Appendix C: Metric Conversion Tables   422**
**Appendix D: Center of Gravity Mannikin   429**

**Indexes   433**

## TABLES

1  Fundamental Movements

2  Bones of the Human Skeleton

3  Classification of Joints by Action

4  Muscle Fibers and Motor Units in Human Muscles

5  Average Speeds and Velocities for Selected World Records in the Track Events

6  Multiplication Factors to Determine the Horizontal and Vertical Velocities of Release

7  Distances a Projectile Falls Due to Gravity in Selected Time Intervals

8  Angles of Takeoff and Release Used in Selected Activities

9  Effect of 5% Increases in Speed, Angle, and Relative Height of Takeoff for a Long Jump

10  Limiting Friction for Different Shoe Soles and Different Surfaces

11  Limiting Friction and Coefficient of Friction for a Composition Sole on "Astroturf" Surface Under Different Loads

12  Angular Momentum of the Human Body During Airborne Phases of Selected Athletic Activities

13  Masses and Moments of Inertia of Golf Clubs About Transverse Axes 10 cm from the Grip End

14  Classification of Results for Selected Skills

15  Muscles Acting on the Neck

16  Muscles of the Abdominal Wall (Anterior Trunk)

17  Muscles of the Trunk (Posterior)

18  Muscles Used in Respiration

19  Muscles Acting on the Shoulder Girdle

20  Muscles Acting on the Upper Arm (Shoulder Joint)

21  Muscles Acting on the Forearm

22  Muscles That Move the Wrist and Hand

23  Muscles Acting on the Thigh (Hip Joint)

24  Muscles Acting on the Leg (Knee Joint)

25  Muscles That Move the Foot

26  Conversion of Force Units: Newtons and Pounds

27  Conversion of Mass Units: Kilograms and Slugs

28  Conversion of Length Units: Meters and Feet

29  Conversion of Length Units: Kilometers and Miles

30  Conversion of Length Units: Centimeters and Inches

31  Conversion of Speed Units: Meters/Second and Miles/Hour

# Preface

During the last 10 to 15 years, many departments of physical education have developed programs leading to graduate degrees with a specialization in anatomy and biomechanics—two of the component disciplines in the broad field of kinesiology. The development of these programs has led to a great disparity in the nature of the instruction currently offered in undergraduate kinesiology courses. In those departments where strong graduate programs have been developed, there have generally been concomitant developments at the undergraduate level. These have included (a) the replacement of the introductory kinesiology course with separate courses in applied (or functional) anatomy and biomechanics—a process that has usually meant both an increase in the total time devoted to the area and a compounding of existing difficulties in integrating and applying knowledge derived from the two sources—and (b) the shifting of the instructional emphasis away from traditional, practice-oriented, qualitative method toward research-oriented, quantitative methods, now much in use. In contrast, in those departments with limited or non-existent graduate programs in kinesiology, anatomy, or biomechanics, undergraduate instruction in kinesiology has generally been little affected by overall development in the field.

In response to a growing concern among its members for the numerous, and often serious, problems implicit in this situation, the Kinesiology Academy of the American Alliance of Health, Physical Education, and Recreation prepared a position paper[1] that set out what it considered the principal objectives of the undergraduate course in kinesiology and the ways in which these objectives might best be attained. Briefly stated, this paper made the following key points:

1. The principal objectives of the undergraduate course in kinesiology are to provide students with (a) knowledge of a systematic approach to the analysis

[1]"Guidelines and Standards for Undergraduate Kinesiology", *Journal of Physical Education and Recreation*, 51 (1980), 19–21.

of human movements and (b) experience in applying that knowledge to the evaluation of both the performer and the performance.
2. The course should consist of three units (or sections)—Anatomical Considerations, Mechanical Considerations, and the Application of Kinesiological Concepts—and no less than one-third of the total time should be devoted to the last of these three.
3. The use of some quantitative methods to increase understanding of basic concepts is desirable, but a command of the qualitative method of analysis must be the primary goal of all introductory courses in kinesiology at the undergraduate level.

One serious obstacle to compliance with these guidelines (and, thus, to the attainment of the stated objectives) is the lack of textbooks suited to the purpose. There are several textbooks available that cover the anatomical aspects well; a few that provide comparable coverage of the mechanical aspects; a very few that consider the qualitative analysis of human movement; and none that cover all three aspects thoroughly. This book has been written with the objective of remedying this situation and, in so doing, providing a basis for the systematic attainment of the objectives set down in the guidelines of the Kinesiology Academy.

The book is divided into three main parts. After the first chapter, which defines kinesiology and its anatomical and mechanical components, as well as defining the role of kinesiology in physical eduction, athletics, and rehabilitation, Parts I and II contain discussions of the basic anatomical and mechanical concepts relevant to the analysis of human motion. Part III introduces a system for the qualitative analysis of human movements. This system differs from other systems previously designed for the same purpose in that it permits human movements to be analyzed in a logical step-by-step fashion and in such a way that only those factors that have a direct and demonstrable causative relationship with the performance are considered. Further, it ensures that all of the important factors in determining the quality of the performance are taken into account.

To provide a concise reference source, those elementary anatomical and mathematical concepts that are considered essential prerequisites for an introductory course in kinesiology are presented in summary form in the Appendices. As a further aid to students, a set of exercises and a list of recommended readings are included at the end of each chapter.

Anatomical nomenclature based on a modified Basle Nomina Anatomica (1895) and Birmingham Revision (1933) terminology is used in the anatomical sections of the book.

The metric system of units is used throughout the book. While this may create some difficulties initially—linear and metric equivalents are provided in Appendix C in an attempt to minimize these—it is the authors' conviction that the use of the metric system in kinesiology texts is not only warranted but long overdue. The metric system is a part of the language of science, and since kinesiology is undeniably a science, it is only fitting that it be used in a book of this kind.

Finally, the authors would like to extend their sincere appreciation to the many people who contributed in one way or another to the writing of this book—Stan Morton and Charmaine Dapena, who prepared line drawings for Chapters 2–7 and 8–16, respectively; Bryan Dumouchel, who took anatomical photographs for Chapters 2–7; Eleanore Grefe, Betty Schieck, Carol Boyle, Jean Graham, and Sandi Dillon who typed the manuscript; Jack Engsberg, Antonio Guimaraes, Walter Herzog, Robert Nadler, Carol Putnam, Decio de Souza, Alex Stacoff, and Christopher Vaughan who assembled and prepared material for the figures and tables and assisted in a variety of other ways; Jack Griffith, of *Athletic Journal*, who provided the 35 mm camera with which the photo sequences of Chapters 15 and 16 were taken; the subjects who gave of their time to pose for photographs; and Hilary Hay and Patricia Reid who provided much-needed support and encouragement throughout the many months it took to write this book.

*James G. Hay*
*J. Gavin Reid*

# Chapter 1

# Why Study Kinesiology?

A knowledge of the anatomical and mechanical bases of human movement—a field of study commonly referred to as *kinesiology*—is widely considered essential to the professional preparation of physical education teachers, athletic coaches, and trainers. Although not always obvious to those required to study kinesiology, the reason for this is not hard to find. Movements of the human musculoskeletal systems, movements that are the central focus of the fields mentioned, are produced by the contraction of muscles and are moderated by environmental factors such as gravity and friction. The resulting movements thus depend on two sets of factors—those that influence the muscular contraction itself and those that influence the moderating effects of the environment. The first includes the structural and functional characteristics of the muscles involved and of the nervous system that initiates and controls their activity. The study of such characteristics lies within the field of *anatomy*. The second includes gravity, friction, fluid resistance, and the so-called normal reaction forces evoked through contact with other bodies such as the ground, an opponent, or a prosthesis. The study of these quantities lies within that branch of physics known as *mechanics*. In view of these relationships between human movement and the sciences of anatomy and mechanics, physical educators, coaches, and trainers should derive many benefits from the study of these disciplines.

What are these benefits? The answer to this question is perhaps best stated in terms of the practical situations in which a knowledge of kinesiology might be applied. In the normal course of their work, physical education teachers, coaches, and trainers are repeatedly faced with certain fundamental questions: "Which technique is the best?" "Should I teach this technique to my students or is it suitable only for top-class athletes?" "What is wrong with this individual's performance and how can I correct it?" "What exercises should I prescribe to improve this individual's physical condition?"

Reduced to their essential elements, these are primarily kinesiological questions—that is, questions that are most likely to be solved via appropriate anatomical and mechanical analysis.

In many cases, a brief consideration of the anatomical and/or mechanical factors contributing to the performance is all that is needed to obtain a satisfactory answer to the question. Consider, for example, the case of a diver who hits the water in a tucked position after completing $1\frac{1}{4}$ of a planned $1\frac{1}{2}$ front somersault dive. A brief mechanical analysis reveals that there are only three factors that directly influence the number of somersaults a diver can complete before hitting the water—the "amount of rotation" acquired prior to takeoff, the length of time spent in the air (a function of the maximum height attained), and the tightness of the body position adopted while in the air. Under such circumstances, a failure to complete the dive as planned is clearly due to one or more of these factors being insufficient for the purpose. Once the factors responsible for the unsatisfactory performance have been identified, and a simple observation of the movement is usually sufficient for the purpose, the teacher's task is clear. The teacher must work to eliminate the cause of the problem.

In other cases, simple qualitative analyses are unlikely to yield much of real value. This is especially true, for example, when the subject of the analysis is a highly skilled performer whose flaws in technique and weaknesses in physical makeup are very subtle or well concealed. Identifying these flaws and weaknesses requires a thorough quantitative analysis involving the use of sophisticated methods for gathering data (for example, electromyography, cinematography, and dynamography) and the use of a computer for processing and analyzing that data. Although quantitative analyses of this kind are used quite often to evaluate the performance of top-class athletes, the sheer volume of work involved precludes their use on a routine basis.

Finally, in still other cases, the complexity of the questions involved makes it unlikely that satisfactory answers will be obtained in the near future. Such questions, for example, as "What is the optimum technique for performing a given skill?" "What is the optimum way in which to load muscles to produce maximum strength gains?" and "What is the optimum design for a specific prosthesis?" are questions for which the answers are still being sought. It is highly likely, however, that kinesiological analysis will play a prominent role in their eventual solution.

In summary, a knowledge of the anatomical and mechanical bases of human motion provides physical education teachers, coaches, and trainers with the means to perform valid qualitative analyses to resolve recurring problems in their everyday work. Should they choose to proceed further, it also permits them access to the even more powerful tool of quantitative analysis—a tool that can be used to resolve the problems of top-class athletes and, perhaps too, the more general practical problems that face their professions.

# Part I

# Basic Anatomical Concepts

# Chapter 2

# Introduction

Recent scientific and technological advances have had a profound effect on practices in medicine: cures for many types of cancer, drugs to relieve pain and cure disease, techniques for organ transplant and replacement, and well-designed orthoses and prostheses for those who need them are examples. All these advances, and many others like them, have resulted in large part from the use of previous knowledge in the areas concerned.

To understand the movement capabilities of humans is to understand the gross structure of the human body itself. Without a knowledge of this structure—the basic foundation—it is almost impossible to study the normal and abnormal motor behavior of humans.

The study of the structure of human beings is the science of anatomy. The word *anatomy* comes from the Greek word to "cut," in today's terms, to dissect. Dissection was the term used by the Greeks for the study of anatomy because it was the main technique used. The basic study of the human body is termed *gross anatomy*; however, there are many divisions of study within the science of anatomy, such as histology, cytology, organology, and functional, developmental, and comparative anatomy. The principal focus of this section of the book will be on *functional anatomy*, with specific reference to human movement during the performance of physical activity.

## SURFACE ANATOMY

Of all subjects to be studied, it would seem that human anatomy should be the easiest. All students have observed the growth, development, and motor behavior of the human body for almost as many years as they are old. Not only do students have their own models close and available for study, but living human models are in plentiful supply and may be studied with sincere academic fervor as well as for aesthetic and pleasurable purposes.

The physical education instructor, the coach, and the trainer should be aware of acceptable structural differences and deviations among human bodies. It is also important that the student learn to detect deformities and abnormalities in body alignment that reflect poor posture and have a thorough understanding of those differences crucial to anyone involved in the prescription of exercise. The student must recognize telltale differences in skin condition, muscle bulk or tone, and joint movement capabilities. Pallid skin, for example, may indicate nutritional deficiency and/or illness; abnormal lack of muscle bulk or tone may reflect atrophy of muscle and indicate neuromuscular disorder, which may in turn result from injury or disease. The detection, however, is the important factor and should be followed by referral to a physician. Finally, the ability of joints to move through their normal range is an important consideration in physical fitness, in posture, and in the performance of physical skills. It is imperative that the physical therapist be able to recognize normal and abnormal joint mobility for manipulative therapy and for ameliorative purposes.

Surface anatomy involves the study of the shape, form, and marking of the body surface. Visual inspection provides the opportunity for the student to study landmarks; however, the use of palpation accompanying visual study enables the student to feel with the hands the form of the body, the size, position and movement capabilities of many structures, and the texture of the body tissues both superficial and subdermal. When muscles contract, they feel as though they have increased their firmness, or tonus. The ability to differentiate between active and passive muscles and degrees of activity by feeling is easily mastered and becomes a useful tool in the study of human movement.

A study of changes in the form of a shirtless subject's respiration patterns at rest and immediately following vigorous exercise, using both observation and palpation techniques—and assisted, perhaps, by markings placed on the surface of the thorax—can provide much information. This information can serve as the basis for an interesting discussion on the roles of the respiratory muscles and the movements of the ribs during inspiration and expiration. Students should make such observations while many of the calesthenic exercises and physical activities that are discussed in this book are being performed.

The surface features illustrated on Figs. 1, 2, and 3 should be noted and studied on a human subject. It may be advisable, at this stage, to review the bones that constitute the human skeleton (Appendix A).

## SPATIAL TERMS

It should become clear that, in the anatomical study of human movement, not only does the student require a thorough knowledge of the names of the parts of the skeleton as well as the muscles that provide or restrict movement to the

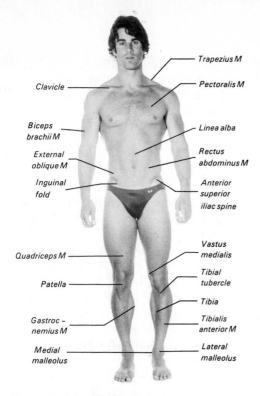

Clavicle

Biceps
brachii M

External
oblique M

Inguinal
fold

Trapezius M

Pectoralis M

Linea alba

Rectus
abdominus M

Anterior
superior
iliac spine

Quadriceps M

Patella

Gastroc -
nemius M

Medial
malleolus

Vastus
medialis

Tibial
tubercle

Tibia

Tibialis
anterior M

Lateral
malleolus

**Fig. 1.** *Surface features (anterior view).*

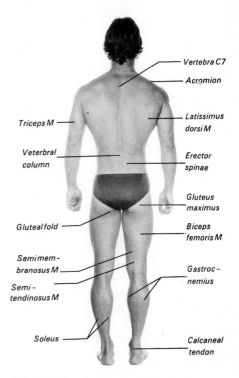

Triceps M

Veterbral
column

Gluteal fold

Semimem-
branosus M

Semi -
tendinosus M

Soleus

Vertebra C7

Acromion

Latissimus
dorsi M

Erector
spinae

Gluteus
maximus

Biceps
femoris M

Gastroc -
nemius

Calcaneal
tendon

**Fig. 2.** *Surface features (posterior view).*

**Fig. 3.** *Surface features (lateral view).*

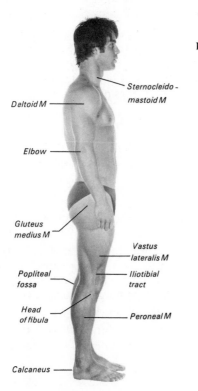

Deltoid M

Elbow

Gluteus
medius M

Popliteal
fossa

Head
of fibula

Calcaneus

Sternocleido -
mastoid M

Vastus
lateralis M

Iliotibial
tract

Peroneal M

skeletal parts, but also the types of movement that are possible at each joint. It is necessary, therefore, to familiarize oneself with the basic terminology used in relating to the *standing position*, the *directional terms* for the body, the *planes and axes* for movement, and the nomenclature of joint *movements*.

## Standing Position

The *anatomical position* is the reference position used to describe the location of anatomical parts and to describe and explain human movement. The position is illustrated in Fig. 4. The body is erect and facing the observer. Arms are at the side with the palms of the hands facing forward. To locate or describe a position of a part of the body, the body must be referenced to a single position—the anatomical position. The importance of this position will become clear as the reader proceeds through this discussion. It is necessary, also, to use basic anatomical references, as common terms such as "above" and "below" are relative terms only. For example, in Figure 4, the head is above the feet; however, if the subject were balanced upside down on his or her head, the feet would be above the head. It is necessary, therefore, to use directional terms to avoid any resulting confusion.

## Directional Terms

In studying anatomy, it is often necessary to describe the position of a body structure; for instance, it may be important to locate the radial pulse for checking heart rate during a physical conditioning period or to monitor heart beat if an athlete were injured and possibly in a state of shock. The radial pulse is found on the anterior aspect of the forearm, superficial to the distal end of the radius, which is the lateral bone of the forearm. This permits the use of a simple standard guide, or reference, to locate the exact position of the radial pulse for counting heart beats. These directional terms and others are defined in the following paragraph. Reference should be made to Fig. 5, which shows a subject in the anatomical position.

"A is *superior* to B" means that A is nearer the head (or the upper part of a structure—usually a structure of the trunk) than B. "B is *inferior* to A" states the same thing—that is, that B is farther away from the head than A. "A is *lateral* to B" means A is farther away from the midline of the body or structure. "B is *medial* to A" means that B is closer to the midline than A. "C is *proximal* to D" means that C is nearer than D to the attachment of a limb or extremity to the trunk; conversely, "D is *distal* to C" means that D is farther away than C from the attachment. The directional terms *anterior* and *posterior* are also important. Anterior is defined as nearer to or in front of the body; posterior refers to that

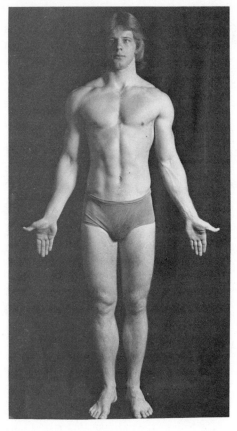

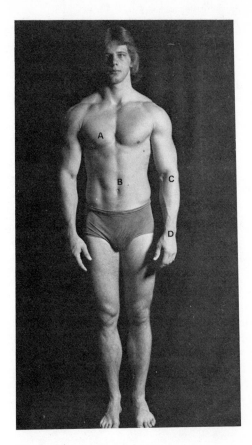

**Fig. 4.** *The anatomical position.*

**Fig. 5.** *Illustration for directional terms.*

which is nearer to or at the back of the body. Reference is often made to *superficial* and *deep* structures. Those referred to as superficial are on or near the surface of the body; those that are deep are farther away from the surface of the body.

## Planes and Axes

A further structural plan for the body involves the cardinal orientation planes. The three cardinal planes correspond to the three dimensions of space, each plane being perpendicular to each of the other two (Fig. 6). The point at which the three cardinal planes or midplanes make contact while the body is standing in the anatomical position is the center of gravity of the body (pp. 202–216).

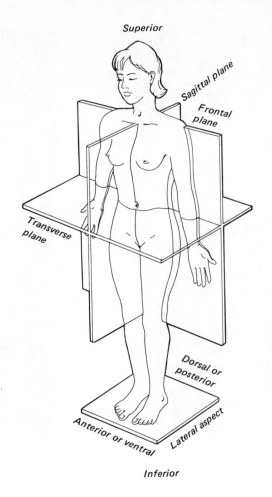

**Fig. 6.** *Cardinal orientation planes.*

**Planes.** The *sagittal* plane is the vertical plane that divides the body into left and right parts. If this plane passes through the body so that the masses of the halves are equal, it is called the midsagittal plane. Line A–B in Fig. 7 defines the midsagittal plane whereas line C–D defines a sagittal plane through the subject's right hip joint. The *frontal* plane is the vertical plane passing through the body so that it divides the body into anterior and posterior parts. If this plane passes through the whole body so that the masses of the halves are equal, it is called the midfrontal plane. The *transverse* plane is the third plane, and it divides the body into superior and inferior parts. As with the other planes, if the masses of the two parts are equal, this plane is referred to as the midtransverse plane.

These planes, though imaginary, are useful in describing basic movements. The planes are always the same regardless of the orientation of the body to the earth. One merely refers back to the anatomical position for a description of a

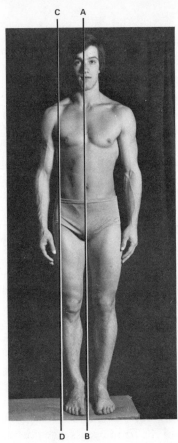

**Fig. 7.** *Sagittal planes of the body; AB = mid-sagittal plane, CD = sagittal plane.*

movement performed in one or more planes. An example is the bending and straightening of the elbow. The plane in which the movement is occurring is the sagittal plane; similarly, if the subject were to lie on his or her side maintaining a position similar to the previous one and perform the same movement, the motion would still be in the sagittal plane. The subject's head would still be superior to his or her feet despite the fact that in this position they would be on the same level with reference to the earth. The importance of relating body positions back to the anatomical position cannot be overemphasized.

***Anatomical Axes.*** There are three imaginary anatomical axes, each associated with a plane of motion and perpendicular to that plane.

1. The *transverse axis* passes through the body from side to side. The plane perpendicular to this axis, therefore, is the sagittal plane.
2. The *anteroposterior axis* passes through the body from front to back and is associated with movement in the frontal plane.
3. The *longitudinal axis* passes through the body from top to bottom and thus lies perpendicular to the transverse plane.

The movement capabilities of the joints of the body are complex and variable. Many joints permit simple movements in one plane only, and the axis of rotation remains essentially constant. Other joints permit movement in as many as three planes and thus involve continual changes in the location of the axis. Although the study of the complexity of joint motion is an interesting academic exercise, the concern of this section will be simply to explain the nomenclature of joint motion necessary to permit one to describe adequately the movements common in exercise prescription.

It should be recalled that the starting position for defining the movements is the anatomical position. It should be remembered, also, that in this position the joints of the subject are fully extended while in the upright position with the palms of the hands facing forward.

The fundamental movements discussed in the next section are presented in Table 1 for easy reference.

**Movements in the Sagittal Plane**    *Flexion* is movement of a body part (or parts) so that the movement performed is in the sagittal plane and the axis around which the movement occurs is transverse. (Flexion at a joint results in a decrease of the angle between the two segments that meet at that joint.)

*Extension* is the opposite of flexion; however, it occurs in the same plane also using a transverse axis and increases the joint angle. Flexion of the phalanges (fingers) of the hand occurs when the open (extended) hand is moving into the position of a clenched fist. When moving in the opposite direction, the action being performed is extension. From the extended position of the arm, in the anatomical position, flexion of the wrist and elbow results in decreasing the angles at the respective joints. Extension occurs when the joints return to the extended position. If movement occurs beyond the extended position, the action is called *hyperextension*. Movement of the extended arm in a forward and upward direction is flexion at the shoulder joint. A shoulder joint is in full flexion when the arm is directly superior to the shoulder joint. Extension at the shoulder joint occurs during the movement back to the anatomical position. Movement from this position in a posterior direction is hyperextension of the shoulder joint. Flexion of the knee and hip joint poses no difficulty in understanding; however, when considering the ankle joint. In the anatomical position of full extension, the foot is almost at a right angle to the leg. Movements of the foot in both directions is termed flexion. When the top (dorsum) of the foot moves towards the shin (tibia), the action is called *dorsiflexion*. Movement of the sole of the foot away from the tibia, as in "pointing toes," is called *plantar flexion*.

Confusion often arises when considering movement of the trunk. Flexion of the trunk is the action that produces an anterior concavity of the spinal column, whereas extension occurs while moving back from this flexed position to the anatomical position. Hyperextension of the trunk involves a backward arching of the trunk. Trunk flexion should not be confused with hip flexion and should be measured independently—movements such as "toe touching" involve flexion at the hip joint as well as the trunk and therefore must not be regarded as a measure of trunk flexion.

**Movements in the Frontal Plane**  *Abduction* occurs when a body part is moved away from the midline of the body in the frontal plane about an anteroposterior axis. Examples of this action are the raising of the arm sideways from the anatomical position and likewise raising the leg sideways. *Adduction* is the opposite movement in which the body part moves toward the midline of the body. Exceptions from this rule occur in the hands and the feet, in which cases the fingers or the toes are spread away from the midline of the hands or feet, respectively—as, for example, in spreading the fingers. A further exception occurs in the movements of the thumb, which will be outlined later.

**TABLE 1**  **Fundamental Movements**

| Action | Definition |
|---|---|
| Flexion | Decreasing the angle between two bones |
| Extension | Increasing the angle between two bones |
| Abduction | Moving away from the midline (of body or part) |
| Adduction | Moving toward midline (of body or part) |
| Elevation | Moving to superior position |
| Depression | Moving to inferior position |
| Rotation | Turning about the vertical axis of the bone |
| Supination | Rotating wrist or hand laterally from the elbow |
| Pronation | Rotating wrist or hand medially from elbow |
| Inversion | Lifting the medial border of the foot |
| Eversion | Lifting the lateral border of the foot |
| Dorsiflexion | Moving the top of the foot toward the shin |
| Plantar flexion | Moving the sole of the foot downward (pointing toes) |
| Circumduction | Circumscribing a conical area, involving flexion, abduction, extension, and adduction in sequence |

Adduction of the fingers and toes occurs when they move toward the midline of their respective parts. When the humerus is abducted in the frontal plane to shoulder level and is then moved toward the midline of the body in the transverse plane, this action is called *horizontal adduction*. Movement back to the side position is termed *horizontal abduction*. It should be noted, therefore, that abduction and adduction may take place in more than one plane. *Elevation* is a movement of the shoulder girdle in the frontal plane to a superior position, whereas *depression* is a movement in the opposite direction. Bending of the spinal column in the frontal plane is referred to as *lateral bending* to the left or right.

**Movements in the Transverse Plane** *Rotation* to the left and right are directional terms used to describe rotation of the head and neck and the trunk. When describing the motion of the limbs of the body, *inward* or *medial* rotation is used to indicate the movement of the anterior surface of the limb toward the midline of the body. The opposite movement is *outward* or *lateral* rotation.

*Supination* and *pronation* are actions of the forearm. In the anatomical position with the hand facing forward, the forearm is in the supinated position. A movement of the hand inward or in a medial direction with the shoulder fixed is called *pronation.* The opposite action is termed *supination* of the forearm. These motions occur at the radioulnar joints of the forearm. Although the terms pronation and supination are used to describe similar movements of the foot, more commonly the term *inversion* is used instead of supination and *eversion* in place of pronation. An everted or pronated position of the foot is commonly termed a "flat-footed" position.

### Complex Movements

*Circumduction* is a combination of movements in the sagittal and frontal planes and the oblique planes between them, which results in the moving segment outlining a cone. An example of this movement is the sequential motion of the hand at the wrist joint during the tracing of a circle by the fingers while the forearm is stationary.

Human movement is extremely complex and variable and to describe exactly the position of body parts during movement of the body is most difficult. Attempts have been made to record joint motions and to describe and measure the orientation of body parts with precision; however, the difficulty of these methods far outweighs their practical value. The verbal description of the position and motion of the joints during movement provides a practical and adequate means of describing human motion.

### EXERCISES

1. Assume a prone-lying position with hands beside the shoulders, palms flat, fingers directed forward, ready to perform a push-up. Moving the arms in a sagittal plane, execute a push-up and return to a prone-lying position.

   Using a copy of the Exercise Sheet (p. 15) to concisely summarize your conclusions, determine
   a. the positions of the joints listed at the start of the exercise,
   b. the joint actions that take place as the body is raised,
   c. the positions of the joints when the push-up action has been completed, and
   d. the joint actions that take place as the body is lowered.

In the course of this analysis, list also the planes and axes associated with the movements that take place.

How would your conclusions have differed if the push-up had been performed incorrectly by first straightening the arms and then lifting the abdomen from the floor? People who do not have the strength to perform a push-up correctly often perform it in this manner. Why is an incorrect push-up like this easier to perform?

How would your conclusions have differed if a kneeling push-up had been performed instead of a regular push-up?

2. Repeat the same form of analysis used in Exercise 1 for a bent-knee sit-up with hands held behind the head. Consider both the lifting and lowering phases of the exercise.

3. Repeat Exercise 1 for an arm swing sideward and upward to clap the hands overhead as is seen in the "jumping jack" (or astride-jumping) exercise. Consider both lifting and lowering phases of the exercise and assume that the arms are kept straight throughout.

4. Locate on a subject each of the surface features indicated in Figs. 1, 2, and 3.

## RECOMMENDED READING

Steindler, Arthur, *Kinesiology of the Human Body: Under Normal and Pathological Conditions.* Springfield, Ill.: Charles C Thomas, 1955, pp. 110–121.

**EXERCISE SHEET**

| Exercise: | | | | | | | | |
|---|---|---|---|---|---|---|---|---|
| | Start Position | Lifting Phase | | | End Position | Lowering Phase | | |
| | | Movements | Plane | Axis | | Movements | Plane | Axis |
| Neck | | | | | | | | |
| Trunk | | | | | | | | |
| Shoulder girdle | | | | | | | | |
| Shoulder | | | | | | | | |
| Elbow | | | | | | | | |
| Radioulnar | | | | | | | | |
| Wrist | | | | | | | | |
| Fingers | | | | | | | | |
| Hip | | | | | | | | |
| Knee | | | | | | | | |
| Ankle | | | | | | | | |
| Toes | | | | | | | | |

# Chapter 3
# The Skeletal System

The human skeleton is comprised of bones, related cartilages and ligaments, and joints. *Osteology* is the study of bone whereas *arthrology* is the study of joints. Both these topics will be discussed in this chapter. The human skeleton is situated within the soft tissues of the body. It is alive and capable of growth and regeneration. It has the human capabilities of adaptation. There are a number of important functions of the human skeleton. The skeletal system gives support to the soft tissues of the body in that it provides attachments for the muscles, ligaments, and fascia. It provides protection to many of the vital organs of the body enclosing, for instance, the brain within the skull, the lungs and heart within the thoracic cage, and an important system of nerves within the spinal column. Protection is also provided for organs such as the bladder and uterus within the bony pelvis. Blood cell formation occurs in the red marrow of many bones; moreover, the bones have storage capabilities for calcium. The bones serve as levers and, in conjunction with muscles, cartilage, and ligaments, provide the bases for movement.

## BONES

The distribution of the 206–210 bones that constitute the skeleton is presented in Table 2. The bones may also be classified into four groups according to their shapes (Fig. 8):

1. *Long bones* such as the femur, tibia, and fibula of the legs and the humerus, radius, and ulna of the arms comprise the limbs of the body. In movement, these bones are generally used as levers.
2. *Short bones* are usually about equal in length and width. They include the

## TABLE 2   Bones of the Human Skeleton

| Skeleton | Number of Bones* |
|---|:---:|
| Axial | |
|     Vertebral column | 26 |
|     Cranium and face | 22 |
|     Sternum, ribs, and hyoid | 26 |
| Appendicular | |
|     Upper extremities | 64 |
|     Lower extremities | 62 |
| Total | 200 |

*In addition, there are auditory ossicles (6) and a number of sesamoid bones.

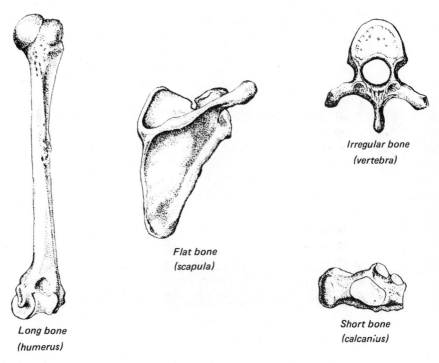

Long bone
(humerus)

Flat bone
(scapula)

Irregular bone
(vertebra)

Short bone
(calcanius)

**Fig. 8.** *Classification of bones according to shape.*

    bones of the wrist and foot where strength is required and mobility is less necessary.

3. *Flat bones* make up the vault of the skull, the scapula, and the sternum and ribs of the thorax. They are generally used for protection; however, the ribs do provide a certain amount of movement during respiration.

4. *Irregular bones* are named according to their complex and varied shapes.

They are used for protection, for support, and for leverage in movement. Irregular bones are those comprising the vertebral column, the pectoral and pelvic girdles, the patella, and the small bones of the face.

The student should refer to Appendix A for a review of the bones of the skeletal system.

### Structure of Bone

The hardness of bone is due to the mineral salts that are incorporated within it. Calcium carbonate and calcium phosphate form a large proportion of the inorganic matter that makes up about two thirds of the weight of the bone; the remaining one third is organic. In young children, organic matter is more plentiful than in adults and therefore the bone is softer and less brittle. With advancing age, the bones become less pliable, less flexible, and more brittle, and, because they are not able to absorb forces over an appreciable range, they are more susceptible to breakage. There are two types of bony substance in the body: *compact* tissue and spongelike *cancellous* tissue. The amounts of these two tissues vary throughout the body and throughout life. The hard outer layer of bones is the compact substance; the spongelike interior is the cancellous substance. Cancellous tissue is more common in large flat bones and the expanded ends of the long bones. Figure 9 shows a frontal section through the proximal end of the femur.

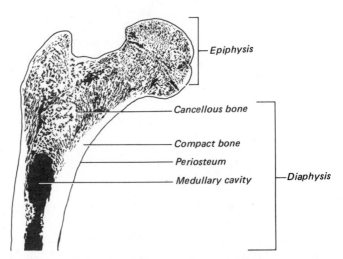

**Fig. 9.** *Frontal section through the proximal end of femur.*

The shaft of the femur consists almost entirely of compact bone surrounding a *medullary cavity* that contains marrow. The head or proximal end of the femur is cancellous bone covered by a thin layer of compact bone. It has no true marrow cavity. Figure 9 shows the long bone with its end or *epiphysis* being joined by the shaft, the *diaphysis*. Articular cartilage, a special kind of hyaline cartilage, covers the articular ends of the bones. With the exception of these ends, living bone is covered by a two-layered fibrous membrane, the *periosteum*. The outer of the two layers of the periosteum is fibrous and contains blood vessels, whereas the inner layer consists of white and more elastic fibers. The white fibers are continuous with the matrix of the bone to give strong support.

### Radiological Anatomy

An X-ray picture enable one to see the size, shape, and varying density within and between bones and the relationship of one bone to another. To use this aid effectively, one should have a simple but clear understanding of radiography.

Radiography is based on a differential absorption of X-rays; this simply means that certain structures are easily penetrated by X-rays while others are not penetrated or are penetrated less easily.

X-rays were discovered at the end of the nineteenth century. Similar to infrared, ultraviolet, and radio waves, they are called electromagnetic waves, which are wavelike packages of energy that arise in association with an acceleration of electrons. Although all these waves travel at the same speed, they differ in frequency and wavelength. The wavelengths of X-rays are very short. X-rays thus have the power to penetrate dense matter. The density of the matter is proportional to the amount of X-rays absorbed. Bone, for example, absorbs X-rays more easily than do soft body tissues.

The most dense tissues or structures (very often foreign bodies) appear the whitest in an X-ray. An example of this is the metal ring on the finger of the hand in Fig. 10.

The next most dense tissue is the enamel of the teeth, the density of which is greater than that of bone. Soft tissues such as muscles and internal organs are less dense and therefore appear darker than the preceding structures. The least dense tissue is fat, which appears very dark, while the black of an X-ray represents air as it is found, for example, in the lungs and stomach and the area that surrounds the body part shown in the X-ray.

The cancellous and compact substances of bone are easily observed in Fig. 10. Note that the greater thickness of compact bone is in the diaphyses rather than in the epiphyses of the long bones and that the compact bone surrounds the cancellous bone. Because compact bone is more dense than the cancellous bone, it appears whiter in the X-ray.

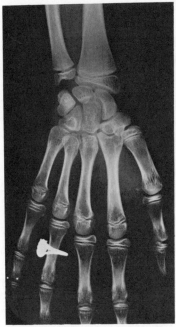

**Fig. 10.** *X-ray of the right wrist and hand of a teenager.*

## Growth and Development of Bone

The bones of the body develop from *mesenchyme*, which is embryonic connective tissue. They either develop as *intramembranous* bones, which form directly in a fibrous membrane with the primary center of ossification in the middle, or as *endochondral bones*, which form by the replacement of hyaline cartilage.

The intramembranous bones are the flat bones of the cranial vault, which, prior to birth, are membranous to enable the head of the fetus to pass through the birth canal during delivery. Ossification (formation of fibrous tissue into bone) usually commences after birth and the process continues for many years until completion.

Endochondral bone forms from a cartilaginous model that exists at birth and represents the skeleton in shape and relationship. The growth and development of endochondral bone follows a fairly typical pattern.

The cartilaginous model is invaded by osteoblasts resulting in a thin layer of bone around the diaphysis (shaft). Cartilage is then replaced by bone (endochondral ossification) at the center of the bone, and ossification spreads from this *center of ossification* toward the ends of the bone. At one end (epiphysis), an epiphyseal center of ossification occurs and bone development commences. This is followed by a similar process at the other end of the long bone. An *epiphyseal disc* or plate is formed at each end of the bone between the diaphysis and the epiphysis. Figure 11 shows the epiphyseal disc of the radius next to the elbow joint; Fig. 10 shows many of these discs in the hand. These growth zones consist of cartilage cells that grow rapidly and are then replaced by bone, thereby increasing the length of the shaft of the bone. The center last formed is the first to fuse, which results in stoppage of growth at that end. Longitudinal growth ceases when both epiphyseal discs become ossified. Although growth stops, there is a continual replacement of old tissue by new.

Epiphyses have been classified in two types, traction and pressure, by Larson in his comprehensive paper on physical activity and the growth and development of bone and joint structures.[1] *Traction epiphyses* contribute to the shape of bone and are found at the bony origins and insertions of certain muscles—at the point on the tibia to which the patellar tendon of the quadriceps muscle is attached, for example. Figure 11 shows a traction epiphysis on the proximal end of the ulna to which the triceps muscle is attached. Traumatic epiphysitis or traumatic separation of traction epiphyses, such as occurs at the tibial tuberosity and the medial epicondylar epiphysis of the humerus (Little Leaguer's elbow), are related to a single traumatic strain or excessive repetitive stress to the epiphysis.[2] *Pressure epiphyses* are located at the ends of long bones and are subjected to compression as in weight bearing.

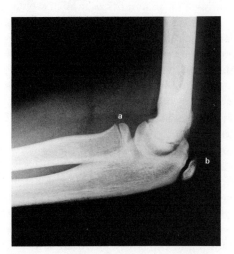

**Fig. 11.** *(a) Epiphysial disc of radius (b) Traction epiphysis of proximal end of ulna.*

[1] Robert L. Larson, "Physical Activity and the Growth and Development of Bone and Joint Structures, in *"Physical Activity Human Growth and Development*, ed. G. L. Rarick (New York: Academic Press, 1973), pp. 32–59.

[2] Larson, "Physical Activity and the Growth and Development of Bone and Joint Structures," p. 45.

It has been well documented that immobilization results in a loss of mineral salts and a decrease in the cortical substance in bone. These changes, which have been attributed to the absence of functional mechanical stimulation,[3] result in the bone's becoming less dense.

Appearance and closure of the growth plates and the ossification of bones vary according to factors such as sex, hormonal influences, health of the individual, and bone injury. Normally, however, there is a similarity in the time of ossification of the bones of the wrists, and this consistency has provided a fairly accurate means of determining skeletal and physiological maturity. Figure 12 shows the ossification of the carpal bones at three stages of growth. Two of the X-rays are of prepubertal children; the third is of a teenager in his upper teens.

The use of X-rays for the purpose of indexing age is not of value to the physical education teacher or coach. However, the importance of the knowledge is in the awareness of the injury potential for the growing child.

It should be noted that the closure or ossification dates for long bones range from six to twenty years, with most of the plates fusing by the late teens. One must bear in mind that growth plates, which consist of growing cartilage cells, are susceptible to injury due to traumatic impact, as well as to excessive or repetitive shearing and torsion forces, any of which could result in premature closure.

The effects of impact on the human body as a result of car crashes and ejection from airplane seats have been studied; however, there is a paucity of literature dealing with human impact tolerance levels in sporting events. The body of the athlete often experiences vary rapid changes in velocity. Because rapid changes in velocity that accompany impacts are an integral part of many sports, an understanding of the effects on the body of the athlete is essential. Too often, the coach or teacher ignores this aspect of athletic competition and fails to teach the athlete how to minimize the effects of impact. As a result, injury to the athlete may occur.

Davidovits[4] calculated that a 70 kg person could generate an impact force capable of fracturing bone in a jump from only 41.6 cm if the landing was with legs perfectly rigid on a hard floor. On the other hand, there have been reports of individuals who escaped injury in a fall from an airplane when they landed in deep snow, in a forest, or in water. These illustrations demonstrate that great variation in the injury potential associated with falling from various heights, depending upon the nature of the landing. Davidovits has pointed out that the important factors in determining whether an injury occurs following a fall from a

---

[3] J. A. Gillespie, "The Nature of the Changes Associate with Nerve Injuries and Disuse," *Journal of Bone and Joint Surgery*, 36-B (August 1954), 471.

[4] Paul Davidovits, *Physics in Biology and Medicine* (Englewood Cliffs, N.J.: Prentice-Hall, 1975), p. 61.

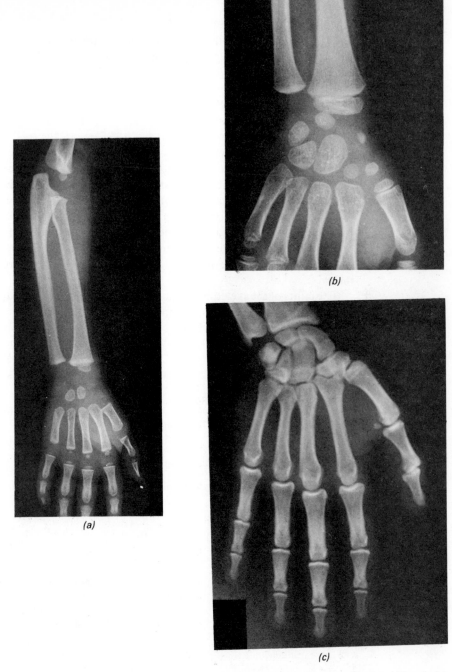

**Fig. 12.** *Ossification of carpal bones: (a) normal 2-year-old female, (b) normal 6-year-old female, (c) normal 19-year old male.*

given height are the pliability of the landing surface, the area of contact, the use of flexion of the joints to gradually decrease velocity, and the duration of the impact.

## JOINTS

Most bones of the skeleton act as levers when muscles pull on them thereby causing movement; the junctions of the levers or skeletal parts are called joints or *articulations*. Movement of the system at some joints is extensive; at others it is limited, and in certain cases it is prevented by the union of bones or cartilage. Extensive mobility is permitted at the shoulder joint, while at the other extreme immobility is demonstrated at the bony joints or sutures of the skull. It is necessary, therefore, that the functional skeleton have joints that provide mobility for a wide range of movement as well as stability and strength to prevent injury.

The joints of the skeleton are classified according to the amount of movement that they permit. Table 3 is a summary of joint classification; Fig. 13 presents examples of the different joints.

**TABLE 3    Classification of Joints by Action**

| Class | Common Name | Technical Name | Movement | Examples |
|---|---|---|---|---|
| Immovable (synarthrodial) | Fibrous | Suture Syndesmosis | None | Sutures of skull Distal tibiofibula |
| Slightly movable (amphiarthrodial) | Cartilaginous | Synchrondosis Symphysis | Negligible | Sternocostal Epiphyseal plates Pubic symphysis |
| Freely movable (diarthrodial) | Synovial | | | |
| | Ball and socket | Enarthrosis | Triaxial | Hip and shoulder |
| | Condyloid | Ellipsoid | Biaxial | Wrist |
| | Gliding | Arthrodial | Nonaxial | Intercarpal and intertarsal |
| | Hinge | Ginglymus | Uniaxial | Elbow |
| | Pivot | Trochoid | Uniaxial | Atlanto-axial |
| | Saddle | Sellar | Triaxial | Carpometacarpal of the thumb |

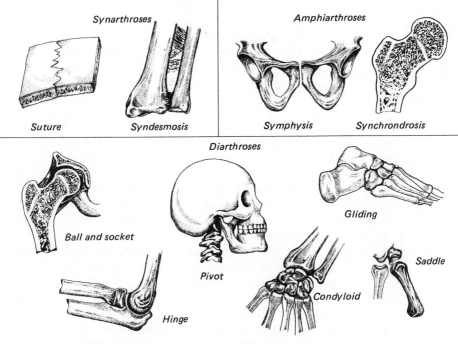

**Fig. 13.** *Classification of the joints of the body.*

## Classification of Joints

1. *Immovable (synarthrodial)*—Fibrous joints of which there are two types:
   a. *Sutures* are the immovable joints found between the bones of the skull in which the fibrous tissue is continuous with the periosteum and the bones are in close proximity with each other.
   b. *Syndesmoses* are joints consisting of dense fibrous tissue that bind the bones together and permit very limited movement. An example is the articulation between the radius and ulna.
2. *Slightly movable (amphiarthodial)*—Cartilaginous joints of which there are two types:
   a. *Synchrondroses* are the slightly movable sternocostal joints of the thorax and the epiphyseal plates that separate the epiphyses from the diaphyses in the long bones. The cartilaginous growth plates ossify on completion of growth and thereafter are not considered functional joints.

      b. *Symphysis* joints are usually composed of fibrocartilage that is separated from the bones by thin plates of hyaline cartilage. Examples of fibrocartilaginous joints are the symphysis pubis and the joints between the bodies of the vertebrae where movement is limited.
3. *Freely movable* (*diarthrodial*)—This joint category provides a variety of movements, many over a wide range of motion. Some of these joints are specially constructed to permit a considerable degree of movement and are thus of special interest.

    The articulating surfaces of the bones are covered with articular cartilage and are connected by a capsular ligament of fibrous tissue that is taut enough to provide stability and yet lax enough to permit the necessary movement at the joint.

    Where there is movement, there will be friction; however, this is reduced by the synovial membrane that supplies synovial fluid to the joint surfaces, thereby lubricating the surfaces and providing less friction on the articulating surfaces. The synovial membrane is attached to nonarticular bone. This membrane covers the deep surface of the capsular ligament and intra-articular structure such as fibrocartilaginous discs, tendons, ligaments, and fat pads; however, it does not cover the articular discs and cartilage.

    Although a synovial joint is said to have a joint cavity, it is normally only a potential cavity that contains a small amount of synovial fluid. Injury such as rotational wrenching or a direct blow to a joint is often accompanied by increased secretion of synovial fluid that usually results in swelling.

    The range of motion of synovial joints varies considerably as does the structural design of the joints. Due to the range of motion capabilities, the joints must be able to provide a certain amount of strength and stability to minimize the effects of dislocative forces that could result in injury. This strength and stability depends on (a) the architecture of the bones, (b) the ligaments, fascia, tendons, and muscles crossing the joint, and (c) the articular fibrocartilage within the joint.

**Bony Architecture.**    If one bone fits snugly into another as in the hip joint, there is less chance of dislocation than when the socket is shallow as in the shoulder joint. The film of synovial fluid on the articular surfaces of such bones assists in the maintenance of the integrity of the joint. One common example is the analogy of two sheets of wet glass that are placed one on top of the other; they slide easily but they are extremely difficult to separate.

**Ligaments, Fascia, Tendons, and Muscles.**    The ligaments and articular capsules are made of a type of connective tissue composed of tough, inelastic, white fibers. They are extremely strong and are used to stabilize the joint, limit excessive movement, and prevent injury.

When white fibrous tissue is found in sheaths around muscle, it is called fascia; in addition to its protection of the muscle fibers, it separates muscles from each other to enable each muscle to act independently within its own sheath. It also binds muscles together for protection and efficient movement. An analogy to the fascia sheath is panty hose, which provides support to the muscles of the thigh and which are being recommended by some athletic trainers to assist in the recuperative process for hamstring pulls.

A tendon is also white fibrous tissue. It is used to bind muscle fibers to their bony attachments by means of connection to the periosteum layer of white fibrous tissue on bone. Muscle tendons are often intertwined with the capsular ligaments of joints.

The tone and strength of muscle and its tendons aid the stability of joints. Most muscles, when contracting, increase the stability of the joint over which they act by pulling the bones involved together (Fig. 14(a)). However, they may also decrease the stability of the joint by pulling the bones apart (Fig. 14(b)).

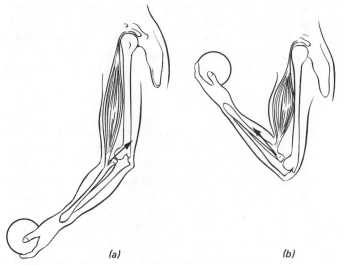

*(a)*                    *(b)*

**Fig. 14.** *Influence of muscle contraction on joint stability.*

***Articular Fibrocartilage.*** A disc of fibrocartilage is found within the joint cavity of certain joints and thereby separates the articulating bones. An example of this is at the sternoclavicular joint. Others such as the menisci, or semilunar cartilages of the knee, are incomplete discs that assist, in addition to their shock-absorbing and stability roles, in facilitating lubrication.

There are six types of freely movable diarthrodial joints (Table 3):

1. *Ball and socket (enarthrodial)*—These joints provide the greatest movement. A somewhat rounded articulating head of a long bone fits into a cup-shaped

cavity permitting flexion, extension, abduction, adduction, rotation, and the combined movement of circumduction. This type of joint is sometimes classified as triaxial. The hip and shoulder are examples.

2. *Condyloid (ellipsoidal)*—This is a modified ball-and-socket joint. Movement is limited somewhat in that it is biaxial, and, although flexion, extension, abduction and adduction are possible, there is no rotation. Examples of this joint are the metacarpophalangeal joints of the fingers. This does not include the thumb.

3. *Gliding (arthrodial)*—These joints permit movement in one plane (uniaxial) in the form of sliding motion. Examples are found in the tarsal and carpal bones and the articular processes of the vertebrae.

4. *Hinge (ginglymus)*—Movement in these joints is in the sagittal plane around a transverse axis only. The flexion and extension occurs at the elbow, ankle, and interphalangeal joints.

5. *Pivot (trochoid)*—Movement at this type of joint occurs in one plane only (transverse) about a longitudinal axis. Examples of this joint are the radioulnar and the atlas rotating around the odontoid process of the axis during rotation (turning) of the head.

6. *Saddle (sellar)*—This joint is shaped like a saddle. It is biaxial permitting movements of flexion, extension, abduction, and adduction. The only true saddle joint is the carpometacarpal joint of the thumb.

## Joint Mobility

Concern for the maintenance and improvement of the range of joint motion was prevalent in ancient Greece. Hippocrates, for example, claimed that all parts of the body that have a function become healthy, well developed, and age slowly if exercised; however, if they are not exercised, they become liable to disease, defective in growth, and age quickly. His claim was made with reference particularly to joints and ligaments.[5]

Today our concern for the maintenance or improvement of range of joint motion is similar. Wright and Johns[6] represented the concern for range of motion in medicine at present in their claim that joint stiffness was a major manifestation of joint disease; however, they also suggested that a lack of scientific knowledge on the subject existed. The desire to achieve excellence in physical performance in the form of recreation, sports events, and entertainment has furthered the concern for knowledge relating to the maintenance and improvement of joint motion.

[5]E. T. Withington, *Hippocrates, with an English Translation* (New York: Putnam, 1927), pp. 339, 341.

[6]Verna Wright and Richard J. Johns, "Observations on the Measurement of Joint Stiffness," *Arthritis and Rheumatism*, 3 (August 1960), 328.

Scientific investigations and writings concerned with the range of human joint motion have resulted in controversy regarding the description of the phenomenon. The term describing the change in angle at a joint has been variously defined as mobilization, freedom to move, and more commonly as flexibility. This last term (flexibility), according to Holland,[7] implies flexion alone and should be replaced with a more appropriate designation.

Range of joint motion has an established role in physical fitness, sports ability, posture, surgery, and physical medicine (Rasch and Burke[8]). In a comprehensive review of the literature dealing with the physiology of flexibility, Holland[9] suggested that there is probably no such phenomenon as general flexibility and that specific measures of flexibility are only indicative of the range of motion in the specific joints measured; they cannot be used to predict flexibility in other body parts. Extremely low relationships have been reported between flexibility and motor ability; however, much of the literature supports the concept that participation in certain activities or kinds of performance result in the development of rather specific patterns of joint flexibility. Holland[10] believed that this suggested that range of motion was modified and that, by observing the movement of patterns of different athletes and performers, the mechanics of joint stress and connective tissue adaptation could be analyzed. Observation of the extreme range of motion attained through practice by high-caliber gymnasts and dancers readily supports the premise that range of motion can be modified.

**Properties of Connective Tissue in Joint Mobility**   The range of movement at a joint is determined by the extensibility of the muscles, elasticity of the articular capsules, fluidity of the discs, extensibility of the longitudinal ligaments, anatomical architecture (structure) of the joint bones (articulations), and resistance of the surrounding tissues.

Kottke, Pauley, and Petak,[11] in describing the rationale for the use of prolonged stretching to correct a shortening of connective tissue, reported that clinical observation has shown that collagenous connective tissue has the property of shortening in the absence of tension as well as plasticity in that it slowly elongates under moderate constant tension. They claimed that connective tissue has a very high tensile resistance to a suddenly applied tension of short

---

[7]George J. Holland, "The Physiology of Flexibility: A Review of the Literature," in *Kinesiology Review* (Washington, D.C.: American Association for Health, Physical Education, and Recreation, 1968), p. 49.

[8]Philip J. Rasch and Roger K. Burke, *Kinesiology and Applied Anatomy*, 5th ed. (Philadelphia: Lea and Febiger, 1974).

[9]George J. Holland, "The Physiology of Flexibility," pp. 49–62.

[10]Holland, "The Physiology of Flexibility," p. 57.

[11]Frederic J. Kottke, Donna L. Pauley, and Rudolph A. Petak, "The Rationale for Prolonged Stretching and Correction of Shortening of Connective Tissue," *Archives of Physical Medicine and Rehabilitation*, 47 (June 1966), 347.

duration; however, under prolonged mild tension, it shows plastic elongation. This plastic elongation was attributed to the separation of the attachments at the point of contact of adjacent collagen fibers in the connective tissue meshwork. They claimed that normal motion in joints and soft tissues is maintained by the normal movement of the parts of the body that elongate and stretch joint capsules, muscles, subcutaneous tissue, and ligaments through the full range of motion many times each day.

***Exercise for Development of Range of Motion.***   Exercise for the development and maintenance of range of motion plays an important role in rehabilitation of the handicapped and is an essential part in treatment of acute and chronic trauma in orthopedic work. Specific exercises such as those used for postural correction are the concern of orthopedists, pediatricians, physical educators, and physical therapists. Various types of exercises are used to restore and recondition the patient who has been bedridden through internal disease; they are used to return the patient to normal life in the shortest span of time after surgery; they are used to prepare athletes for the performance of physical skills, and they are used by most persons in various forms of recreational activities. Wessel and van Huss[12] pointed out that the fundamental questions concerning the effects of exercise apply equally to therapeutic exercise as they do to exercise and sports generally.

The exercises used for the development or maintenance of flexibility are of two types: *passive* and *active*. Passive exercise can be divided into *nonforced* and *forced*. Nonforced is movement within the limits of voluntary range of motion that does not cause pain or spasm. This type of movement has been found to assist in the prevention of contractures, adhesions, capsular tightness, and muscle shortening. Forced passive exercise is movement carried beyond the existing voluntary range of motion.[13]

*Active exercise* may be classified as *static* or *kinetic*. According to Wessel and van Huss,[14] the static type of exercise is performed when the muscles contract isometrically—that is, without producing joint motion. This type of exercise may be performed with or without resistance. Kinetic exercise is performed when the muscles contract isotonically (concentrically or eccentrically) to produce single or multiple joint movement. This type of exercise may also be performed with or without resistance. The literature presents other classifications for these exercises;[15] however, they are basically similar to those suggested previously and differ mainly in semantics.

[12] Janet A. Wessel and Wayne van Huss, "Therapeutic Aspects of Exercise in Medicine," in *Science and Medicine of Exercise and Sports*, ed. Warren R. Johnson, (New York: Harper, 1960), p. 665.

[13] Wessel and van Huss, "Therapeutic Aspects of Exercise and Sport," p. 672.

[14] Wessel and van Huss, "Therapeutic Aspects of Exercise and Sport," p. 673.

[15] H. A. Rusk, *Rehabilitation Medicine: A Textbook on Physical Medicine and Rehabilitation*, 2nd ed. (Saint Louis, Mo.: C. V. Mosby, 1964).

Techniques advocated and practiced for the development of range of joint motion used in sport activities vary considerably. Investigations of the influences of different exercises on the development of range of joint motion in humans have produced contradictory results.[16,17,18,19] The techniques used in research to increase the range of joint motion have differed in terms of the force creating stretch and, specifically, in terms of the magnitude of the force, the length of time the force was applied, and the frequency of application of the force.

Active exercises, those designed to stretch muscles crossing a joint, have been found to increase the range of joint motion significantly. A slow, controlled, intermittent force of a given intensity was found to develop range of motion more rapidly than was a passive or constantly applied force of similar intensity.[20] From a safety standpoint, it would be advisable to advocate the use of passive stretching or a *very* slow stretching action to increase range of joint motion.

**EXERCISES**

1. Construct a goniometer, similar to that illustrated in Fig. 15, out of cardboard or wood. Measure the maximum range of motion of a number of joints on a partner and compare corresponding left and right joints. Consider the factors that make this form of measurement merely an approximation of the true range of motion of a joint. What is inhibiting the range of motion? What would be the influence of a thorough warm-up on the measurements?
2. Consider the range of joint motion an athlete requires to permit optimum performance in the following physical activities (1 = great, 2 = moderate, 3 = little):

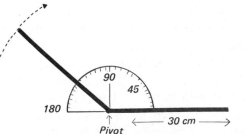

Fig. 15. *Simple goniometer.*

[16]S. Weber and H. Kraus, "Passive and Active Stretching of Muscles," *Physical Therapy Review*, 22 (September 1949), 407–410.

[17]Herbert A. de Vries, "Evaluation of Static Stretching Procedures for Improvement of Flexibility," *Research Quarterly*, 33 (May 1962), 229.

[18]Laurence E. Holt, Thomas M. Travis, and Ted Okita, "Comparative Study of Three Stretching Techniques," *Perceptual and Motor Skills*, 31 (October 1970), 611–616.

[19]J. Gavin Reid, "Effects of Constant and Intermittent Forces on Range of Joint Motion," in *Biomechanics V-A, International Series on Biomechanics*, ed. P. V. Komi, Vol. 1A, (1976) pp. 461–467.

[20]Reid, "Effects of Constant and Intermittent Forces on Range of Joint Motion," pp. 461–467.

| Joints | Archery | Dance | Gymnastics | Pool | Swimming |
|---|---|---|---|---|---|
| Intervertebral | | | | | |
| Shoulder | | | | | |
| Elbow | | | | | |
| Wrist | | | | | |
| Hand | | | | | |
| Hip | | | | | |
| Knee | | | | | |
| Ankle | | | | | |
| Foot | | | | | |

Which activity did you find required the greatest overall range of joint motion? Which required the least? Compare your estimate with those of other students.

3. What are the dangers of performing strenuous physical activities when an athlete possesses
   a. excessive joint suppleness or
   b. excessive joint stiffness?

## RECOMMENDED READINGS

American Academy of Orthopedic Surgeons, *Measuring and Recording of Joint Motion.* Chicago: The Academy, 1963.

Barnett, C. H., D. V. Davies, and M. A. MacConaill, *Synovial Joints: Their Structure and Mechanics.* London: Longmans, Green, 1961, pp. 159–164.

Hamilton, W. J., G. Simon, and S. G. Ian Hamilton, *Surface and Radiological Anatomy.* Baltimore, Md.: Williams and Wilkins, 1971, pp. 34–36.

Harris, Margaret L., "Flexibility: Review of the Literature," *American Physical Therapy*, 49 (June 1969), 591–601.

Holland, George J., "The Physiology of Flexibility: A Review of the Literature," *Kinesiology Review.* Washington, D.C.: American Association for Health, Physical Education and Recreation, 1968, pp. 49–62.

Larson, Robert L., "Physical Activity and the Growth and Development of Bone and Joint Structures," in *Physical Activity, Human Growth and Development*, ed. G. L. Rarick. New York: Academic Press, 1973, pp. 32–59.

Malina, R. M., "Exercise as an Influence upon Growth: Review and Critique of Current Concepts," *Clinical Pediatrics.*, 8 (January 1969), 16–26.

# Chapter 4

# The Muscular System

The bones and joints provide the framework of the body; however, the addition of functional muscular tissue is necessary for the body to reach its movement potential. Almost forty percent of the total body weight is comprised of skeletal muscles of which there are approximately 500. Most of these contain many thousands of multinucleated, single-cell, muscle fibers that are stimulated by means of a gigantic network of nervous tissue.

## CHARACTERISTICS

Each muscle fiber, many of which are grouped to comprise the muscle, has four distinct characteristics:

*Irritability.* Muscle has the capability of receiving and responding to various stimuli. It may respond to an action potential propagated along a nerve to an electrical stimulus applied directly to the surface of the muscle or to a sharp blow to the muscle or one of its tendons.

*Contractility.* When a stimulus is received, the muscle has the capability of shortening.

*Extensibility.* Muscle has the characteristic of being able to lengthen, either when it is in a passive or active state.

*Elasticity.* Whenever a muscle has been shortened or lengthened, it has the ability to return to its normal or resting length or shape.

## TYPES OF MUSCULAR TISSUE

There are three types of muscular tissue:

*Smooth muscle* is often termed unstriated or involuntary which are descriptive of its structure and in most cases its action. The muscle lines the hollow viscera and vessels of the body and is regulated by the autonomic nervous system. Involuntary action is apparent during erection of hair and the closing and opening of the iris of the eye.

*Cardiac muscle* is found in the heart and is similar to smooth muscle in that it is controlled by the autonomic nervous system. Although cardiac muscle consists of individual cells, the muscle reacts synchronously in that the cells contract and release at approximately the same time.

*Skeletal muscle* is so named because the vast majority of muscles within this group is attached to bone, although a limited number of skeletal muscles are attached to fascial sheaths, aponeuroses, integument, and cartilage. Skeletal muscle is referred to also as striated muscle, because, when observed under a microscope, it appears to be striated or striped. Finally, it is sometimes called voluntary muscle because it is generally under control of the will. Skeletal muscle is the type of muscle with which physical education teachers, coaches, and others are primarily concerned. The function of skeletal muscle, therefore, provides much of the focus of this book. There has been an attempt to avoid discussion of factors such as the microstructure of the muscle, the cellular response to excitation, and the energetics of muscular contraction, which are dealt with in the study of the physiology of muscular contraction. Suggested readings in this area are included at the end of the chapter for the interested student.

## MUSCLE STRUCTURE AND ATTACHMENTS

Observation of the movement of the human body in normal everyday activities such as walking, standing, and sitting and in precise movements such as typewriting, combing one's hair, or picking up an article with the fingers or toes makes one aware of the complexity of the neuromusculoskeletal system. This awareness is heightened to levels approaching disbelief when we observe athletes performing physical skills at high levels of proficiency. Although the nervous system provides the stimulus for muscle contraction, it is the contraction and relaxation of the muscles and the associated lever systems provided by the skeleton that permit the complex movements observed.

This ability to move depends largely on the form of muscles and the architecture of the skeletal system. Muscles vary in size, shape, and structure according to the functions they have to perform. Some are designed predominantly for *power*, others to provide a wide *range of motion*, many to permit *rapid movement*, and some are structured for the performance of delicate, *precise movements*.

**Structure.**  The muscle cell or fiber is multinucleated and resembles a long, straight cylinder. The fibers range from 0.01 mm to 0.1 mm in diameter and from a few centimeters to over 30 cm in length, often traversing the full length of the muscle. The nuclei (as many as several hundred in a single fiber)[1] are located in a cell membrane (plasma) called the sarcolemma. Each fiber can be stimulated by nervous impulses and has the ability to propagate an action potential and to contract.

The fiber is surrounded by a connective tissue sheath called the endomysium. In turn, fibers are grouped together into fasciculi that are enclosed by perimysium. In large numbers, these fasciculi are bound together by epimysium, which is tissue similar to fascia and sometimes binds with it. At the ends of the muscle, the fibrous epimysium and perimysium sheaths blend with the white fibers of the tendons with which the muscle is attached to the periosteum of the bones.

## Fast-Twitch and Slow-Twitch Fibers

Muscle fibers have, for simplicity's sake, been classified into two types: the first, when stimulated, responds with a rapid contraction and is called a *fast-twitch (or white) fiber*; the other contracts much more slowly and is referred to as a *slow-twitch (or red) fiber*. Although this classification simplifies discussion, it is not always appropriate; some red muscles have been found to possess fast-twitch capacities, but this is seen mainly in animals.

Fast-twitch (or white) fibers are high in glycolytic enzymes, possess a low oxidative (endurance) capacity, and have a high myofibrillar ATPase activity. That is, in strenuous power–speed work, the fast-twitch white fibers fatigue rather quickly because they depend upon a finite anaerobic (glycolytic) energy source, and such rapid mechanical activity depletes this source relatively quickly.

In contrast, slow-twitch (or red) fibers differ from fast-twitch ones in that they have low glycolytic, high oxidative capacities and low myofibrillar ATPase activity. Such fibers are thus highly dependent upon the body's capacity to deliver oxygen, and this to a great extent limits the energy production capacity of red muscle; however, it should be appreciated that the greater the oxygen transport capacity of an athlete, the greater the ability to perform sustained tonic activity. (A marathon runner is an obvious example.)

The distribution of fast- and slow-twitch muscle fibers varies considerably throughout the skeletal muscles of the body as well as from individual to individual. According to Saltin and his coworkers,[2] in a most comprehensive article concerning fiber types, 52 percent of the total fibers in the vastus lateralis

---

[1] P.-O. Astrand and K. Rodahl, *Textbook of Work Physiology*, 2nd ed. (New York: McGraw-Hill, 1977), p. 39.

[2] Bengt Saltin, Jan Henriksson, Else Hygaard, and Per Andersen, "Fiber Types and Metabolic Potentials of Skeletal Muscles in Sedentary Man and Endurance Runners," *Annals of the New York Academy of Sciences*, 301 (October 1977), 9.

for both sexes were slow-twitch fibers. They claimed that available evidence suggests that no difference exists between the fiber types of males and females; however, for both sexes, a wide variation of fiber composition exists among individuals. The vastus lateralis, rectus femoris, gastrocnemius, deltoid, and biceps muscles contain approximately 50 percent slow-twitch and 50 percent fast-twitch fibers, the soleus has 75–90 percent more slow-twitch than do other leg muscles, and the triceps have 60–80 percent more fast-twitch than do the other arm muscles.[3]

The high slow-twitch fiber composition of the soleus muscle suggests that the muscle should have the ability to perform sustained tonic activity. While standing in a relaxed position, the soleus has been found to show very marked activity and continued contraction.[4,5] This activity, therefore, is consistent with the fiber composition of the muscle.

The muscle fiber composition of athletes differs from that of sedentary persons. Sprinters' legs have a marked predominance of fast-twitch fibers; long distance runners have the opposite.[6] A similar finding of a greater percentage of slow-twitch fibers in endurance athletes over a sedentary group was reported by an independent group of Swedish researchers.[7] While these differences are probably due, in the first instance, to heredity, the performance capacities of athletes can be maximized by the type of training they undertake and by the way in which their muscles adapt as a result.

## Fiber Arrangement and Direction

The bundles of muscle fibers (or *fasciculi*) are grouped within a muscle so that they are parallel to each other; however, groups of fasciculi within a muscle may not always be parallel to each other (Fig. 16).

When all the fibers and groups of fibers within a muscle are parallel with the long axis of the muscle, as in the sartorius and rectus abdominis muscles, the muscle is said to be a *longitudinal muscle*. If the muscle fibers are in the form of a spindle (as is the case with the biceps brachii), the muscle is called a *fusiform*

---

[3] Saltin et al., "Fiber Types and Metabolic Potentials of Skeletal Muscles in Sedentary Man and Endurance Runners," p. 11.

[4] J. Joseph and A. Nightingale, "Electromyography of Muscles of Posture: Leg Muscles in Males," *Journal of Physiology*, 117 (August 1952), 490.

[5] J. Joseph, A. Nightingale, and P. L. Williams, "A Detailed Study of the Electric Potentials Recorded over Some Postural Muscles while Relaxed and Standing," *Journal of Physiology*, 127 (March 1955), 617–625.

[6] Saltin et al., "Fiber Types and Metabolic Potentials of Skeletal Muscles in Sedentary Man and Endurance Runners," p. 13.

[7] A. Thorstensson, L. Larson, P. Tesch, and J. Karlsson, "Muscle Strength and Fiber Composition in Athletes and Sedentary Men," *Medicine and Science in Sport*, 9 (1977), 26–30.

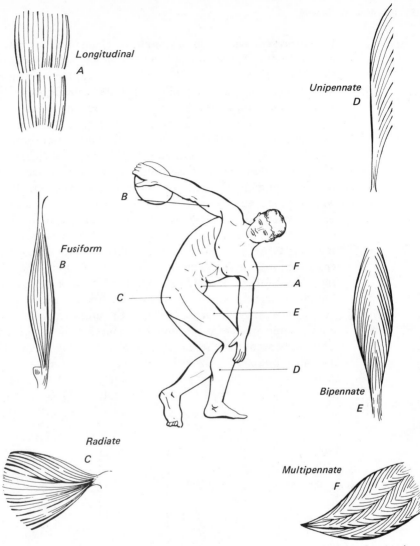

**Fig. 16.** *Fiber arrangement and direction: (a) rectus abdominus, (b) biceps brachii, (c) gluteus medius, (d) tibialis posterior, (e) rectus femoris, (f) deltoid.*

*muscle*. If the fibers fan out from a single attachment, the muscle is called a *radiate muscle*. The gluteus medius and minimus and the pectoralis major and minor are radiate muscles.

When the groups of fibers within a muscle are attached to the side of a tendon that runs the length of the muscle, a featherlike appearance is apparent. These muscles are called *penniform muscles* of which there are three types. *Unipennate muscles* have an origin from a large area of bone and run obliquely to a single tendon. The flexor pollicis longus and the tibialis posterior are examples of this type. *Bipennate muscles* have oblique fibers that are attached to both borders of a central tendon. The rectus femoris is an example. *Multipennate muscles*, such as the deltoid, appear as a combination of several bipennate muscles in that there are a number of tendons to which the fibers converge and attach.

The magnitude of the force that the muscle can produce is proportional to the product of the size and the number of muscle fibers within a muscle. Estimates of the amount of force that can be generated by a muscle range from about 30–60 Newtons (N) for each square centimeter ($cm^2$) of their cross-section (30–60 $N/cm^2$). To measure the so-called physiological cross-sectional area of muscle in the living human is difficult due to the variation in the direction of the fibers within muscle and the inaccessibility of muscle; however, this can be done on anatomical specimens, and, with a loss of a certain amount of accuracy, the data can be extrapolated to the living body. To determine the cross-sectional area of a muscle, it is necessary to observe the direction of the fibers and consider the total cross-sectional area of all the fibers in the calculations. The area perpendicular to the fibers must be considered. Figure 17 shows models representing a longitudinal muscle and a bipennate muscle. Although the circumference of the two models is exactly the same (14 cm), the cross-sectional area of the fibers is considerably greater for the bipennate muscle than for the longitudinal muscle. This can be seen from the perpendicular slices in Fig. 17 and the calculated areas, which are presented. The cross-sectional area of the fibers for the longitudinal muscle is 6 $cm^2$ while that of the bipennate is 16 $cm^2$. Assuming that the amount of force that the muscles could generate is 50 $N/cm^2$, the potential force

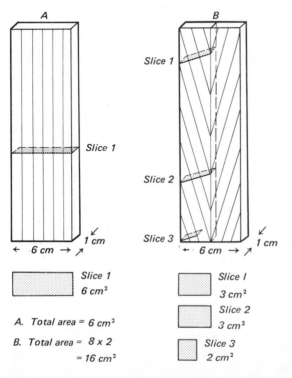

**Fig. 17.** *Cross-section area of muscle models: (a) longitudinal, (b) bipennate.*

production would be 300 N for model A and 800 N for model B. This bipennate muscle model, therefore, would be capable of producing about 250 percent more force than would the longitudinal muscle.

Penniform muscles, therefore, provide a greater number of fibers within the physiological cross-sectional area than do longitudinal muscles, and, assuming that the force capability of the muscle is constant and that the overall muscle length and the mechanical system are similar, the penniform muscle is the stronger muscle. Examples of such strong muscles are the tibialis posterior (unipennate), soleus (bipennate), and deltoid (multipennate).

It has been claimed that muscles with longitudinal fiber arrangements are designed for speed movements rather than for strength. This claim has been associated with the fact that muscle fibers may shorten as much as 50 percent of their resting length and that their fiber arrangement, often the full length of the muscle, permits a wide range of motion; however, most pennate muscle fibers are attached to their tendons at angles of 30° or less, and, therefore, their ability to shorten the whole muscle is no less than that of longitudinal muscle fibers. There is a tendency, however, for longitudinal, fusiform, and radiate muscles to have a greater muscle-to-tendon length ratio than penniform muscles. Two muscles of equal length (bone to bone) may differ considerably in the amount that they can shorten if their muscle-to-tendon length ratio is different (Fig. 18). This can affect the range of motion of a joint. It can be seen from the figure that a muscle with a long tendon, such as the gastrocnemius or flexor hallucis longus would be able to shorten less than a muscle like the pectoralis major or deltoid with the fibers from bone to bone.

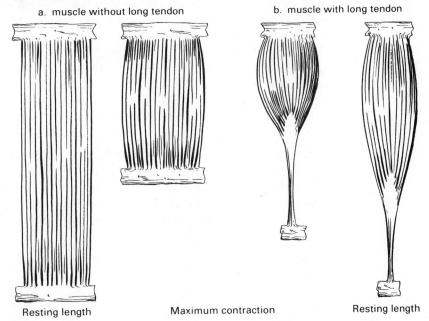

a. muscle without long tendon    b. muscle with long tendon

Resting length        Maximum contraction        Resting length

**Fig. 18.** *Influence of muscle-tendon length ratio on range of motion during maximum contraction. Contractile component (muscle) has shortened 50% of its resting length. Note the difference in range of motion.*

*Fasciae*.   Fasciae are membranous sheaths that enclose structures of the body. They are a part of the structural and functional system and may be classified into three categories: the superficial fascia, deep fascia, and subserous fascia. The first two types are of importance to physical education teachers and coaches. The subserous fascia, which lines the body cavities alongside the serous membrane, is not of major interest.

The *superficial fascia* is found immediately beneath the skin and covers the entire body. It consists of two layers that may be separated during dissection. Found within this fascia are blood vessels, lymphatics, nerves of the skin, fat deposits, and in specific areas, such as the face and neck, muscles that attach to the skin. In many areas of the body, the superficial fascia permits a considerable amount of movement over the deep fascia and deeper structures. This is very noticeable on the posterior surface of the hand where the skin and the superficial fascia may be lifted and moved back and forth especially when the wrist and fingers are extended.

The *deep fascia* consists of a series of dense, membranous sheaths that vary in shape, size, and strength depending upon their function. It is deep to the superficial fascia and supports the muscles and other structures in position so that they may effectively function in producing or restricting movement. Fascial sheaths between muscle groups, called *intramuscular septa*, aid in permitting independent muscle group activity. Some deep fascia is thick and very strong and, although covering a muscle, may also provide a means of attachment for muscle. An example of this is in the extensive fascial sheath of the iliotibial tract (the fascia lata) to which the tensor fasciae latae and the gluteus maximus muscles are attached.

The fascia lata is extremely strong. Gratz[8] reported that, weight for weight, it was almost as strong as soft steel. He found that the fascia had an average tensile strength of approximately 4800 N/cm$^2$. Because of its strength, highly elastic property, and abundancy in the body and its accessibility to the surgeon, it is commonly used for surgical work such as hernia and ligament repair.

*Tendons and Aponeuroses*.   A tendon is composed of bundles of white collagen fibers and serves to connect muscle fibers to their bony attachment. At the attachment, the fibers fan out into the periosteum of the bone. The shape of tendons varies according to their function. Some are cordlike as in the hamstring muscles that can be felt behind the bent knee. Others are broad, flat aponeuroses

[8]Charles Murray Gratz, "Tensile Strength and Elasticity Tests on Human Fascia Lata, *Journal of Bone and Joint Surgery*, 13 (1931), 338.

such as the aponeurosis in the abdominal area. Here, the external oblique muscles insert into a broad, thin, yet strong, aponeurosis that is attached to the linea alba.

***Origins and Insertions.*** The attachments of a muscle are called the *origin* and *insertion*. Traditionally, anatomists have described the origin of a muscle to be the attachment that is closer to the center of the body (proximal), whereas the insertion is the attachment further away (distal). This definition of terms is most apt, as the explanation often used by nonanatomists can cause confusion. Many have claimed that the attachment that is usually stationary during an action of the muscle is called the origin, whereas the attachment that moves is the insertion. This definition results in a change of the origin and insertion of the biceps brachii muscle when a person performs arm curls and then pull-ups. If the skill necessitated equal movement at the joints near both attachments, further confusion would arise.

## MUSCULAR CONTRACTION

***Motor Units.*** The motor unit is regarded as the functional unit of skeletal muscle. Muscles are composed of many thousands of motor units. The motor unit consists of a number of fibers, each innervated separately by a nerve branch from one motor neuron. The number of muscle fibers in a motor unit varies considerably within the body.

Each fiber within a motor unit contracts according to the *all-or-nothing* principle when its motor neuron provides an adequate stimulus. When the stimulus is strong enough to elicit a response, the resulting response will be maximal. A muscle such as the gastrocnemius (2,000 fibers per motor unit) is structured so that when a large number of motor units are activated, the powerful movement of plantar flexion will occur. In contrast, the muscles that produce movements of the eyeball, that is, movements that require great precision and little force, have few muscle fibers per motor unit. The number of fibers that are innervated by each motor neuron is related to the type of movement that the muscle performs. Table 4 presents examples of motor unit sizes for certain muscles of the body as reported by Feinstein and coworkers.[9]

Observation of a contracting muscle suggests that it is a smoothly working machine despite the barrage of up to 2,000 fibers firing at the same time. There are two reasons for the smooth contraction of muscle. First, there is usually a considerable number of motor units active during the same time period to

[9] Bertram Feinstein, Bengt Lindegord, Eberhard Numan, and Gunnar Wohlfart, "Morphological Studies of Motor Units in Normal Human Muscles," *Acta Anatomica* 23 (1955), 128.

produce movement, and they are being stimulated independently and are contracting and relaxing up to 50 times per second. Second, the muscle fibers are spread out within a muscle and intermingle with the fibers of other motor units.

**TABLE 4\*   Muscle Fibers and Motor Units in Human Muscles†**

| Muscle | Number of Muscle Fibers | Number of Motor Units | Mean Number of Fibers Per Motor Unit |
|---|---|---|---|
| Platysma | 27,100 | 1,100 | 25 |
| Brachioradialis | >129,200 | 330 | >410 |
| First lumbrical | 10,000 | 100 | 110 |
| Tibialis anterior | 250,000 | 450 | 600 |
| Gastrocnemius (median head) | 1,120,000 | 580 | 2,000 |

\*Bertram Feinstein, Bengt Lindegord, Eberhard Nyman, and Gunnar Wohlfart, "Morphological Studies of Motor Units in Normal Human Muscles," *Acta Anatomica*, 23, no. 2 (1955), 128.

†Numbers are rounded off for convenience.

***Types of muscular contraction.***   If an action potential traveling along a motor neuron to a muscle fiber is of sufficient strength to elicit a response, the muscle fiber will contract. *Contraction* is a term used to describe the tension developing response of a muscle to a stimulus. Muscle can only pull, and it does so toward the middle of the muscle. The tension at its attachments is usually of equal magnitude. There are two types of contractions; isometric and isotonic.

An *isometric contraction* occurs when tension within the muscle does not result in joint movement; the muscle tension is equal to the resistance. This is often referred to as static contraction and an example is illustrated in Fig. 19.

An *isotonic contraction* involves muscular activity and joint movement. When a muscle becomes active and the resulting tension causes the muscle to create movement by shortening, it is called *concentric contraction*. An example of concentric contraction is when the elbow flexors shorten and the elbow bends as in doing a curl with weights (Fig. 19). The controlled lowering of the weight is a function of *eccentric contraction* in that the elbow flexors, although active, permit the resistance (weight) to lengthen the muscle as it is lowered. It is possible to lower the weight by completely relaxing the elbow flexors and relying on gravity to do the work. Gravity acting on the mass of the forearm and hand is the force responsible for causing the extension. This would make the action very quick and may well result in injury to the elbow and/or shoulder joint and related musculature. One should consider a number of movements such as slow abduction and adduction of the arm. The muscle groups active during both these

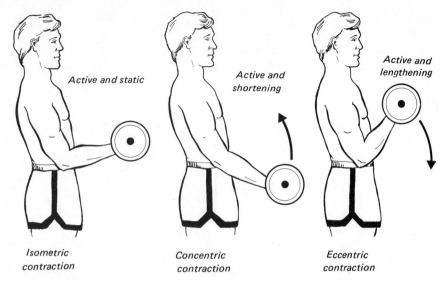

Active and static

Active and
shortening

Active and
lengthening

Isometric
contraction

Concentric
contraction

Eccentric
contraction

**Fig. 19.** *Types of muscle contraction.*

actions are the shoulder abductors. They contract concentrically during abduction and eccentrically during slow adduction. The eccentric contraction of the abductor muscle group provides a counterforce to slow the rate of motion caused by gravity. Thought should be given to the types of contraction necessary for muscle groups during the performance of more complex movements such as sit-ups, pull-ups, push-ups, and squats.

### Stimulus Response of Muscle

When a muscle fiber is stimulated by an impulse at or above threshold level, there will be a short period of no visible response or tension (latent period) followed by contraction and relaxation. The muscle fiber contracts according to the all-or-nothing law of muscle contraction. This response to a single impulse is called a muscle fiber twitch (Fig. 20), the duration and tension level of which varies from muscle to muscle as well as with temperature change. In the slow-twitch fibers of the soleus muscle, the fiber twitch may last as long as 100 msec, while, in the fast-twitch fibers of the gastrocnemius muscle, it may last only 30 msec.

Single-twitch contractions are uncommon in the intact muscle as muscles are usually activated by volleys of impulses. If a second stimulus arrives at a muscle when it is still in a tension-producing state, then the muscle will increase its tension further. This is called *summation* and is shown in Fig. 20. A barrage of high-frequency stimuli at or above threshold level will cause the tension to be held at a higher level than any one twitch would produce, until fatigue occurs. This condition is called tetanus.

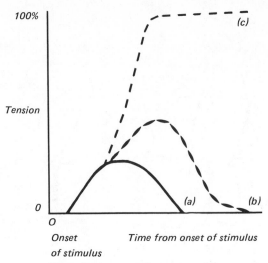

Fig. 20. *Summation from repeated stimuli: (a) muscle fiber twitch from a single stimulus, (b) response to a second stimulus, (c) summation from repeated stimuli.*

## Length-Tension Relationship

In a given muscle, there is a relationship between the length of the muscle at the start of contraction and the amount of tension that the muscle can produce.

If one were to measure the maximum tension that could be generated at the wrist during isometric elbow flexion through a range of elbow joint angles, one would obtain a graph similar to that presented in Fig. 21. It should be noted that when the elbow joint is extended or flexed to its maximum, there is little tension

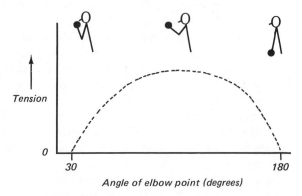

Fig. 21. *Relationship between muscle length and tension in intact skeletal muscle.*

produced by the elbow flexor muscles. The greatest tension is obtained when the elbow is bent to approximately 90°. Basically, this can be attributed to two factors: the mechanical advantage associated with the change in the angle of pull of the functioning muscles and the length of the muscle at the start of contraction. With respect to the latter, a muscle generally exerts maximum tension when its starting length is that found in the normal resting muscle.

It is important, therefore, that, during testing for muscle strength, the joint angle, and thereby the muscle lengths, should be predetermined or standardized.

### Force-Velocity Relationship

Muscle is capable of *producing* its maximum force when it is contracted isometrically although greater forces can be *sustained* in eccentric contraction. Once the contraction becomes isotonic (concentric), the amount of force produced decreases. The force decreases exponentially as shown in Fig. 22. The faster the movement (contraction), the smaller the force generated.

This relationship can be illustrated by a simple example. A person performing elbow flexion (curl) while holding a large weight (500 N) will raise the load slowly; however, if a smaller weight were used, the elbow flexion could be performed with greater velocity. This interesting phenomenon is yet to be fully explained physiologically; however, it is interesting that the shape of the curve can be associated with the rate of energy production of the cell.[10]

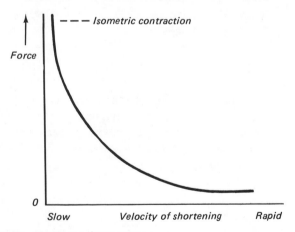

**Fig. 22.** *Force-velocity curve.*

[10] A. V. Hill, *First and Last Experiments in Muscle Mechanics* (New York: Cambridge University Press, 1970).

The contraction of a muscle may effect a joint or joints in one or more ways. The resulting action of the muscle on the joint(s) classifies the muscle into a group. Consider, for example, the biceps brachii that crosses both the shoulder and elbow joint. This muscle is regarded as an *elbow flexor* in that it is very active during resisted flexion of the elbow and as a *supinator* because of its contribution to forearm supination. In addition, the biceps brachii assists a number of movements at the shoulder joint and is even found to be active during rapid elbow extension; however, it cannot be regarded as an elbow extensor.

Whether simple or complex, movements are the result of the coordinated and controlled action of a number of muscles. This group action may well require the cooperation of flexors, extensors, abductors, adductors, and rotators, for it is the coordination of prime movers, antagonists, and synergists that permits efficient movement.

Muscles are further named to indicate their function in the performance of specific movements.

## Movers

There are two types of movers: prime movers and assistant movers. Muscles classified in these groups are responsible for performing the specified movement. A prime mover is a major contributor to the movement. If it does not provide a major contribution to the movement, it is called an assistant mover. The biceps brachii is a prime mover in elbow flexion; it is debatable whether it is a prime mover or an assistant mover in supination; and it is an assistant mover in a number of shoulder movements.

## Antagonists

An antagonist is a muscle whose action on a joint is opposite to that of the prime mover. The triceps is the antagonist to the biceps brachii during elbow flexion. Generally antagonists are relaxed to permit efficient movement; however, with rapid movement they are sometimes active to control the movement and thereby prevent injury.

## Stabilizers

These muscles contribute to movement by stabilizing the attachment(s) of one end of the prime mover(s) so that the contracting muscle(s) can work effectively. An example is the stabilizing of the shoulder girdle, in particular the scapula, to enable one to pull an object toward the body. Without such stabilization, the force exerted by the arms would tend to displace the shoulder girdle as well as the object being pulled.

### Synergists

It should be remembered that when a muscle contracts it will cause motion at all the joints it crosses if they are free to move. Muscles classified as synergists increase the efficiency of the prime movers by preventing inhibiting movements while the movement is being performed. An example of synergistic action is found in tightly clenching the fist. The wrist is held in slight hyperextension to achieve maximum power by the finger flexors. To experience this one should squeeze a finger of one's other hand while moving the wrist slowly from extreme hyperextension to full flexion—the power of the grip varies considerably through the range. The extensors of the wrist act as synergists by holding the wrist in the optimum position for the production of maximum grip strength.

## THE STUDY OF THE ACTION OF MUSCLE

The actions of muscles may be studied in a number of ways. One simple method is to observe the attachments of a muscle on a skeleton or a cadaver and to deduce what movement(s) will occur when the muscle shortens. This method is not very accurate but may act as a supplement to other techniques.

Another approach is by electrical stimulation of the muscle. If the stimulus is sufficient, the contracting muscle will cause motion at the joint(s) that can be observed. This method also has its inaccuracies as muscles seldom act independently.

The most practical and simplest method is by observation and palpation of the muscle during activity. Use of this method to determine which muscles are active during the performance of specific exercises should be encouraged.

Although these methods can be of value, one must be aware that they are subject to error on the part of the observer. Fortunately, the use of sensitive electronic equipment permits one to study, with accuracy, the activity within muscles while they are contracting. The technique used is called *electromyography*, which is the recording of the electric potential of muscle.

The muscle is a source of electrical activity and is surrounded by bodily fluids that are good conductors of electricity, thereby enhancing the recording of the activity. Needle electrodes and/or embedded fine-wire electrodes inserted into the belly of muscles can assist in the production of reliable information regarding the activity of muscles. Surface electrodes for the study of large superficial muscles can be used if the outer layer of skin (dead cells), which has a high resistance to electrical flow, is removed by thoroughly cleaning the skin.

The electric potential received by the electrodes is amplified by a special electromyographic amplifier and may be recorded in many different ways. The data can be studied in "raw" or integrated form and/or processed directly by computer.

The data collected by use of electromyography can add objectivity to the understanding of single muscle activity during motion and the interplay or coordination of various muscles during complex motor activities.

## EXERCISES

1. Discuss the various arrangements of muscle fibers within muscles and consider their advantages and disadvantages.
2. Select a number of superficial muscles on your body and, by palpation, observation, and rationalization, attempt to classify them according to their fiber arrangement and direction. Discuss your conclusions with your fellow students. Check both the fiber arrangement and direction of the selected muscles on dissected specimens in an anatomy laboratory or museum or in anatomy texts or atlases in the library.
3. Attempt to perform a curl (flexion at the elbow joint) as fast as possible while holding a heavy weight. Repeat the action with half the weight and then without a weight. Which action was (a) fastest and (b) slowest? How can you rationalize the outcome?

## RECOMMENDED READINGS

Astrand, P. O. and K. Rodahl, *Textbook of Work Physiology*. New York: McGraw-Hill, 1977.

Basmajian, John V., *Muscles Alive: Their Functions Revealed by Electromyography*. Baltimore, Md.: Williams and Wilkins, 1978.

Cavanagh, Peter R., "Electromyography: Its Use and Misuse in Physical Education," *Journal of Health Physical Education Recreation*, 45 (May 1974), 61–64.

Feinstein, Bertram, Bengt Lindegord, Eberhard Nyman, and Gunnar Wohlfart, "Morphological Studies of Motor Units in Normal Human Muscles," *Acta Anatomica*, 23 (1955), 127–142.

Hill, A. V., *First and Last Experiments in Muscle Mechanics*. New York: Cambridge University Press, 1970.

Hubbard, A. W., "Homokinetics: Muscular Function in Human Movement," in *Science and Medicine of Exercise and Sports*, ed. W. R. Johnson. New York: Harper, 1960, Chap. 1.

O'Connell, A. L., and E. B. Gardner, "The Use of Electromyography in Kinesiological Research," *Research Quarterly*, 34 (May 1963), 166–184.

Winter, D. A., G. Rau, and R. Kadefors, "Units, Terms and Standards in the Reporting of Electromyographical Research," *Proceedings of the 4th Congress of the International Society of Electrophysiological Kinesiology*, Boston, August 1979, pp. 100–101.

# Chapter 5

# The Neck and Trunk

The bones of the vertebral column, the thorax, the pelvic girdle, and the shoulder girdle will be discussed in this chapter together with their joint structure and movement capabilities. In addition, the associated muscle groups and surface anatomy will be considered. Muscles will be discussed within muscle groups such as spinal flexors and neck extensors; however, it must be remembered that, during physical activity, muscles seldom act independently. These discussions point out major surface landmarks that should be observed on oneself or preferably on a partner. The muscles being studied should be observed during contraction that results in movement as well as during isometric contraction.

Reference should also be made to the tables in Appendix A to review the location, origin, insertion, and action of muscles in the respective muscle groups.

## THE VERTEBRAL COLUMN

The vertebral column is an important supporting structure of the human body. The anatomic structure and functional capability of the spinal column and, in particular, its potential to aid in the performance of complex motion, and its limitations and susceptibility to injury, should be clearly understood.

The vertebral column normally consists of thirty-three vertebrae—seven cervical, twelve thoracic, five lumbar, five sacral, and four coccygeal. The first twenty-four vertebral bodies are separated by fibrocartilaginous intervertebral discs, whereas those of the sacrum and coccyx are fused together. Variations in vertebrae, usually congenital, occur occasionally. These include differences in the number, the position, and the shape of the vertebrae.

The vertebral column of the articulated skeleton has four curves (Fig. 23)—the thoracic and sacral curves are concave forward; the cervical and lumbar curves are convex forward. The first two curves mentioned are *primary* because they are present at birth and do not alter their shape. On the other hand, *secondary curves* develop after birth: the lumbar curve develops when the child begins to sit up.

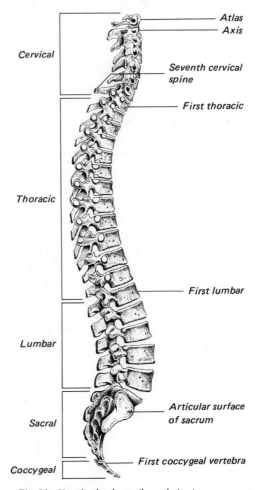

**Fig. 23.** *Vertebral column (lateral view).*

## The Vertebrae

Generally it can be stated that the vertebrae are constructed similarly. Figure 24 is an illustration of the sixth thoracic vertebra. It has a *body* from which two bars of bone, called the *pedicles of the vertebral arch*, project backward. Below each pedicle, which is slightly notched, is an intervertebral foramen that permits

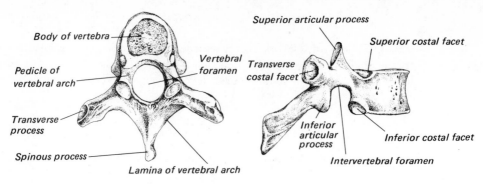

Fig. 24. *The sixth thoracic vertebra.*

passage of spinal nerves. The *laminae of the vertebral arch* pass from the pedicles medially and posteriorly to unite at the *spinous process*, which projects posteriorly. The *transverse processes* project laterally from the junctions of the pedicles and the laminae. The *superior* and *inferior articular processes* also project from this junction for articulation with their respective vertebra above or below. The *vertebral foramen* is bounded by the vertebral body, pedicles, and laminae and provides passage for the spinal cord.

**The Cervical Vertebrae.** The general form of the cervical vertebrae (Fig. 25) is like that of the typical vertebra shown in Fig. 24. However, the following characteristics make the third, fourth, and fifth cervical vertebrae distinctive: the body is small and the lateral margins of its upper surface are turned upward, whereas the lower surface is concavoconvex; each transverse process contains a foramen; the spinous process is bifid; and the vertebral foramen is triangular. The first (*atlas*) and second (*axis*) vertebrae (Fig. 26) differ further from the typical vertebrae.

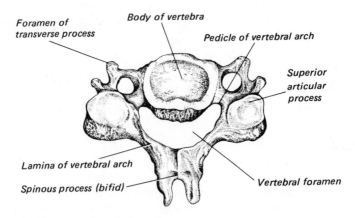

Fig. 25. *A typical cervical vertebra.*

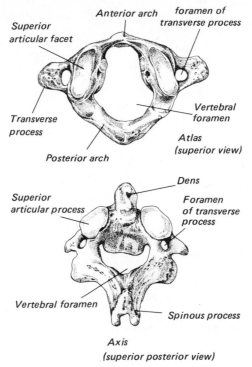

**Fig. 26.** *The atlas and the axis.*

The atlas is a thick bony ring formed by an anterior and posterior arch that joins two bulky masses containing two superior and inferior articular facets for articulation with the articular bone above and the axis below. The transverse processes are fairly long in relation to those of the axis below. The special features of the axis are the stout *dens* lying between the anterior arch of the atlas in the articulated skeleton and the pair of large superior facets for articulation with the atlas.

**The Thoracic Vertebrae.**  Differences in these vertebrae are mainly in *costal* (rib) attachment. Note the half facets on the body of the vertebra and a whole facet on the transverse processes in Fig. 24. The first thoracic vertebra has a whole superiolateral facet for the first rib. The tenth, eleventh, and twelfth thoracic vertebrae have only one facet on each side of the body. A final difference is that the eleventh and twelfth vertebrae have no facets on their transverse processes.

**The Lumbar Vertebrae.**  Lumbar vertebrae (Fig. 27) can be differentiated from other vertebrae in that they do not have facets for ribs, as occurs in the thoracic region, and by the absence of transverse foramina. In addition, they are larger,

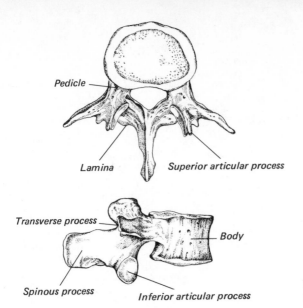

Pedicle

Lamina        Superior articular process

Transverse process

Body

Spinous process

Inferior articular process

**Fig. 27.** *Superior and lateral view of third lumbar vertebra.*

have a large kidney-shaped body, short pedicles, thick laminae, long thin transverse processes, a triangular vertebral foramen, and a horizontal, almost quadrangular-shaped spinous process.

## Joints of the Spine

*Occipito-atlantal.*   The articulation of the atlas with the occipital bone is a double condyloid joint that permits flexion and extension as in nodding the head and, in addition, a very slight lateral bending. The range of motion in the sagittal plane of about 10° flexion and 25° hyperextension[1] is fairly extensive in comparison with other joints of the cervical or thoracic vertebrae.

*Atlanto-axial.*   This is undoubtedly the most mobile of the vertebral joints. Rotation occurs about the odontoid process and involves the axis, atlas, and the occiput. The pivot (odontoid process) is enclosed in the anterior arch of the atlas and is supported by the transverse ligament. About 45° of rotatory motion is possible at this joint; this is about half of the total rotation possible by the cervical vertebrae.[2] About 30° of motion in the sagittal plane is possible at this joint (10° flexion and 5° hyperextension).[3]

*Cervical Spine.*   The joints of the cervical region, below the second vertebra, permit movement in the three planes: sagittal (flexion–extension), transverse (rotation), and frontal (lateral bending).

[1] J. William Fielding, "Cineroentgenography of the Normal Cervical Spine," *Journal of Bone and Joint Surgery*, 39A (December 1957), 1282.

[2] Fielding, "Cineroentgenography of the Normal Cervical Spine," p. 1284.

[3] Fielding, "Cineroentgenography of the Normal Cervical Spine, p. 1286.

***Thoracic Spine.***    The bony architecture, facet positions, and close proximity of spinous processes of these vertebrae permit a fair range of rotation (40° to each side).[4] They also allow a little flexion but limited hyperextension and a slight lateral bending that, due to the primary curve of the column, is accompanied by rotation.

***Lumbar Spine.***    The positions of the articulating facets in the lumbar region permit motion in the sagittal and frontal planes and little to no movement in the transverse plane. The range of motion within the lumbar vertebrae increases downward and varies considerably from one individual to another. In a fully flexed spine the lumbar spine may assume a concave position forward in one individual and a straight position in another; however, many persons, though they may reduce the degree of lordosis, are unable to even straighten their lumbar spine let alone flex it. There is also a wide range in hyperextension of the spine. The young gymnast in Fig. 28 shows an alarming degree of hyperextension, which, although useful in the performance of competitive gymnastics, may well be hazardous to the health of the performer.

**Fig. 28.** *Gymnast displaying considerable hyperextension of the lumbar vertebrae (Photo courtesy Andy Roberts).*

[4]Arthur Steindler, *Kinesiology of the Human Body Under Normal and Pathological Conditions* (Springfield, Ill.: Charles C Thomas, 1955), p. 159.

Although the twenty-four vertebrae of the spine do not provide a large range of motion individually, the total range of motion is considerable in the healthy, physically fit individual. Steindler,[5] for example, has reported that the excursion field of the entire spine in the living is

Sagittal plane 245°

Frontal plane 182°

Transverse plane 210°

The intervertebral discs are of prime importance in the movement of the spine as the other joints permit gliding and rotary motion only if the discs provide the elasticity to accommodate movement. The shape of the discs alters to permit a certain amount of twisting and tilting of the vertebrae.

The intervertebral disc consists of an outer fibrous ring that unites the vertebral body above with that below and is called the *annulus fibrosus*. The fibers are arranged in a diverse pattern so that they permit a fair amount of expansion and yet provide considerable strength. Within the annulus fibrosus is a gellike mass called the *nucleus pulposus* that is moved under compression, tilting, and twisting to alter the shape of the intervertebral disc, thereby providing movement. The nucleus pulposus complies with hydrostatic laws (applicable to fluid under pressure) in that it is incompressible but subject to deformity. The force is transmitted by the nucleus pulposus in all directions and affects the cartilage end plates as well as the annulus fibrosus, which represent the limiting structures of the nucleus.

The disc is very strong and can resist considerable vertical loading forces. It is usual that the vertebra fails and not the intervertebral disc.[6] It has been estimated that the normal disc can sustain a load of up to 6000 N before damage occurs to the vertebral end plate.[7]

With age, the disc experiences a decrease in water content and becomes progressively more fibrous, thereby decreasing the intervertebral space and limiting the range of motion. The narrowing of the disc progresses until the cartilage end plates of vertebral bodies lie almost in opposition.

The movements of the spinal column include flexion and extension, lateral bending, and rotation.

---

[5]Steindler, *Kinesiology of the Human Body*, p. 161.

[6]Olof Perey, "Fractures of the Vertebral End-Plate in the Lumbar Spine: An Experimental Biomechanical Investigation," *Acta Orthopaedica Scandinaviac*, 27 (1958), 237.

[7]Sven Carlsöö, *How Man Moves: Kinesiological Methods and Studies* (London: Heinemann, 1972), p. 74.

*Flexion* of the spine is easily attained by bending at the hips and pulling the top of the head in toward the straight knees; however, it can be achieved in the upright or seated positions by lowering the chin to the chest and shortening the rectus abdominis as much as possible. Spinal flexion does not include hip flexion and should not be measured by toe touching.

*Extension* is the movement from a flexed position to the original extended spinal column as in the anatomical position. Bending of the spine in a backward direction, as in arching the back, is called *hyperextension*.

*Lateral bending* is the bending of the spine to one side or the other. The direction to which the spine bends is included in the description of the movement—lateral bending to the left. Lateral bending of a curved rodlike structure such as the spine is usually accompanied by some rotation; however, in the living it is seldom observable except in cases of scoliosis.

*Rotation* of the spine is a twisting movement in the transverse plane around a longitudinal axis such as turning the head and shoulders to look behind. As with lateral bending, the direction is included in the description.

## Muscle Groups and Surface Anatomy

**The Neck.**   *The posterior aspect of the neck* has muscles that originate on the trunk and insert onto the cervical vertebrae or the posterior surface of the base of the skull. The muscles form the *neck extensor group* and part of the *neck rotators*.

The following landmarks can be located by simple palpation: *occipital protuberance*, which is at the base of the skull and can be located by tracing the *nape ligament* upward from the thoracic vertebrae while the neck is flexed; *seventh cervical vertebra*, which is the first prominent vertebra to be felt at the base of the neck; and the *semispinalis capitis*, which is a ridge of muscle on either side of the midline of the neck—this deep muscle of the neck is covered by the thin upper part of the trapezius muscle. The semispinalis capitis is the prime mover for extension and hyperextension of the neck. When only one side contracts, it assists in lateral bending of the neck.

*The lateral and anterior aspect of the neck* has a prominent landmark called the *posterior triangle*. The apex is above and just behind the mastoid process (posterior to the ear) while the base is formed by the middle part of the clavicle. The anterior border is formed by the *sternocleidomastoid* muscle (Fig. 29) and the posterior border by part of the lateral margin of the *trapezius* muscle. The sternocleidomastoid muscle is prominent when the jaw is held and rotation of the neck is attempted. Note that the contracting muscle is on the side opposite to the direction of the rotation. Both the fleshy clavicular head and the tendinous sternal head of the sternocleidomastoid are visible in Fig. 30. It is a prime mover for flexion of the neck and singly is a lateral bender and rotator of the neck. The

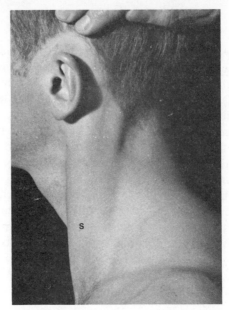

**Fig. 29.** *The sternocleidomastoid muscle (S).*

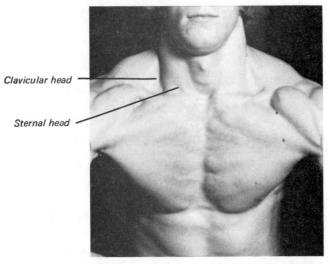

Clavicular head

Sternal head

**Fig. 30.** *Clavicular and sternal heads of the sternocleidomastoid muscle.*

*platysma* muscle, which is subcutaneous, consists of strands of muscle fibers and covers the area of the chin to the chest. By forcibly contracting the muscles around the mouth and protruding the chin, this muscle can be observed on many persons.

**The Trunk**   The *posterior aspect of the trunk* consists of a large number of pairs of muscles. These muscles are grouped according to their function into one or more muscle groups—*trunk extensors, trunk rotators,* and *lateral flexors*—of the trunk. In addition, a number of muscles located on the trunk act upon the shoulder girdle or cross the shoulder joint or hip joint and are involved in movement of the limbs. These muscles will be discussed individually.

The prominent surface features (Fig. 31) are the full length of the *vertebral column* (note the spinous processes during trunk flexion on the living) and the *erector spinae* (sacrospinalis muscle).

The erector spinae is the prime mover for extension and hyperextension of the trunk. It also assists in bending the spine laterally and rotating the vertebral column. This muscle consists of a large number of groups of fibers situated along both sides of the spine. Although the muscle is active during forward bending as in toe touching, it is more active while the trunk is being extended back to the upright position. During complete trunk flexion, the erector spinae becomes inactive (the position being maintained by the spinal ligaments). To prevent injury to the lumbar region, it is wise, during the lifting of heavy loads, to keep the spine extended and as close to the upright position as possible. The load should be lifted by extension at the hips and knees.

**Fig. 31.** *The vertebral column and the erector spinae muscle (ES).*

*The anterior aspect of the trunk* has paired muscles that may be grouped as follows: *trunk flexors, rotators,* and *lateral benders* as well as the muscles involved in limb movement and those responsible for respiration. Respiratory muscles will be discussed shortly. Refer to the tables in Appendix A for the location, origin, insertion, and action of the muscles of the abdominal wall.

The following surface landmarks and muscles should be noted in Fig. 32 and observed firsthand: *clavicle, suprasternal fossa, xiphoid process, inferior margin of ribs, linea alba, rectus abdominis* muscle, *external oblique* muscle, *linea semi lunaris, anterior superior iliac spine, iliac crest,* and *inguinal fold.*

The paired muscles involved in trunk flexion include the rectus abdominis and the external obliques; both lateral bending and trunk rotation are assisted by the internal and external obliques.

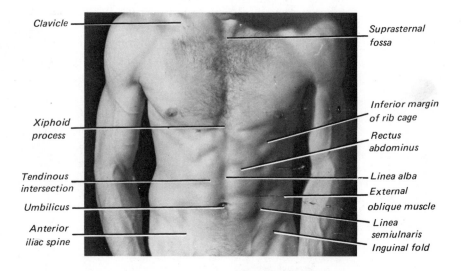

**Fig. 32.** *Surface landmarks on the anterior aspect of the trunk.*

The *rectus abdominus* is a straight, flat muscle found on both sides of the midline of the abdomen. The two halves of the muscle are separated by the *linea alba,* a tendinous cord extending from the xiphoid process to the pubic symphysis. The lateral borders of the muscle form the *linea semi lunaris,* which are easily noticeable in a lean, well-muscled individual. There are usually three horizontal tendinous creases that give the muscle a corrugated appearance. This muscle is a prime mover in trunk flexion (flexion of the spinal column) and may be strengthened by performing various types of exercises in which the body is raised from the supine position to a seated position ("sit-up" exercises). Exercises can be prescribed to provide minimum stress on the abdominal muscles through to very strenuous abdominal exercises; however, care should be taken, in performing abdominal exercises, to avoid the sharp forward pelvic tilt that is accompanied by hyperextension of the lumbar vertebrae during these movements. This action, which usually initiates the sit-up exercise, is caused by the strong psoas muscles that pull on the lumbar vertebrae. To minimize this hyperextension, sit-up exercises should be performed with the knees well bent as shown in Fig. 33 and actively utilize hip extensors to control pelvic tilt. A more strenuous abdominal exercise in which full stress is placed on the abdominals (no activity is necessary by the psoas) is illustrated in Fig. 34. This exercise should be performed by athletes who possess fairly strong abdominal muscles and wish to further increase their strength. The position shown is held statically for about five seconds and repeated a number of times.

**Fig. 33.** *Sit-up on inclined bench with knees bent.*

**Fig. 34.** *Hip lift: a strenuous abdominal exercise.*

The *external oblique* muscle, which is located on the lateral aspect of the trunk above the iliac crest, becomes an aponeurotic sheet anteriorly and fuses with its opposite at the linea alba. The inferior border is a free edge called the *inguinal ligament*. The muscle assists trunk flexion and singly aids trunk rotation and lateral bending. Rotation of the trunk during sit-ups increases the stress placed on this muscle.

The *internal oblique*, deep to the external oblique, has fibers that run medially and upward at right angles to the external oblique, on the same side of the body—they, too, form an aponeurosis. They assist in trunk flexion and lateral bending when acting singly.

## THE THORAX

The thorax is a cagelike structure composed of the sternum and ribs that attach posteriorly to the thoracic vertebrae (Fig. 35).

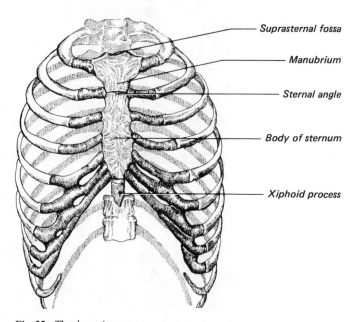

Suprasternal fossa

Manubrium

Sternal angle

Body of sternum

Xiphoid process

**Fig. 35.** *The thoracic cage.*

### The Sternum

Commonly referred to as the "breast bone," the sternum consists of the *manubrium*, which is superior, the body of the sternum below it, and the *xiphoid process*, which is the inferior protuberance. The *sternal angle* is found at the junction of the manubrium with the body. The *suprasternal fossa* or the jugular notch is on the superior border of the manubrium. The lateral borders of the sternum provide articulation with the clavicle and the costal cartilages.

### The Ribs

There are twelve pairs of ribs. The upper seven are *true ribs* as they are attached by their costal cartilages to the sternum. The lower five are *false* ribs and have no direct articulation with the sternum. Although they vary in shape and size, each of the ribs has an anterior and posterior extremity and a long, thin, curved shaft.

### Joints of the Thorax

The joints of the thorax include the joints of the twelve ribs with the twelve thoracic vertebrae and the anterior attachments of the ribs with the sternum and costal cartilages. The true ribs articulate with the sternum and are arthrodial joints. The costal cartilages of the upper three false ribs attach to the cartilage immediately above. Only the inferior two ribs, sometimes called "floating ribs," do not have double attachments; they articulate with the thoracic vertebrae only. The *costovertebral* attachments are supported by a number of ligaments, are strong, and being arthrodial, permit a fair range of movement.

### Movement of the Thorax

Movement of the thoracic cage to enable one to inspire and expire air is essential for survival. The process of normal respiration involves movement of the ribs and the sternum. Observation of a partner's or one's own breathing patterns will enhance the understanding of the mechanics of the respiration process.

*Inspiration* is the act of breathing in and necessitates the expansion of the rib cage so that air is forced into the lungs. This expansion is achieved in three directions:

1. Contraction of the external *intercostal muscles* causes elevation of the ribs, which causes the sternum to move forward as well as upward.
2. The raising of the rib results, also, in a widening of the thorax laterally due to their bony architecture.
3. The descent of the dome of the *diaphragm* pushes down on the abdominal organs and flattens the floor of the thoracic cage; in so doing, it increases the vertical capacity of the thorax.

The results of these actions is a decrease in the intrapleural pressure causing air to be drawn into the lungs.

*Expiration* is due mainly to the elastic recoil of the lungs and the relaxation of the external intercostal muscles and the diaphragm. It is aided by contraction of the *internal intercostal* and the abdominal muscles, in particular the *transversus abdominis*, which is the deepest and thinnest of the abdominal muscles.

## THE PELVIC GIRDLE

Anteriorly and laterally the pelvis consists of two hip bones each formed by three separate bones: the ilium, the pubis, and the ischium. These bones are separate during childhood and become fused during the early teens (Fig. 36). On the lateral surface is a cuplike structure, the *acetabulum*, for articulation with the head of the femur. Each of the three bones contributes to it. Below the acetabulum is the *obturator foramen*, bounded by the ischium and the pubis. The flat bone superior to the acetabulum is the ilium. The pubis is inferior and anterior to the acetabulum and joins with the pubis of the opposite side at the *pubic symphysis*. The ischium is posterior and inferior to the pubis.

Sexual differences in the bony pelvis appear during the rapid growth period of puberty. In the female, the bones are usually lighter, the bony prominences for muscular attachment less defined, and adaptations for childbearing are readily noticeable when a comparison is made with the male pelvic girdle.

### The Ilium

The *iliac crest* begins at the *anterior superior iliac spine*, which is easily felt, and ends at the *posterior superior iliac spine*, which can also be found on oneself or a partner. It is located under a small dimple above the buttock 3–5 cm lateral to the spine. The posterior medial surface is rough due to its articulation with the sacrum. The *greater sciatic notch* is inferior to this region. The interior surface of the ilium is called the *iliac fossa*.

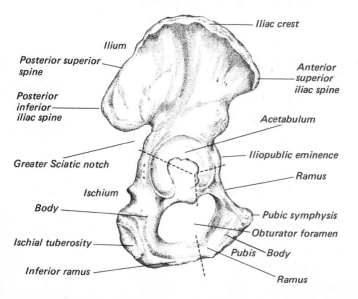

**Fig. 36.** *Pelvic bones (right side).*

### The Pubis

The pubis, which is the smallest of the three bones, fuses with the ilium at the *iliopubic eminence*. The bone consists of a *body* and two *rami*. The superior ramus extends up to the acetabulum, whereas the inferior ramus is continuous with the inferior ramus of the ischium. On the body of the pubis where the two rami join is an articulating surface for union with the other pubis. This articulation consists of a disc of fibrocartilage and is called the *pubic symphysis*.

### The Ischium

The upper part of the ischium forms the lower part of the acetabulum; the lower part forms the *ischial tuberosity*, which supports the trunk while in the sitting position. Both the body of the ischium and the *inferior ramus* of the ischium provide the inferior border of the obturator foramen.

The sacrum and coccyx, which are part of the axial skeleton, form an integral part of the structure and permit it to be called a girdle. The tremendous strength of the pelvic girdle is due to the fusion of the three pelvic bones with each other on both sides and the strength of the articulation of the sacrum to the pelvic bones at the strong sacroiliac joint; furthermore, the anterior pubic symphysis joint is tightly bound.

### Joints of the Pelvic Girdle

The two pairs of three pubic bones become firmly articulated by late puberty.

The pubic symphysis, an amphiarthrodial joint, experiences slight separation due to hormonal influences during pregnancy as does the sacroiliac joint; otherwise, the pelvic girdle, in itself, becomes a rather immobile unit.

### Movements of the Pelvic Girdle

Prior to discussion of the movement of the pelvic girdle it is necessary to establish the normal position of the pelvis in the anatomical position. A plane touching both of the anterior superior spines and the upper edge of the pubic symphysis is a frontal plane that is parallel to the midfrontal plane of the body. A second plane that is necessary to establish the orientation of the pelvis passes through the two anterior superior spines and the two posterior superior spines—this plane is almost horizontal and represents the transverse plane (Fig. 37).[8] All movements of the pelvic girdle occur as a result of motion at one or both hip joints and/or the lumbosacral junction.

[8]Steindler, *Kinesiology of the Human Body*, p. 181.

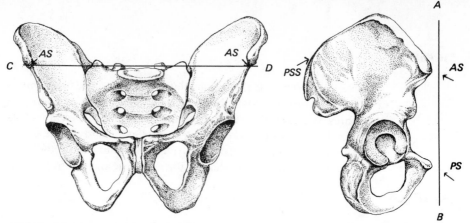

**Fig. 37.** *Orientation of the pelvic girdle: 1. lateral view, AB = vertical plane through pubic symphysis (PS) and anterior superior spine (AS), 2. anterior view, CD = transverse plane through both anterior superior spines (AS) and the posterior superior iliac spines (PSS).*

*Forward rotation or tilt* occurs when the anterior superior spines are moved or displaced forward and downward. This action is usually accompanied by an increased lordotic curve and therefore is common in people who have lumbar lordosis and/or weak, slack abdominal musculature.

*Backward rotation or tilt* is the opposite of forward pelvic tilt in that the anterior superior spines are behind the frontal plane that passes through the pubic symphysis. This position is associated with a flat (straight) lumbar spine and is achieved by simultaneous isometric contraction of the gluteal and the abdominal muscles.

*Lateral tilt* is the raising or lowering of only one side of the pelvis. This action occurs when walking and running and when standing with one leg supporting the mass of the body while the other is slightly bent or placed forward as a common form of stance.

*Rotation* is the twisting or turning of the pelvic girdle in the transverse plane around a longitudinal axis.

### Muscle Groups and Surface Anatomy

Muscle groups are not generally assigned to the movement of the pelvic girdle per se due to the continuation of the spinal column (sacrum and coccyx) into the pelvic girdle. The trunk muscles crossing the lumbar–sacral junction normally affect the rotation or tilting of the hip while acting on the spinal column. Shortening of the erector spinae muscle in the lumbar region causes extension or hyperextension of the lumbar vertebrae and forward rotation or tilt of the pelvis. Backward rotation or tilt of the pelvis is achieved by contraction of the abdominal musculature and relaxation of the erector spinae. While standing erect, this position can be further enhanced by a strong isometric contraction of the gluteus maximus. This exercise should be performed frequently by people who have lumbar lordosis (excessive curvature of the lumbar vertebrae).

# THE SHOULDER GIRDLE

The shoulder girdle consists of two clavicles and two scapulae. It is not a complete girdle—that is, it does not form a complete ring—as the scapulae do not articulate posteriorly. Its only attachment to the axial skeleton is at the sternum. The shoulder girdle has an additional skeletal attachment, laterally on both sides, with the head of the humerus of the arm. It has to support these limbs and this increases its insecurity further.

## The Scapula

The scapula is a flat bone that is roughly triangular in shape (Fig. 38). The triangular bone has a *medial*, *lateral*, and a *superior* border. The three angles are *superior*, *lateral*, and *inferior*. Its *costal surface*, that closer to the rib cage, is slightly concave to correspond with the shape of the rib cage. The *dorsal surface* has a prominent ridge—the *spine of the scapula*—that extends laterally and ends in the *acromion process*, the point at which the clavicle articulates. Inferior to the acromion is a shallow concave articular surface, the *glenoid fossa*, for articulation with the head of the humerus. The *coracoid* (or beaklike) *process* projects forward under the clavicle and toward the head of the humerus from the superior border just medial to the glenoid fossa.

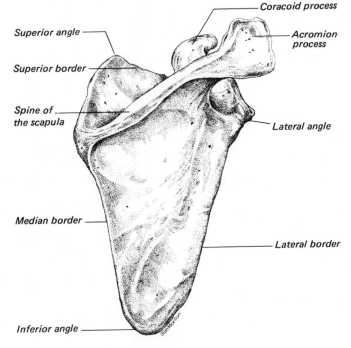

Coracoid process

Superior angle

Superior border

Spine of the scapula

Acromion process

Lateral angle

Median border

Lateral border

Inferior angle

**Fig. 38.** *The right scapula (posterior view).*

### The Clavicle

The clavicle has the appearance of an elongated S and articulates with the acromion process of the scapula laterally and the sternum medially. The medial half of the bone is anteriorly convex; the lateral half is concave.

### Joints of the Shoulder Girdle

The shoulder girdle consists of two joints: the sternoclavicular and the acromioclavicular. On each lateral side is a glenoid fossa for articulation with the head of the humerus.

The *sternoclavicular joint*, which possesses an interarticular fibrocartilage disc, is a synovial joint between the medial end of the clavicle and the superior lateral corner of the manubrium of the sternum and the cartilage of the first rib. A *fibrous capsule* that covers the articulation is aided in providing strength to the joint by an *anterior* and *posterior sternoclavicular ligament*, an *interclavicular ligament*, and a *costoclavicular ligament*. This joint is strong and dislocation uncommon. When the acromion of the scapula is struck or when a force is transmitted from an outstretched arm when the hand strikes the ground on falling, the clavicle may break but the sternoclavicular joint will rarely dislocate.

The *acromioclavicular joint*, also an arthrodial joint, is the union between the lateral end of the clavicle and the acromion process of the scapula. The *superior acromioclavicular* and *inferior acromioclavicular* ligaments support the joint. The *coracoclavicular ligament*, not part of the joint, helps to maintain the integrity of the joint. Dislocation of this joint is common in contact sports such as football, rugby, and ice hockey as when the athlete falls on his or her shoulder; this injury is often incorrectly called a "shoulder separation."

### Movement of the Shoulder Girdle

All movements of the scapula depend on the combined motion capability of the sternoclavicular and acromioclavicular joints. The sternoclavicular joint permits movement in almost all directions including circumduction, whereas the acromioclavicular joint permits a gliding motion of the articular end of the clavicle on the acromion and some rotation of the scapula forward and backward on the clavicle.[9]

The following movements of the scapula, combined with the clavicle, are possible:

1. *Adduction and abduction.* Adduction of the scapula occurs when the medial

---

[9] Henry Gray, *Anatomy of the Human Body*, ed. C. M. Goss rev. ed. 28th (Lea & Fibeger) 1966, p. 326.

border of the scapula moves toward the spine; abduction is in the opposite direction. The act of sticking out the chest and pulling back the shoulders results in adduction in the scapula. This can easily be seen and felt on a partner.

2. *Elevation and depression.* Movement of the scapula upward, with no rotation, as in raising the shoulders, is called elevation; the opposite movement is depression. Elevation and depression of the scapula may be felt by placing a hand on the scapula, the clavicle, or both, simultaneously.

3. *Rotation.* Upward rotation is the movement of the inferior angle of the scapula outward and upward. The axis of rotation may be at the sternoclavicular or the acromioclavicular joints. Downward rotation is in the opposite direction.

## Muscle Groups and Surface Anatomy

The muscles that originate on the trunk and insert into the shoulder girdle form, in part, the following muscle groups: shoulder girdle adductors, abductors, rotators, elevators, and depressors. Additional muscles that cross the shoulder joint and contribute to shoulder girdle movement are discussed in the next chapter.

Locate the following surface landmarks on yourself or a partner: the *acromion*, *coracoid*, *spine of scapula*, *inferior angle of scapula*, *medial border of scapula*, *trapezius* muscle, and the *serratus anterior* muscle.

There are four important pairs of muscles on the posterior aspect of the trunk that act on the shoulder girdle: trapezius, levator scapulae, rhomboid major, and rhomboid minor.

The *trapezius* (Fig. 39) is a large triangular-shaped muscle that is divided into four parts each with its own innervation. The upper part is a thin sheetlike muscle attached from the base of the skull to the neck of the clavicle. Its prime action is elevation of the scapula, and therefore it is very active during weight bearing of the upper limb such as carrying a suitcase. The second part (immediately below) is attached to the acromion and is involved in elevation and upward rotation and assists in adduction. Next below is the prime mover for adduction of the scapula, while the fourth part is involved with upward rotation and depression and assists in adduction. This muscle is easily observed and should be studied on oneself or a partner during shoulder girdle movement.

The *levator scapulae* is a small muscle situated deep to the upper part of the trapezius, its main action being elevation.

The *rhomboid major* and *minor*, located below the trapezius, are strong adductors of the scapula and also contribute to downward rotation of the scapula.

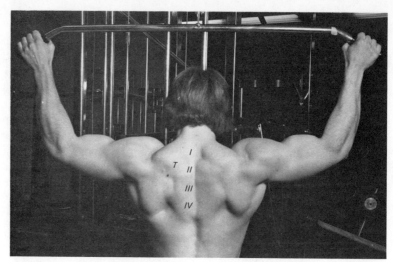

**Fig. 39.** *The trapezius (T) in action. Note the four heads.*

On the anterior aspect of the trunk there are two important pairs of muscles that act on the shoulder girdle: serratus anterior and pectoralis minor.

The *serratus anterior* is a broad muscle on the lateral side of the chest where it originates on the upper eight or nine ribs (Fig. 40). It has a serrated appearance and is easily visible on a well-muscled person. The muscle is inserted into the medial border of the scapula, and, when it contracts, during a forward-pushing

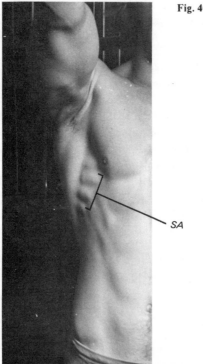

**Fig. 40.** *The serratus anterior (SA).*

action, it abducts the scapula while holding it in close proximity to the thoracic cage. A strengthened and shortened serratus anterior reduces the condition of the protruding inferior angle of the scapula common referred to as *winged scapula*.

The *pectoralis minor* is a small muscle deep to the pectoralis major. It is inserted into the coracoid process and during contraction depresses the superior lateral angle of the scapula, thereby causing the inferior angle to protrude if it is not supported by the serratus anterior.

## EXERCISES

1. Observe and discuss the movements of the different parts of the thorax during regular breathing as well as the mechanical changes in respiration following a short vigorous bout of exercise.
2. Present a simple exercise that would develop the strength of each of the following: (a) the trunk extensors, (b) the trunk flexors, and (c) the shoulder adductors. Consider the plane, axis, range, and speed of movement, the type of muscle contraction, and, of course, the safety factors of the exercise.
3. In the toe-touching exercise that starts from an erect position (legs straight),
   a. Which muscle group(s) initiate(s) the move? Is the muscle action concentric or eccentric?
   b. What other muscle groups are active once the motion is started? How do they contract when the person is moving downward to touch the toes?
   c. Which groups are active while coming up toward the standing position? How do they act this time?
4. While holding the push-up position (elbows extended), which of the extensor and/or flexor muscle groups of the trunk and neck are active statically? Prod the appropriate muscles of a partner while the position is being held to detect activity or relaxation of the muscles.

## RECOMMENDED READINGS

Asmussen, Erling, "The Weight-Carrying Function of the Human Spine," *Acta Orthopaedic Scandinaviac*, 29 (1960), 276–290.

Basmajian, J. V., and R. K. Greenlaw, "Electromyography of Iliacus and Psoas with Inserted Fine-Wire Electrodes," (abstract), *Anatomical Record*, 160 (February 1968), 310–311.

de Sousa, Odorico Machado, and José Furlani, "Electromyographic Study of the M. Rectus Abdominis," *Acta Anatomy*, 88 (1974), 281–298.

Flint, M. Marilyn, and J. Gudgell, "Electromyographic Study of Abdominal Muscular Activity During Exercise," *Research Quarterly*, 36 (March 1965), 29–37.

Floyd, W. F., and P. H. S. Silver, "Electromyographic Study of Patterns of Activity of the Anterior Abdominal Wall Muscles in Man," *Journal of Anatomy*, 84 (April 1950), 132–145.

Godfrey, K. E., L. E. Kindig, and E. J. Windell, "Electromyographic Study of Duration of Muscle Activity in Sit-up Variations," *Archives of Physical Rehabilitation*, 58 (March 1977), 132–135.

Jonsson, Bengt, "Morphology, Innervation and Electromyographic Study of the Erector Spinae," *Archives of Physical Medicine and Rehabilitation*, 50 (November 1969), 638–641.

Joseph, J., *Man's Posture: Electromyographic Studies*. Springfield, Ill.: Charles C Thomas, 1960.

Keagy, Robert D., Joel Brumlik, and John J. Bergan, "Direct Electromyography of the Psoas Major Muscle in Man," *Journal of Bone and Joint Surgery*, 48A (October 1966), 1377–1382.

Klausen, Klaus, "The Form and Function of the Loaded Human Spine," *Acta Physiologicia Scandinaviac*, 65 (1965), 176–190.

Morris, J. M., D. B. Lucas, and B. Bresler, "Role of the Trunk in Stability of the Spine," *Journal of Bone & Joint Surgery*, 43A (April 1961), 327–351.k

Vitti, Mathias, Makoto Fujiwara, John V. Basmajian, and Masuru Iida, "The Integrated Roles of Longus Colli and Sternocleidomastoid Muscles: An Electromyographic Study," *Anatomical Record*, 177 (December 1973), 471–484.

# Chapter 6

# The Upper Extremity

The bones of the upper extremity, from proximal to distal, are the humerus, the ulna and radius, the carpals, the metacarpals, and the phalanges.

## The Humerus

The largest bone of the upper extremity is the humerus (Fig. 41). Its proximal end consists of a large smooth articular surface, the *head of the humerus*, which

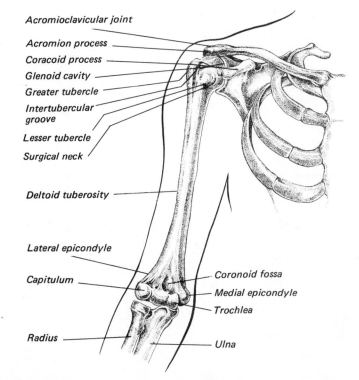

Acromioclavicular joint

Acromion process

Coracoid process

Glenoid cavity

Greater tubercle

Intertubercular groove

Lesser tubercle

Surgical neck

Deltoid tuberosity

Lateral epicondyle

Capitulum

Radius

Coronoid fossa

Medial epicondyle

Trochlea

Ulna

**Fig. 41.** *Shoulder joint and the humerus.*

73

articulates with the glenoid fossa of the scapula to form the shoulder joint; a *surgical neck*, which is a slightly thinner region below the head; and the *greater* and *lesser tubercle*. The greater tubercle is lateral to the head; the lesser tubercle is slightly anterior. The *bicipital groove* lies between the two tubercles and runs down the proximal third of the shaft. The *deltoid tuberosity* is midway on the lateral aspect of the shaft.

The upper half of the shaft is almost cylindrical, but toward its distal end it flattens anteroposteriorly. The distal end consists of two rounded condyloid articular surfaces—the *medial* and *lateral epicondyles*—with their respective *medial* and *lateral supracondylar ridges* extending up the shaft. Between the condyles is the distal articular surface of the humerus, the *trochlea*. A deep depression, the *olecranon fossa*, is located on the posterior surface while anteriorly is the shallow *coronoid fossa*.

### The Ulna

The ulna is on the medial or little finger side of the forearm (Figs. 42 and 43). The length of the bone can be felt through the skin on the posterior aspect from the tip

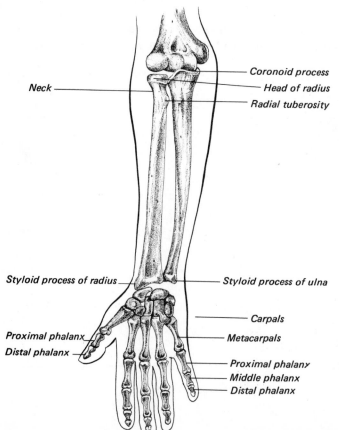

**Fig. 42.** *The forearm and hand (anterior view).*

Neck

Coronoid process
Head of radius
Radial tuberosity

Styloid process of radius
Styloid process of ulna

Carpals

Proximal phalanx
Metacarpals

Distal phalanx

Proximal phalanx
Middle phalanx
Distal phalanx

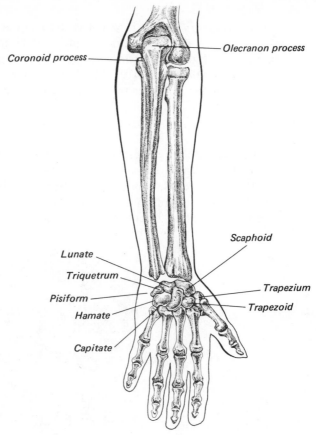

Coronoid process

Olecranon process

Scaphoid

Lunate

Triquetrum

Trapezium

Pisiform

Trapezoid

Hamate

Capitate

**Fig. 43.** *The forearm and hand (posterior view).*

of the elbow to a small projection at the wrist. It is the larger of the forearm bones illustrated in the figures and has a sizeable proximal end for articulation with the humerus to form the elbow joint. Its size decreases distally where it ends in a smooth rounded end and a pointed *styloid process*. Striking features of the ulna include the large, curved *olecranon process* on the posterior aspect of the proximal end and a triangular eminence of bone called the *coronoid process* anterior to it. Between these two processes is the concave *semilunar notch*.

**The Radius**

Lateral to the ulna is the shorter radius bone that has a small proximal end, a shaft, and a large distal end for attachment to the carpal bones of the wrist (Fig. 188 in Appendix A). The proximal *head* is cylindrical with a concave end and is supported by a *neck*. The concave end articulates with the humerus, while the head is held against the *radial notch* of the ulna by the *anular ligament*. A prominent *radial tuberosity* is clearly visible a few centimeters below the head. This tuberosity is due to the attachment of the biceps muscle.

The larger distal end has two articular surfaces. The small notch on the medial aspect is for articulation with the ulna, whereas the larger, smooth, and somewhat concave surface is for articulation with the carpal bones.

## The Carpal Bones

The carpal bones consist of eight individual bones that vary from one another in shape and size. They are arranged in two rows of four (Fig. 44). The proximal row, medial to lateral, consists of the *pisiform*, *triquetrum*, *lunate*, and the *scaphoid*; the distal row includes the *hamate*, *capitate*, *trapezoid*, and the *trapezium*. The radius articulates with the lunate and the scaphoid bones.

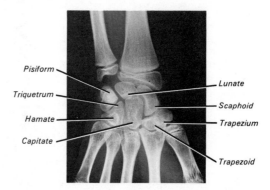

**Fig. 44.** *X-ray of the bones of the hand.*

## The Metacarpals

The bony framework of the body of the hand consists of the five long, cylindrical *metacarpal* bones. The proximal end or *base* of each bone is cuboidal in shape and articulates with the wrist and adjacent bones. The rounded distal ends form the knuckles and articulate with the first (proximal) phalanges.

## The Phalanges

The bones of the fingers number fourteen on each hand and are called the phalanges. Each of the fingers has three phalanges; the thumb does not have a middle phalanx.

## THE SHOULDER JOINT

This typical ball-and-socket joint is formed by the union of the shallow glenoid cavity of the scapula and the hemispherical head of the humerus (Figs. 45 and 46). The glenoid cavity is deepened by a low rim of fibrocartilage called the *glenoid labrum*. The lax *articular capsule* provides little strength or stability to the joint and does not inhibit the large range of motion of the joint. Within the fibrous membrane the *inferior, middle,* and *superior glenohumeral* ligaments fan out from the superior glenoid margin to the humerus to provide strength to the joint, while externally the *coracohumeral* and the *transverse humeral* ligaments reinforce the joint. The ligaments of the shoulder joint do not provide a great deal of strength or stability; however, additional strength is possible due to the two layers of muscles crossing and covering the joint. The relatively small muscles internally (teres minor and infraspinatus posteriorly, supraspinatus superiorly, and subscapularis anteriorly) are covered by the large deltoid muscle externally. These "rotator cuff" muscles are largely responsible for maintaining the integrity of the joint, due to their angles of pull, coupled with the slant of the glenoid fossa.

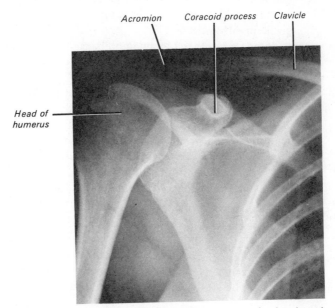

**Fig. 45.** *X-ray of shoulder joint showing humerus, scapula, clavicle, and ribs.*

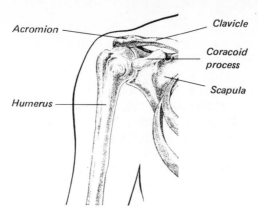

**Fig. 46.** *Shoulder joint showing humerus, scapula, clavicle, and ribs.*

## Movement of the Shoulder Joint

Glenohumeral (shoulder joint) movements should not be confused with movements of the shoulder girdle, although the two usually occur together and should be considered together. Flexion, extension, slight hyperextension, abduction, adduction, circumduction, and medial and lateral rotation may occur at the shoulder joint, but the range of motion is limited if there is no shoulder girdle involvement.

Inman and others[1] in a comprehensive study on the function of the shoulder joint claimed that during all flexion and abduction motion of the glenohumeral joint there is simultaneous shoulder girdle (scapulothoracic) movement. Through the first 30–60°, the scapula may remain fixed, there may be motion at the joint until a stable position is obtained; or the scapula may move on the chest wall. After 30° of abduction or 60° of forward flexion, there is a constant relationship between the humeral and scapula movement with two degrees of humeral motion to every one degree of scapular rotation. This claim has been partly supported by Ito and others[2] in a study of the shoulder joint and girdle using electrical goniometers to measure the glenohumeral and scapulothoracic angles. They found that, during 30–130° of humeral abduction, the ratio for movement of these two angles was 2 to 1. Figure 47 shows the change in glenohumeral and the scapulothoracic angles during abduction of the arm; note, also, the inferior angle of the scapula in Fig. 39. From the anatomical position, the full range of movement in flexion of the arm to above the head is only accomplished if medial rotation of the humerus occurs, whereas full abduction is

[1]Verne T. Inman, J. B. DeC. M. Saunders, and Le Roy C. Abbot. "Observations of the Function of the Shoulder Joint," *Journal of Bone and Joint Surgery*, 26 (January 1944), 9.

[2]N. Ito, R. Suzuki, K. Ishimura, and H. Kuwahara. "Electromyographic study of shoulder joint," *Proceedings of the 4th Congress of the International Society of Electrophysiological Kinesiology*, Boston, August 1979, p. 22.

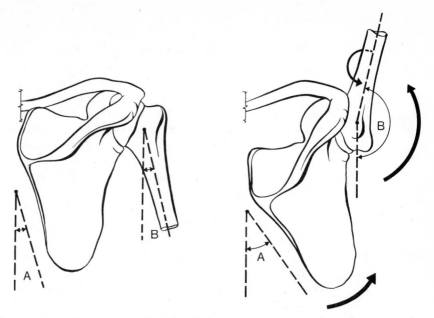

**Fig. 47.** *Posterior view of scapula and humerus during abduction of the humerus: (a) scapulothoracic angle, (b) glenohumeral angle.*

possible from this position. If abduction is attempted with the palm of the hand facing the thigh, it is limited to approximately 90°; lateral rotation at this point will permit further abduction.

### Muscle Groups and Surface Anatomy

Eleven muscles cross the shoulder joint and contribute to motion. The muscle groups of the shoulder joint and their prime movers are as follows:

*Shoulder flexors*—Clavicular pectoralis and anterior deltoid.
*Shoulder adductors*—Sternocostal pectoralis, latissimus dorsi, and teres major.
*Shoulder abductors*—Middle deltoid and supraspinatus.
*Shoulder adductors*—Sternocostal pectoralis, latissimus dorsi, and teres major.
*Inward rotators*—Teres major and subscapularis.
*Outward rotators*—Teres minor and infraspinatus.

Many of these muscles, and others such as the posterior deltoid, coracobrachialis, biceps, and long head of triceps, contribute to different movements at the shoulder joint; their contribution to movement may be vital in many instances.

*Shoulder flexion* is achieved mainly by the action of the clavicular pectoralis and the anterior deltoid. The *clavicular pectoralis* is the upper portion of the large fan shaped muscle of the chest, which is commonly called the pectoralis major. (The two parts of the pectoralis muscle are clearly shown in Fig. 48.) Increased activity occurs in this head of the muscle during flexion with maximum activity being reached at 115° of flexion.[3] The *anterior deltoid* (Figs. 48 and 49) is very active during resisted flexion. It is a superficial muscle that may be observed and palpated when the arm is abducted to 90°.

In addition to these muscles, shoulder flexion is assisted by the coracobrachialis and the short head of the biceps.

*Shoulder extension* (resisted) is accomplished by active contraction of the sternocostal pectoralis, the latissimus dorsi, and the teres major muscle. Actions in which this occurs are rope-climbing and pull-up exercises. The *sternocostal pectoralis* is the lower and larger portion of the pectoralis muscle, as shown in Figure 48. The *latissimus dorsi* is the very broad muscle located on the back that can be clearly seen on a partner. It is superficial except for a small part that is covered by the lower part of the trapezius. It is a very powerful extensor and becomes very prominent in trained swimmers who use the shoulder extensor

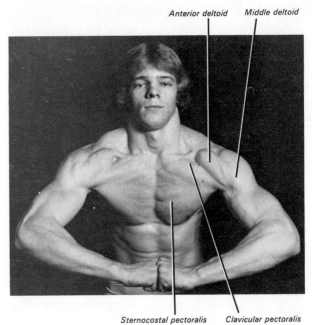

Anterior deltoid      Middle deltoid

Sternocostal pectoralis      Clavicular pectoralis

**Fig. 48.** *Surface muscles active in shoulder joint motion (anterior view).*

[3]John V. Basmajian, *Muscles Alive: Their Functions Revealed by Electromyography,* 4th ed. (Baltimore, Md.: Williams and Wilkins, 1978), p. 190.

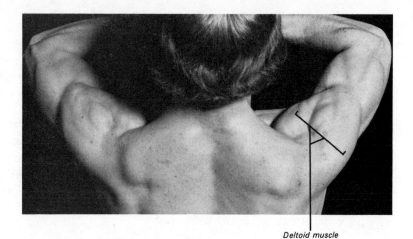

Deltoid muscle

**Fig. 49.** *Surface muscles active in shoulder joint motion (posterior view).*

muscles frequently in the propulsive phase of swimming. (Note the large *latissimus dorsi* muscles of the swimmer shown in Fig. 31 and the "lats exercise" used for developing the muscle in Fig. 39.) The *teres major* appears to be active during extension against resistance but inactive during motions without resistance.[4] The posterior deltoid and the long head of the triceps assist during shoulder extension.

*Shoulder abduction* is performed mainly by the *middle deltoid* and the supraspinatus. The greatest activity in the deltoid during abduction occurs between 90° and 180°. The middle deltoid is a multipennate muscle that is well located for abduction at the shoulder joint and may be felt just lateral to the acromion when the arm is abducted to about ninety degrees. It can be clearly seen in Figs. 48 and 49.

The *supraspinatus* is found superior to the spine of the scapula and is deep to the trapezius and deltoid. It is called an initiator of abduction but has been found to assist the deltoid through 110° of abduction.

Abduction may be achieved through the assistance of the anterior deltoid, clavicular pectoralis, and the long head of the biceps.

For shoulder adduction the prime movers are the *sternocostal pectoralis*, *latissimus dorsi*, and *teres major*. The gymnastic move called the "iron cross" is held by very vigorous contraction of these adductor muscles to prevent the further abduction that would occur if gravity were allowed to pull the gymnast downward. Adduction may be assisted by the short head of the biceps and the long head of the triceps, whereas the subscapularis and the corocobrachialis assist when the arm has been abducted to a position above 90°.

[4]Basmajian, *Muscles Alive*, p. 195.

*Medial rotation* is achieved by the action of *subscapularis* and *teres major* with assistance from the anterior deltoid, clavicular and sternocostal pectoralis, latissimus dorsi, and the short head of the biceps. The prime movers of *lateral rotation* are *teres minor* and *infraspinatus* with assistance from the posterior deltoid.

*Horizontal adduction* is performed by the contraction of the clavicular and sternocostal pectoralis, the anterior deltoid, and the coracobrachialis, with assistance from the short head of the biceps. The bench-press exercise, as illustrated in Fig. 50, is an excellent exerciser of these muscles.

*Horizontal abduction* requires action by the middle and posterior deltoid muscles, infraspinatus, and teres minor. These muscles may be assisted by the latissimus dorsi and teres major.

**Fig. 50.** *Bench-press exercise for development of shoulder joint horizontal adductors.*

## THE ELBOW AND RADIOULNAR JOINTS

The elbow joint is uniaxial; the *humeroulnar* is a hinge joint that provides flexion and extension in the sagittal plane. The *articular capsule*, with *pads of fat* and *synovial membrane* within, covers all three bones at the joint. The capsule is strengthened laterally by the *radial collateral ligament* and on the medial side by the *ulna collateral ligament*. The joint is reinforced further by additional smaller ligaments and the *anterior* and *posterior ligaments*.

There are two *proximal radioulnar joints:* the head of the radius is cupped to articulate with the *capitulum* of the humerus to constitute the first, while, immediately below the head of the radius, the anular ligament supports it against the ulna. Figure 51 shows the elbow joint and its location within the arm. The *intermediate radioulnar joint* is the articulation of the shafts of the radius and ulna by an *interosseous membrane* for muscle attachment and possibly to transmit forces from one bone to the other. The fibers of the interosseous membrane are directed from the radius to the ulna in a proximal to distal direction. The probable function of this membrane is to prevent upward displacement of the radius when force is taken on the hand. At the end of the radius and ulna is the *distal radioulnar joint*, a pivot joint that contains a triangular cartilaginous articular disc.

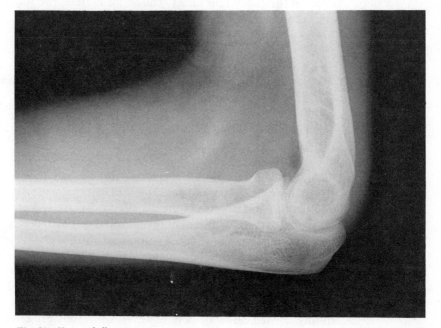

**Fig. 51.** *X-ray of elbow joint.*

## Movement of the Elbow Joint

Flexion and extension and slight hyperextension are possible at the elbow joint. The range of flexion is limited by the muscle bulk on the anterior aspect of the arm. The degree of hyperextension varies from one individual to another; in females, for example, it can be as much as 20° without undue alarm.

### Movement of the Radioulnar Joints

The movements of *pronation* and *supination* are achieved by rotation of the radioulnar joints. In the anatomical position the radius and ulna are side by side, almost parallel, in the supinated position. Pronation of the forearm occurs when the palm of the hand is rotated medially toward the midline of the body. Range of motion for supination-pronation is approximately 130°.

### Muscle Groups and Surface Anatomy

Although some fifteen different muscles cross the elbow joint, there are few regarded as prime movers of the forearm at the elbow joint. The basic movements at the elbow joint include flexion and extension; pronation and supination occur at the radioulnar joints. The major muscles producing these movements are as follows:

*Elbow flexors*—Biceps brachii, brachialis, and brachioradialis.
*Elbow extensors*—Triceps brachii.
*Radioulnar pronators*—Pronator quadratus, pronator teres, and anconeus.
*Radioulnar supinators*—Supinator and biceps brachii.

One should review the location, origin, insertion and action of the muscles of the elbow and radioulnar joints in Appendix A.

**Elbow Flexion.** The *biceps brachii* is one of the best known muscles of the body. Its two heads, one long and one short, originate from the supraglenoid tuberosity and the coracoid process of the scapula, respectively. The long head has a long tendon that passes through the bicipital groove before joining the more anteriorly and medially placed short head. Although the bellies of these heads are separate, they join in a common tendon of insertion into the radial tuberosity and the deep fascia of the forearm. The biceps brachii is easily palpable as is its distal tendon. The biceps brachii acts on three joints: the shoulder, elbow, and radioulnar. When the forearm is supinated, the biceps brachii is very active during resisted elbow flexion; however, when fully pronated, it is less active, as the tuberosity of the radius, to which the muscle tendon is attached, is medially rotated resulting in the tendon's being drawn toward the elbow joint as it wraps around the bone. This movement can be felt at the elbow if a finger is placed against the tendon while the elbow is held at 90° and the forearm pronated and supinated. Figures 52 and 53 show the contracted biceps brachii during pronation and supination, respectively. This muscle is very active during undergrip pull-ups (forearms supinated) and undergrip curls with a weight.

**Fig. 52.** *Contraction of the elbow flexors during pronation.*

**Fig. 53.** *Contraction of the elbow flexors during supination.*

The *brachialis* is the most powerful elbow flexor and is located deep to the biceps brachii on the anterior aspect of the arm (Fig. 54). It is a single joint muscle with its insertion in the tuberosity of the ulna and the coronoid process; there is no change in its line of pull during pronation and supination, and it is, therefore, equally active in both positions.

The *brachioradialis* (Figure 54) is a strong fusiform muscle that flexes the elbow when a quick movement is required or when the elbow is flexed slowly against resistance.[5] This muscle has been called a pronator when the forearm is supinated and a supinator when the forearm is pronated. Basmajian has suggested that it does neither when the elbow is extended unless the movements are performed against resistance and, at best, that it acts as an accessory muscle when strength is required to supinate or pronate the forearm.[6]

[5] J. V. Basmajian and A. Latif, "Integrated Actions and Functions of the Chief Flexors of the Elbow: A Detailed Electromyographic Analysis," *Journal of Bone and Joint Surgery*, 39A (October 1957), 1106–1118.

[6] Basmajian, *Muscles Alive*, p. 204.

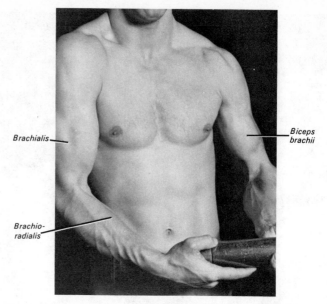

Brachialis

Biceps
brachii

Brachio-
radialis

**Fig. 54.** *The elbow flexors lateral and anterior view.*

**Elbow Extension.**    The *triceps brachii* has three heads, one of which crosses both the shoulder and elbow joints. The other two, the main mass of the muscle, originate on the shaft of the humerus. The attachment of the triceps brachii to the olecranon process of the ulna provides a lever system of the first class and, although it is designed to favor speed, provides considerable power during elbow extension (Fig. 55).

**Radioulnar Pronation.**    The *pronator quadratus* lies deep on the forearm, near the wrist, and as its name suggests is a quadrilateral muscle that acts as the prime mover of pronation of the forearm. The *pronator teres* is a deep fusiform muscle that runs obliquely on the anterior proximal aspect of the forearm. Its origin is on the medial epicondyle of the humerus and the coronoid process of the ulna. Due to its insertion on the lateral side of the middle portion of the radius, it pulls the radius over and in front of the ulna when it contracts, thereby pronating the forearm.

**Radioulnar Supination.**    The prime mover for supination of the forearm is the rather small *supinator* muscle;[7] however, when supination of the forearm is resisted, the biceps brachii becomes active. Supination of the forearm thus becomes an extremely powerful motion. It has been found that, through supination, one can develop a much greater torque than by pronation of the forearm.

[7]Anthony Travill and John V. Basmajian, "Electromyography of the Supinators of the Forearm," *Anatomical Record*, 139 (April 1961), 560.

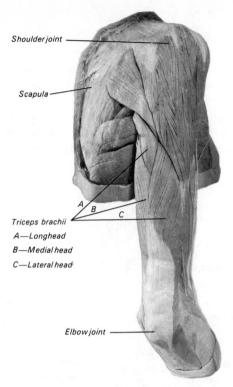

Shoulder joint

Scapula

A
B
C

Triceps brachii
A—Long head
B—Medial head
C—Lateral head

Elbow joint

**Fig. 55.** *The triceps brachii muscle.*

# THE WRIST JOINT

The wrist or *radiocarpal* joint is a condyloid articulation between the distal end of the radius and the proximal two carpal bones (scaphoid and lunate). The arc of the joint has its concavity downward (refer to Fig. 44). The joint is supported by a number of ligaments and many tendons from the muscles of the forearm that act upon the wrist joint.

## Movement at the Wrist Joint

Although there is no rotation at the joint, it is designed to permit a fairly wide range of motion. *Flexion*, in which the palm of the hand approaches the anterior aspect of the forearm, is through almost 90°. Hyperextension of the wrist, commonly referred to as *extension* at this joint, has a similar range. *Abduction* at the wrist—movement in the frontal plane with the thumb leading—is limited due to the medial ligament; however, there is a greater range of *adduction* at the wrist joint. Combined flexion, abduction, extension, and adduction in sequence provides circumduction.

Limited gliding among the carpal bones adds slightly to the range of motion at the wrist joint.

The two muscle groups of the wrist joint are the *wrist flexors*, of which the bulk of the muscles is location on the anteromedial aspect of the forearm, and the *wrist extensors*, which are located on the posterolateral side of the forearm. Both groups can be felt in action by placing the fingers on the extensors and the thumb on the flexors and then, fairly rapidly, flexing and extending the wrist. Figure 56 shows a dissection of the wrist flexors and extensors.

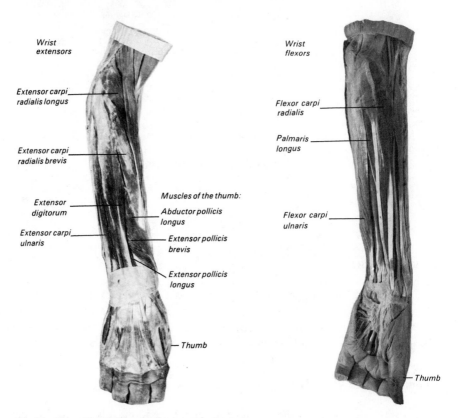

**Fig. 56.** *Dissection of the wrist flexors and extensors.*

*Flexion* is achieved by the simultaneous contraction of *flexor carpi radialis* and *flexor carpi ulnaris*, although neither muscle is a true flexor. Flexor carpi radialis flexes and abducts the wrist; flexor carpi ulnaris flexes and adducts. In a similar manner, wrist *extension* is accomplished by simultaneous contraction of the *extensor carpi radialis* and *extensor carpi ulnaris* (longus and brevis). Figure 57 presents a model of the actions of these muscles at the wrist joint.

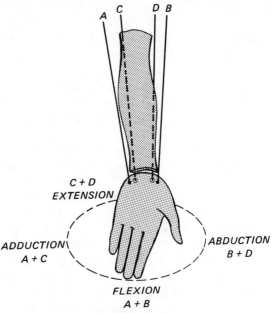

**Fig. 57.** *Model of muscle actions at the wrist joint: (a) flexor carpi ulnaris, (b) flexor carpi radialis, (c) extensor carpi ulnaris, (d) extensor carpi radialis.*

*Abduction*, although limited, is performed by an interplay of contraction between flexor carpi radialis and extensor carpi radialis; *adduction* is produced by flexor carpi ulnaris and extensor carpi ulnaris.

A number of muscles responsible for movement at the carpometacarpal, metacarpophalangeal, and interphalangeal joints cross the wrist and assist in movement at this joint.

## THE CARPOMETACARPAL JOINTS

The *first carpometacarpal joint* is a saddle-shaped joint between the trapezium and the proximal end of the first metacarpal. This joint of the thumb is surrounded by a strong articular capsule and is strengthened further and stabilized by a number of ligaments.

The remaining four joints are arthrodial. The synovial membrane of the carpal bones covers these joints, which are supported by strong ligaments.

### Movement of the Carpometacarpal Joints

The concavoconvex articular surface of the trapezium and *first* metacarpal permits all movement. Terminology traditionally used for describing movements of the thumb is presented in this section because of the complexity of the joint, which is located at an angle to the plane of the other metacarpals. Flexion and extension of the thumb occur in the same plane as abduction and adduction of the fingers while abduction and adduction of the thumb occurs in the same plane as finger flexion and extension.[8] Movement occurs around two axes. A transverse axis allows *abduction* of the thumb away from the index finger and *adduction* back along side it. An anteroposterior axis enables the thumb to be *flexed*, that is, moved across or parallel to the palm, and *extended* back again. The four movements combined, in logical sequence, results in *circumduction*. A limited degree of *rotation* is possible so that during flexion the anterior surface of the thumb can face the other digits. This movement is known as *opposition*. Movement at the other four carpometacarpal joints is limited to slight gliding; the middle two joints have the least movement.

Apart from the *extrinsic* muscles, those coming from or originating outside the hand, which are responsible for movement of the hand, a number of small muscles, *intrinsic* muscles, are situated entirely within the hand. Their actions, though vital to efficient motion of the hand, will not be discussed in this text.

The location, origin, insertion, and actions of the muscles of the hand are presented in Appendix A.

## THE METACARPOPHALANGEAL JOINT

These joints constitute the knuckles of the clenched fist. They are condyloid joints formed by the round distal heads of metacarpals articulating with the concave ends of the phalanges. Flexion, extension, abduction, and adduction are possible at these joints.

## THE INTERPHALANGEAL JOINTS

These are the simple hinge joints of the fingers. Flexion and extension are possible in each of the two joints of the fingers and the single joint of the thumb.

## EXERCISES

1. Why is the clavicle more susceptible to injury (breakage) than the scapula?
2. Why are the spiral grooves of screws always in the same direction? Consider the important anatomical reasons. How about screws for left-handers?
3. Present a simple exercise that would develop the strength of each of the

---

[8]Ellen Neall Duvall, *Kinesiology: The Anatomy of Motion.* (Englewood Cliffs, N.J.: Prentice-Hall, 1959), p. 46.

following: (a) elbow extensors, (b) shoulder adductors, (c) clavicular pectoralis, and (d) latissimus dorsi. Consider the plane, axis, range, and speed of movement, the type of muscle contraction, and, of course, the safety factors of the exercise.

4. Divide a number of students into two approximately equal groups (in terms of strength). Have one group perform a maximum number of pull-ups using an overgrip grasp on a bar while the other uses undergrip. Compare the average number of pull-ups for the two groups. Have each group perform maximum pull-ups again using the opposite hand grip. Compare the averages of the groups again. Discuss the reasons for the differences.

5. A gymnast wishes to do the "iron cross" on the rings and is lacking the necessary strength. Which muscles are primarily responsible for holding the position? Suggest two exercises to develop strength in each of the muscles so as to improve the gymnast's chance of success.

## RECOMMENDED READINGS

Basmajian, J. V., and A. Latif, "Integrated Actions and Functions of the Chief Flexors of the Elbow: A Detailed Electromyographic Analysis," *Journal of Bone and Joint Surgery*, 39A (October 1957), 1106–1118.

Basmajian, J. V., and A. Travill, "Electromyography of the Pronator Muscles in the Forearm," *Anatomical Record*, 139 (January 1961), 45–49.

Bearn, J. G., "An Electromyographic Study of the Trapezius, Deltoid, Pectoralis Major, Biceps and Triceps Muscles, During Static Loading of the Upper Limb," *Anatomical Record*, 140 (June 1961), 103–107.

De Luca, C. J., and W. J. Forrest, "Force Analysis of Individual Muscles Acting Simultaneously on the Shoulder Joint During Isometric Abduction," *Journal of Biomechanics*, 6 (July 1973), 385–393.

de Sousa, O. Machado, F. Berzin, and A. C. Berardi, "Electromyographic Study of the Pectoralis Major and Latissimus Dorsi Muscles During Medial Rotation of the Arm," *Electromyography*, 9 (November 1969), 407–416.

Furlani, J., "Electromyographic Study of the M. Biceps Brachii in Movements at the Glenohumeral Joint," *Acta Anatomica*, 96 (1976), 270–284.

Travill, A. A., "Electromyographic Study of the Extensor Apparatus of the Forearm," *Anatomical Record*, 144 (December 1962), 373–376.

Travill, Anthony, and John V. Basmajian, "Electromyography of the Supinators of the Forearm," *Anatomical Record*, 139 (April 1961), 557–560.

Simons, D. G., and E. N. Zuniga, "Effect of Wrist Rotation on the X Y Plot of Averaged Biceps EMG and Isometric Tension," *American Journal of Physical Medicine*, 49 (August 1970), 253–256.

# Chapter 7

# The Lower Extremity

The bones of the lower extremity from proximal to distal are the femur, the patella, the tibia and fibula, the tarsal bones, the metatarsal bones, and the phalanges.

## The Femur

The longest and strongest bone in the body is the femur (Fig. 191 in Appendix A). It transmits the entire weight of the body onto the tibia. Its shaft is long and almost cylindrical, and its ends are enlarged and designed for the complexity of movements such as ambulation. The shaft of the femur is at an angle of approximately 80° to the vertical in males and 75° in females. This angular difference is due to the relatively wider pelves in females. A round *head*, attached to the shaft by means of a *neck*, fits snugly into the acetabulum. The neck joins the shaft at an angle, which further increases the distance of the proximal ends of the femur from each other due to pelvic width. The angle and the reduced diameter of the neck makes it susceptible to fracture, particularly in older persons when the bone becomes more brittle. This injury is often referred to as a "broken hip." Close to the junction of the neck to the shaft are two protuberances. The larger of these two—the *greater trochanter*—is lateral to the hip joint and can be felt through the skin. The *lesser trochanter* is distal to the neck and on the posterior medial surface of the shaft. Between these two trochanters is the *trochanteric line*.

The *linea aspera* is a prominent ridge along the posterior surface of the shaft. The distal end of the femur is large and consists of two smooth articular surfaces, the *medial* and *lateral condyles*, which articulate with the tibia and patella. The two condyles are easily palpable. Between the condyles on the posterior surface is the *intercondyloid fossa*.

## The Patella

The patella is a small, flat, roughly triangular sesamoid bone that is easily felt through the skin. It is embedded in the extensor tendon and improves the leverage of this knee extensor muscle group and provides protection to the anterior aspect of the knee joint, especially when the knee is in flexion. Dislocation of the patella is uncommon in sport; however, when it does occur, it occurs more often in females than in males due to differences in the angle of the shaft of the femur.

## The Tibia

The medial and larger of the two bones of the leg, the *tibia*, commonly called the shin bone, can be palpated through its full length (Fig. 191 in Appendix A). The proximal end is large, and its *medial* and *lateral condyles* are rather flat. A small facet is located on the lateral side, slightly below the lateral condyle for articulation with the fibula. Very noticeable on the anterior surface of the head is a large bone mass called the *tibial tuberosity* to which the major knee extensors are attached. The shaft has three easily identifiable borders—anterior, internal, and external. The distal end has a smooth, wide, flat articular surface for attachment to the talus bone of the tarsals. Laterally is the *fibular notch* for articulation with the fibula; medially the *medial malleolus* extends to form a large prominence at the ankle joint.

## The Fibula

The fibula is a long, thin bone on the lateral side of the tibia. Its proximal end is slightly enlarged and is attached to the tibia a little to the posterior surface of the lateral condyle. It does not articulate with the femur and therefore is not part of the knee joint. The distal end, called the *lateral malleolus*, is also slightly enlarged and projects downward. It articulates with the tibia, and, with the medial malleolus, they form the sides of the socket of the ankle joint. Although the fibula is not a weight-bearing bone, it has important muscular attachments and adds to the stability of the ankle joint.

### The Tarsal Bones

The seven tarsal bones of the foot are the *talus, calcaneus, cuboid, navicular*, and the three *cuneiform* bones (Fig. 58). The talus is the bone on which the lower extremity rests; it in turn is supported almost entirely by the calcaneus, which is commonly referred to as the *heel bone*.

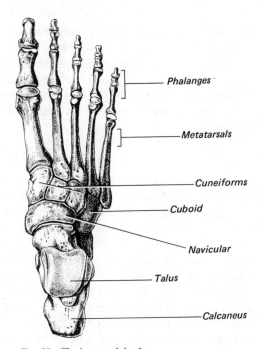

**Fig. 58.** *The bones of the foot.*

### The Metatarsals

Anterior to the tarsus are the five *metatarsal bones*, each cylindrical and slightly curved. They are similar to the metacarpals, having irregular surfaces at their proximal ends and round condylar surfaces distally.

### The Phalanges

As in the hand there are fourteen *phalanges*—two in the great toe and three each for the other four toes.

# THE HIP JOINT

The hip joint is a synovial ball-and-socket joint formed by the articulation of the head of the femur and the acetabulum. The joint is designed for weight bearing, and, although it is very strong, it provides the lower extremity with a fairly substantial range of motion. Its movement capability, however, is considerably less than that of the shoulder joint.

The joint is supported by a number of ligaments. The *capsular ligament*, which encloses the neck of the femur and is attached to the superior margin of the acetabulum, deepens the cavity of the joint and thus increases the suction quality that is present due to the synovial fluid on the smooth joint surfaces. The capsule is strengthened by three ligaments: iliofemoral, ischiofemoral, and pubofemoral (Figs. 59 and 60). The names of these ligaments indicate their attachments.

The *iliofemoral* ligament, on the medial anterior aspect of the joint, is regarded as the strongest ligament in the body. When the leg is extended, the ligament becomes taut; this provides a mechanism for one to stand upright with the hip joint supported by the ligament. This is possible because the line of gravity falls behind the hip joint tending to hyperextend the joint. Most muscles

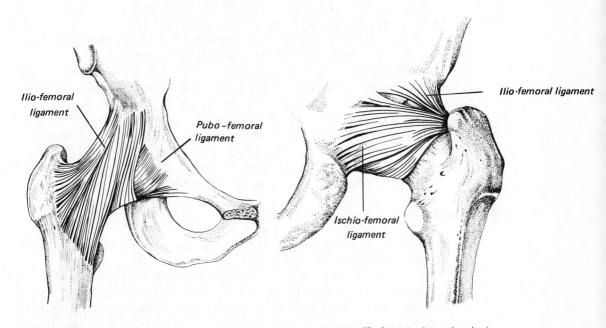

**Fig. 59.** *The hip joint (anterior view).*      **Fig. 60.** *The hip joint (posterior view).*

crossing the hip joint can therefore be relaxed. Electromyographic studies have supported this premise except for the iliopsoas, which remains quite active in the erect posture.[1]

The *ischiofemoral* ligament, which is located on the posterior aspect of the joint, passes laterally and anteriorly from the ischium to the femoral neck, whereas the *pubofemoral* passes from the pubis to the inferior aspect of the femoral neck. Both ligaments are taut during hip extension and relaxed during flexion; they assist the weight-bearing function of the iliofemoral ligament during standing.

### Movement of the Hip Joint

The lower extremity has a fairly extensive range of motion provided by the hip joint.

*Flexion* of the femur at the hip joint is limited by the compression of the muscles of the thigh or the abdomen when the knee joint is flexed. Unless a large range of motion at the hip joint is possible, this action would be limited, if the knees were extended, by the hamstring muscles. The range of flexion, therefore, could be as much as 160°. Extension is movement from a flexed position to the extended position of upright stance. While *hyperextension* is possible, it is limited to approximately 15° by the iliofemoral ligament.

*Rotation* is possible through approximately 90° and may be performed freely when the leg is in any position except during taut hyperextension.

*Abduction* may occur through up to 90° as shown by gymnasts, dancers, and other finely tuned athletes by tilting the pelvis forward, rotating the femur laterally, and stretching the adduct muscles and associated ligaments. It can, however, be limited to some 30° by short adductor muscles. Adduction is the movement from an abducted position back to an upright stance where it is limited by contact with the other leg.

*Circumduction*, the combination of flexion, extension, abduction, and adduction, is possible through a fairly wide range.

### Muscle Groups and Surface Anatomy

The muscles that act on the hip joint may be grouped as follows: hip flexors, extensors, abductors, adductors, and rotators.

The reader should become familiar with the following landmarks and structures: *greater trochanter, gluteus maximus, gluteus medius, tensor fasciae latae* and *iliotibial tract, inguinal ligament* and *femoral artery, pubic symphysis, anterior superior iliac spine*, and the *ischial tuberosity*. (For assistance, refer to Fig. 190 in Appendix A and to Fig. 61.)

---

[1]John V. Basmajian, *Muscles Alive: Their Functions Revealed by Electromyography*, 4th ed. (Baltimore, Md.: Williams and Wilkins, 1978), p. 182.

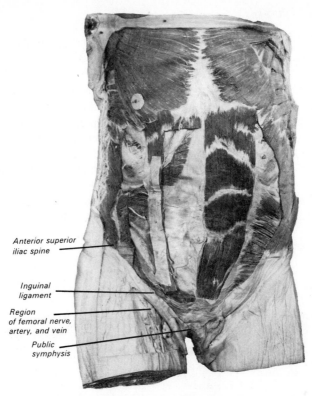

Anterior superior
iliac spine

Inguinal
ligament

Region
of femoral nerve,
artery, and vein

Public
symphysis

**Fig. 61.** *Dissected anterior trunk showing pelvic region.*

## Hip Flexion

The *iliopsoas* (iliacus and psoas major) is the prime mover of hip flexion; in addition, slight activity and bursts of activity at irregular periods have been recorded electromyographically during erect standing, thereby suggesting that it is active in maintaining the erect posture.[2] The iliopsoas is a fusiform muscle running from the lumbar vertebrae (psoas) and the upper part of the iliac fossa (iliacus) to the lesser trochanter of the femur. The two iliopsoas muscles are almost cylindrical in shape and are surprisingly large. Their cross-sectional area is almost four times that of the rectus femoris that assists the iliopsoas in performing most sit-up exercises. The iliopsoas is only palpable close to its distal end. If one locates the femoral artery at the inguinal ligament, the iliopsoas can be felt below and to the lateral side of the femoral artery, vein, and nerve, which are in close proximity to each other.

The *rectus femoris* is a bipennate muscle crossing both the hip and knee joints. It is a prime mover for flexion of the hip and assists in knee extension. It is active in kicking, in which both actions are required, and it can be felt and seen to be active during sit-ups.

[2]Basmajian, *Muscles Alive*, p. 238.

The *tensor fasciae latae* is active during flexion, medial rotation, and abduction of the thigh at the hip joint. Although it is a relatively small muscle it has two functional heads. The *anteromedial fibers* are tendinous and from their insertion at the pelvis run down the thigh ending in fascia distal to the patella. They assist thigh flexion during the initial period of thigh acceleration in walking and are more active during running.[3] The *posterolateral fibers* join part of the iliotibial tract that inserts into the lateral tibial condyle. Pare[4] suggested that their activity during support in gait is related to abduction and medial rotation at the hip joint. The tensor fasciae latae has also been called an extensor of the knee; however, it appears that its main function is to tense the iliotibial tract that blends with the fasciae latae that surrounds the knee and is attached to the femoral condyles. This adds considerable stability to the knee; therefore, consideration should be given to strengthening this muscle during rehabilitation following knee injury.

## Hip Extension

The large, coarsely fibered, quadrilateral shaped muscle that is the major muscle of the buttock is called the *gluteus maximus* (Fig. 62). This muscle is a powerful extensor of the hip joint and is active mainly during power movements such as

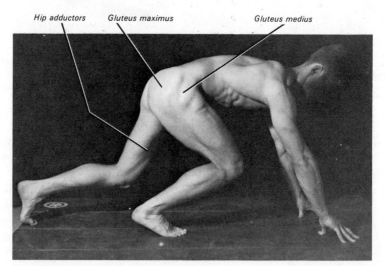

Hip adductors   Gluteus maximus   Gluteus medius

**Fig. 62.** *Hip extensors, abductors and adductors.*

[3]Edmond B. Paré, Jeffrey M. Schwartz, and Jack T. Stern, Jr., "Electromyographic and Anatomical Study of the Human Tensor Fasciae Latae Muscle," *Proceedings of the 4th Congress of the International Society of Electrophysiological Kinesiology*, Boston, August 1979, p. 53.

[4]Pare, "Electromyographic and Anatomical Study," p. 53.

running and climbing stairs. It is inactive during standing and slow walking on the level. Its other prime function is lateral rotation of the thigh at the hip joint.

The *hamstring* muscle group consists of the *biceps femoris, semitendinosus,* and *semimembranosus* (Fig. 63). The muscles of this two joint muscle group (extensors of the hip and flexors of the knee) have a common origin in the ischial tuberosity and constitute the muscle bulk of the posterior thigh. The short head of the biceps femoris has its origin in the lateral lip of the linea aspera. This muscle inserts into the lateral condyle of the tibia and the head of the fibula and therefore can rotate the leg laterally when the knee is flexed. The tendon of the biceps femoris is easily felt behind the knee, on the lateral side, when the knee is flexed. The semitendinosus and semimembranosus are felt on the posterior medial side of the lower thigh.

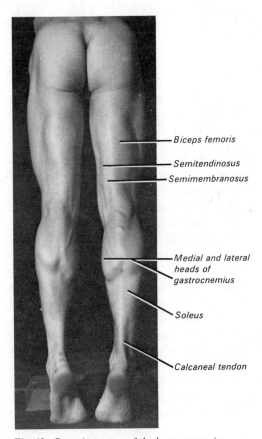

**Fig. 63.** *Posterior aspect of the lower extremity.*

### Hip Abduction

The major abductors of the thigh at the hip joint are the *gluteus medius* and the *gluteus minimus* (Fig. 62). Both muscles are very active at the supporting hip while standing on one leg. These muscles prevent the pelvis from tilting toward the unsupported side. During walking the gluteus medius and minimus of the supporting limb abduct the limb at the hip joint, that is, they hold or tilt the pelvis so that the side of the pelvis of the unsupported (or swinging) leg remains up to enable the foot to swing forward without hitting the ground.

### Hip Adduction

The main adductors are *gracilis, pectineus, adductor magnus, adductor longus*, and *adductor brevis*. The medial muscle of the thigh is the long, slender, two-joint gracilis muscle. The other adductors are lateral to this muscle and insert into the medial-posterior surface of the femur. They are powerful adductors of the thigh at the hip joint and contribute to medial rotation and flexion of the thigh (Fig. 62).

## THE KNEE JOINT

The knee joint is a double condyloid joint between the condyles of the femur above and those of the tibia below; in addition, the patella articulates with the anterior aspects of the condyles of the femur (Figs. 64 and 65).

The main structures that support the knee externally include the tendons of muscles that cross it from above and below, the *patellar ligament* anteriorly, the capsule of the knee, and the lateral and medial ligaments (collateral ligaments) of the knee. Support from within the knee joint is provided by the two *cruciate ligaments* (anterior and posterior) and the two *menisci* (medial and lateral), also referred to as the semilunar cartilages.

The *patellar ligament* is part of the tendon of the quadriceps muscle group and is attached to the patella, which it also covers, and the tubercle of the tibia.

The *capsule* of the knee is strengthened by many ligaments and is inseparable from the aponeurotic expansion of the knee extensor muscles.

Lateral bending of the knee is limited by the two *collateral ligaments*. The lateral ligament is a cordlike structure attached to the head of the fibula below and the lateral condyle of the femur above. It can be felt, in the living, close to the head of the fibula when the knee is flexed. Care must be taken to avoid confusing it with the tendon of the biceps femoris, which has its insertion on the head of the fibula. The medial ligament is longer and wider than the lateral ligament and is attached to the medial condyles of the femur and tibia. This ligament also is attached to the capsule of the knee and thereby to the medial meniscus.

The *cruciate* ligaments are two extremely strong ligaments inside the joint (Fig. 64). They cross each other and are named in accordance with their tibial attachment (anterior and posterior). The anterior cruciate ligament prevents a forward displacement of the head of the tibia and limits hyperextension at the joint; the posterior cruciate ligament restricts displacement of the tibia backward.

The menisci are semilunar-shaped cartilages that tend to increase the concavity of the articular surface of the tibia and thereby increase the stability of the joint; they are not anchored tightly in the joint and so slide during motion of the knee joint (Fig. 65).

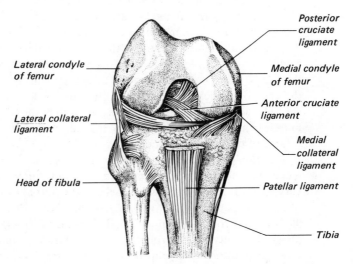

**Fig. 64.** *Anterior view of flexed right knee (patella has been removed).*

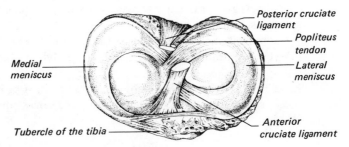

**Fig. 65.** *Superior view of left tibia.*

*Flexion and Extension.* In the upright stance the knee joint is usually in a position of slight hyperextension. Normally, this hyperextension is limited to about 5–10°. The knee joint can be flexed to enable one's heel to be placed against the buttock where further flexion is restricted by the muscle mass of the posterior thigh. The flexion–extension motion of the knee joint is not that of a true hinge joint movement as the femoral condyles glide or slide back and forth on the tibial condyles. The menisci move forward when the knee is extended, and during flexion they glide backward.[5]

During extension, when the knee is almost fully extended and the tibia fixed, as in standing, the femur rotates medially. The lateral condyle is almost stationary during this movement, whereas the medial condyle slides backward. This rotary movement completes the extension of the knee; in fact it places it in a slightly hyperextended position. This rotary phenomenon is commonly termed the "screw-home movement" of the knee and in fact is a normal locking action of the knee joint. Flexion of the knee is initiated by lateral rotation of the femur on the tibia. This "unlocking" action is the result of the contraction of the *popliteus* muscle, which is a lateral rotator of the femur when the tibia is fixed or a medial rotator of the tibia when the foot is free.

*Rotation* of the knee is possible when the knee is flexed, but, when it is in the locked position of slight hyperextension, no rotation is possible. It is therefore very vulnerable to injury, when locked, as external forces applied to the leg are often not dissipated effectively. When the knee is flexed to 90°, approximately 90° of tibial rotation is possible. This rotation is vital in the performance of physical activity such as soccer, ice skating, skiing, and of course many others; however, the knee is susceptible to injury in this position due to a slackness of the supporting medial and lateral ligaments.

## Muscle Groups and Surface Anatomy

The muscle groups that are active at the knee joint include the knee flexors, knee extensors, and knee rotators.

Prominent anatomical features include the patella and its tendon, the tibial tubercle, the head of the femur, and the condyles of the tibia and femur.

*Knee Flexion.* The prime movers for knee flexion are the three hamstring muscles: biceps femoris, semitendinosus, and semimembranosus. These muscles

---

[5]Michael C. Hall, *The Locomotor System: Functional Anatomy* (Springfield, Ill.: Charles C Thomas, 1965), p. 339.

have been discussed in this chapter. A number of muscles assist in knee flexion: the *popliteus*, a short flat muscle posterior to the knee, is known to "unlock" the knee and thereby initiates knee flexion; the *sartorius*, a straplike muscle that is the longest muscle in the body, is a weak flexor and medial rotator of the knee; the *gracilis*, previously mentioned as an adductor of the hip, contributes to knee flexion; and the *gastrocnemius*, which is discussed under muscles acting on the ankle joint.

**Knee Extension.**   The *quadriceps femoris*, consisting of the *rectus femoris, vastus lateralis, vastus medialis,* and *vastus intermedius*, is the prime mover for extension at the knee and is capable of producing tremendous power. The first three muscles are superficial and are easily located on a partner. The rectus femoris has its origin on the pelvis, whereas the three vasti muscles have their origin on the femur. The quadriceps femoris has a common insertion into the tibial tubercle by means of the patellar ligament.

**Knee Rotation.**   When the knee is bent, the *semitendinosus* and *semimembranosus* are the prime movers of medial rotation and the *sartorius* and *gracilis* assist in the motion. It should be remembered that *popliteus* "unlocks" the knee by medial rotation of the tibia when the foot is free or lateral rotation of the femur on the tibia when the tibia is fixed. *Lateral* rotation is achieved by contraction of the *biceps femoris* when the knee is bent.

## THE ANKLE JOINT

The *ankle joint* is a synovial hinge joint between the distal end of the tibia and fibula and the talus bone. The weight of the body is transferred vertically to the talus, which is supported in a mortise on the lateral side by the malleolus of the fibula and on the medial side by the malleolus of the tibia. It has a capsule and is strengthened by the *deltoid* ligament on the medial side and the calcaneofibular ligament and the *anterior* and *posterior talofibular ligaments* on the opposite side.

### Movement of the Ankle Joint

Movement at this hinge joint is limited to the sagittal plane and is called *dorsiflexion* and *plantar flexion*. Raising the toes and the top of the foot toward the shin is termed dorsiflexion; plantar flexion is the opposite action and occurs when one raises one's heel off the ground, commonly referred to in physical activities as "pointing the toes" (Fig. 66). The range of motion possible in plantar flexion is about 40°, whereas only about 20° of dorsiflexion are possible.

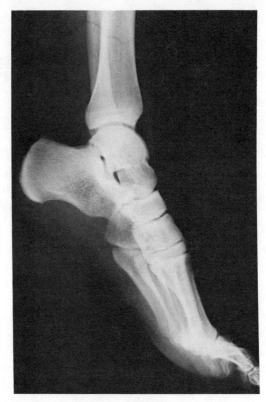

**Fig. 66.** *X-ray of foot in plantar flexion.*

## Muscle Groups and Surface Anatomy

Only two muscle groups act on the ankle joint: the *plantar flexors* and the *dorsiflexors.*

The following landmarks should be observed on the living (refer also to Figs. 63 and 67 and Appendix A): *gastrocnemius, soleus, lateral* and *medial malleoli,* and the *calcaneal tendon.*

*Plantar flexion.*   The prime movers of this action are the *gastrocnemius* and the *soleus.* Assistant movers are the plantaris, peroneus longus, peroneus brevis, and tibialis posterior.

The two heads of the gastrocnemius are shown clearly in Fig. 63. They are next to each other, but separate, and approximately midway down the leg they join a common aponeurosis that becomes the *calcaneal tendon,* commonly referred to as the "Achilles tendon." The short fibers of the gastrocnemius restrict its range of movement but permit it to be a powerful plantar flexor.

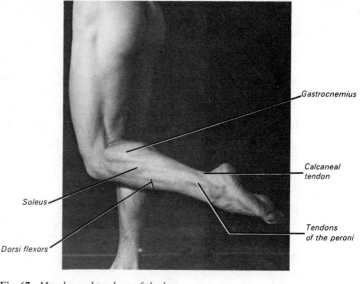

Fig. 67. *Muscles and tendons of the leg.*

The *soleus* is deep to the gastrocnemius and can be seen on both sides distal to it (Fig. 63). Unlike the gastrocnemius, the soleus is a one-joint muscle inserting into the calcaneal tendon that is attached into the posterior aspect of the calcaneus. The predominance of slow-twitch fibers in the soleus suggests that the muscle is able to perform sustained tonic activity. This has been supported electromyographically; the soleus is very active during standing and is, therefore, an important postural muscle.

**Dorsiflexion.**    The prime movers, located on the lateral anterior side of the tibia are the tibialis anterior, extensor digitorum longus, and extensor peroneus tertius.

The large powerful *tibialis anterior* is easily seen on the anterior aspect of the leg, and its tendon can be seen to extend to the base of the first metatarsal and the first cuneiform bone. It is the prime mover for dorsiflexion of the ankle and assists in inversion of the foot. While walking, it assists in raising the foot during the swing forward and thereby prevents stubbing of the toes.

The *extensor digitorum longus*, the extensor of the lateral four toes and an assistant mover in dorsiflexion of the ankle, is a flat penniform muscle located immediately lateral to the tibialis anterior.

*Extensor peroneus tertius* has its origin on the distal anterior aspect of the fibula and part of the interosseous membrane and inserts into the dorsal surface of the base of the fifth metatarsal. It is a prime mover for dorsiflexion and assists in eversion of the foot.

### The Joints of the Foot

The *intertarsal* joints are arthrodial and permit a limited amount of gliding motion. The *tarsometatarsal* joints unite the distal ends of the tarsal bones with the five metatarsals. These joints are also arthrodial and permit some gliding motion. The *metatarsophalangeal* joints are separate synovial condyloid joints between the metatarsals and the phalanges. Flexion and extension and some abduction and adduction occurs at these joints. The last joints of the foot are the hinged *interphalangeal* joints of the toes where flexion and extension is possible.

### Movement of the Foot

The combined movements of the foot permit only a limited range of *flexion* and *extension*. Foot *inversion*, a combination of gliding movements, results in the medial aspects of the foot's being raised while in the movement of *eversion* the lateral side of the foot is raised.

### Muscle Groups and Surface Anatomy

The muscle groups acting on the joints of the foot are the *evertors, invertors, flexors,* and *extensors.* As in the hand there are extrinsic and intrinsic muscles. Although of some importance in movement of the foot, the intrinsic muscles will not be discussed in this text.

**Eversion.**   The prime movers for eversion of the foot are the *peroneus longus*, with its long tendon that runs below the lateral malleolus and can be easily felt; the *peroneus brevis*, whose tendon runs a similar path around the lateral malleolus; and, the *peroneus tertius*, which pulls up the lateral aspect of the foot and assists eversion.

**Inversion.**   The main invertor of the foot, which acts in opposition to the peroni muscles, is the *tibialis posterior* muscle. *Tibialis anterior*, the prominent muscle on the front of the leg immediately lateral to the tibia, is the other prime mover for inversion.

**Flexion.**   *Flexor hallucis longus* and *flexor digitorum longus* are the prime movers for flexion of the toes and assist in plantar flexion.

**Extension.**   The *extensor hallucis longus* and *extensor digitorum longus* are the prime movers of extension and hyperextension of the toes. They are active, also, in dorsiflexion of the ankle. The tendons can be felt and seen on the dorsal surface of the foot, running toward the toes. The ball of the foot is soft and pliable, and the skin can be moved from side to side and proximodistally when the foot is relaxed. Hyperextension or dorsiflexion of the toes results in the ball becoming tense, and mobility of the skin is greatly reduced. This ensures that the shear forces during rapid locomotion are not absorbed by the skin alone but are transferred to the connective tissue and the skeleton.[6]

## Arches of the Foot

The bones of the foot are arranged so that the weight of the body is supported by the heel (calcaneus) and the distal heads of the metatarsal bones. The shape of the bones and their ligamentous and muscular support present two arches. The largest and most noticeable, the *longitudinal arch*, from the calcaneus to the distal heads of the metatarsals, is more pronounced medially, where it forms the instep, than laterally where the foot is usually in contact with the ground during standing (Fig. 68). The *transverse arch* is formed by the bases of the metatarsal bones and the distal tarsal bones.

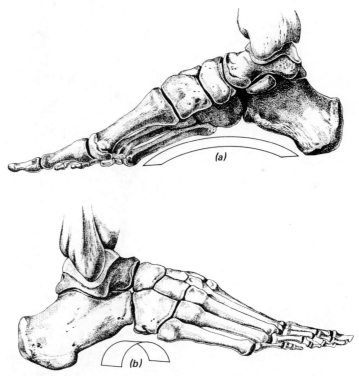

**Fig. 68.** *The arches of the foot: (a) longitudinal, (b) transverse arch.*

[6] Finn Bojsen-Moller and Larry Lamoreux, "Significance of Free Dorsiflexion of the Toes in Walking," *Acta Orthopaedica Scandinaviac*, 50 (1979), 475.

The arches of the foot are maintained, predominantly, by the plantar ligaments—the long plantar, the plantar calcaneocuboid, and the plantar calcaneonavicular. In addition, support is provided by the plantar aponeurosis and to a lesser degree by the short plantar muscles and a number of those long muscles and their tendons that cross the ankle joint.

The arches are very strong and possess springlike qualities. A weighted foot tends to spread, thereby absorbing the impact of the force, and has recoil qualities that assist in locomotion.

When the longitudinal arches collapse, the condition of *pes planus* or "flat feet" occurs. Figure 69 shows a normal footprint with the main supporting areas at the heel and ball of the foot. The large medial longitudinal arch is evident; however, the lateral arch does not show because tissue, other than bone, is in contact with the surface. It is fairly common for this lateral margin to be absent—this condition may still represent a normal arch, although it could be indicative of the condition of *pes cavus*, which is an abnormally high arch.

**Fig. 69.** *A normal footprint.*

## EXERCISES

1. Explain why so few muscles of the lower limbs are active during erect standing.
2. Consider the structural design of the knee joint, its strengths, and its weaknesses. What architectural changes, to improve the efficiency or function of the joint, would you suggest? How does rotation of the knee joint assist a swimmer doing the breaststroke and the soccer player kicking a ball?
3. Describe the movements within the foot.
4. Compare the pelvis and shoulder girdles and the hip and shoulder joints as to their strength, function, and movement capabilities.
5. Present a simple exercise that would develop the strength of each of the following: (a) hip extensors, (b) knee flexors, and (c) soleus. Consider the plane, axis, range, and speed of movement, the type of muscle contraction, and, of course, the safety factors of the exercise.
6. Compare the structure and functions of the feet and hands. Discuss the advantages and disadvantages of replacing the feet by the hands.
7. A high jumper, a ballet dancer, and a weight lifter wish to increase the strength of the muscles of their lower extremities to improve their performances in their respective activities. List the muscle groups and the important muscles within each group that need to be strengthened.
   a. Present two exercises that could be used by all three persons to improve their performances.
   b. Present two exercises for each athlete considering the specific needs of the athlete in the performance of his or her activity.

## RECOMMENDED READINGS

Bierman, W., and H. J. Ralston, "Electromyographic Study During Passive and Active Flexion and Extension of the Knee of the Normal Human Subject," *Archives Physical Medicine*, 46 (January 1965), 71–75.

Brandell, B. R., "Functional Roles of the Calf and Vastus Muscles in Locomotion," *American Journal Physical Medicine*, 56 (April 1977), 59–74.

Carlsöö, S., and A. Nordstrand, "The Coordination of the Knee-Muscles in Some Voluntary Movements and in the Gait in Cases With and Without Knee Joint Injuries," *Acta Chirurgica Scandinavica*, 134 (1968), 423–426.

Fischer, F. J., and S. J. Houtz, "Evaluation of the Function of the Gluteus Maximus Muscle," *American Journal Physical Medicine*, 47 (August 1968), 182–191.

Fujiwara, M., and J. V. Basmajian, "Electromyographic Study of Two-Joint Muscles," *American Journal Physical Medicine*, 54 (October 1975), 234–242.

Herman, R. and S. Bragin, "Function of Gastrocnemius and Soleus Muscles," *Physical Therapy*, 47 (February 1967), 105–113.

Lieb, F. J., and J. Perry, "Quadriceps Function: An Electromyographic Study Under Isometric Conditions," *Journal of Bone and Joint Surgery*, 53A (June 1979), 749–758.

Lovejoy, F., and T. P. Harden, "Popliteus Muscle in Man," *Anatomical Record*, 169 (April 1971), 727–730.

Paré, Edmond B., Jeffrey M. Schwartz, and Jack T. Stern, Jr., "Electromyographic and Anatomical Study of the Human Tensor Fasciae Latae Muscle," *Proceedings of the 4th Congress of the International Society of Electrophysiological Kinesiology*, Boston, August 1979, pp. 52–53.

Walmsley, Roy P., "Electromyographic Study of the Phasic Activity of Peroneus Longus and Brevis," *Archives Physical Medicine and Rehabilitation*, 58 (February 1977), 65–69.

# Part II

# Basic
# Mechanical
# Concepts

# Chapter 8

# Human Motion

There are three forms of human motion—*linear motion, angular motion* and combinations of these known as *general motion*.

## LINEAR MOTION

Linear motion (also known as *translation*) occurs when all the parts of a body[1] move the same distance, in the same direction, in the same time. A ski jumper experiences linear motion as he begins his downward approach to the takeoff (Fig. 70), and downhill skiers, figure skaters, and tobogganists often experience linear motion for short periods during the course of their events.

A simple test can be used to determine whether a given motion is linear or nonlinear. This test consists of arbitrarily selecting two points on the body and visualizing what happens to a line joining these points during the course of the motion. If this line remains the same length throughout and is always oriented so that it is parallel to every other position it has occupied, the motion is linear. If it does not, the motion is nonlinear. Consider, for example, the case of the ski jumper. As he begins his approach (Fig. 71(a)), a line joining his right shoulder and right ankle joints (or, indeed, any other two points that one might choose) remains the same length and parallel to its previous positions throughout. The motion is, therefore, linear. However, when he enters the curved portion of the approach (Fig 71(b)), these two conditions are no longer satisfied. The length of this line remains the same (at least, initially), but, due to the curvature of the approach, it no longer moves so that it is parallel to all of its previous positions. The motion is thus no longer linear.

[1] In this context, the term *body* can refer to the whole human body, to some part of it (such as the head, thigh or foot), or to some object (such as a tennis racket, javelin, or crutch) that is being used.

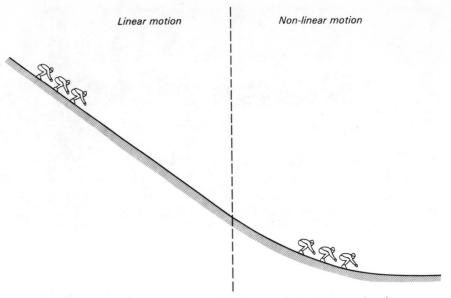

Linear motion          Non-linear motion

**Fig. 70.** *A ski-jumper experiences linear and non-linear motion during his approach to the takeoff.*

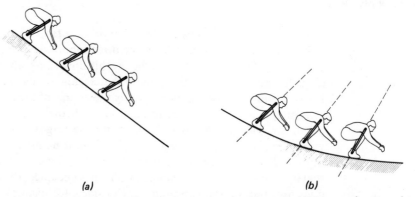

(a)                          (b)

**Fig. 71.** *Consecutive positions of a straight line between shoulder and ankle (or any other two points on the body) can be used to determine whether the motion of the body is (a) linear or (b) non-linear.*

At first glance, one might be tempted to conclude that linear motion can only occur when a body moves in a straight line. This, however, is not correct for, although exceedingly rare in human motion, it is possible for a body to move along a curved path and still satisfy the conditions of linear motion (Fig. 72).

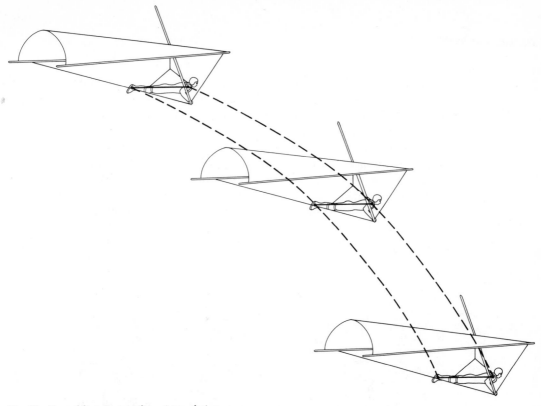

**Fig. 72.** *Curved-line (or curvilinear) translation.*

## ANGULAR MOTION

Angular motion, (also known as *rotation*) occurs when a body moves on a circular path about a central line so that all parts of the body move through the same angle, in the same direction, in the same time (Fig. 73). The central line, which lies at right angles to the plane of motion, is known as the *axis of rotation*. (It is important to emphasize here that an axis of rotation is a line and not a point. This distinction is important because, although axes can generally be treated as points when dealing with motions that are confined to one plane, the extension of this practice to more complex motions often leads to erroneous conclusions.)

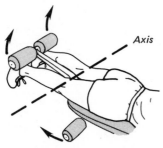 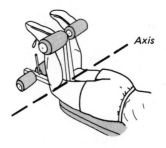

*Axis*        *Axis*

**Fig. 73.** *Angular motion about a transverse axis.*

Linear and angular motions are much less common in physical education, athletics, and rehabilitation than are motions that are either a combination of these two or a combination of several linear or angular motions. Such motions are referred to here as *general motions*.

A girl in a wheelchair imparts linear motion to the rest of her body via the angular motions of her arms (Fig. 74); a pole vaulter imparts linear motion to his pole as the result of the angular motions of his legs (Fig. 75); and a gymnast executing a Thomas Flair on the pommel horse combines many angular motions to produce the desired result (Fig. 76). All these, and many others, are examples of general motion.

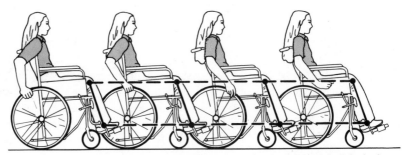

**Fig. 74.** *The angular motions of the arms produce a linear motion of the rest of the body.*

**Fig. 75.** *The angular motions of the legs produce a linear motion of the pole.*

**Fig. 76.** *The Thomas Flair—a complex general motion.*

## EXERCISES

1. Draw up a list of five examples of (a) linear motion, (b) angular motion, and (c) general motion in physical education or athletics. Try to find examples that differ from each other and from those presented in Chapter 8. Be sure to describe each example in sufficient detail that the form of motion involved can be recognized with ease.
2. There are very few examples of curvilinear motion in physical education and athletics. Can you name five? [Note: To qualify as an example of curvilinear motion, all parts of the body involved must move along a curved path and travel the same distance, in the same direction, in the same time.]

## RECOMMENDED READING

Kelley, David L., *Kinesiology: Fundamentals of Motion Description.* Englewood Cliffs, N.J.: Prentice-Hall, 1971, pp. 58–69 (The Classification of Whole-Body Motion).

# Chapter 9

# Describing
# Linear Motion
# (Linear Kinematics)

The science of mechanics can be divided conveniently into two parts—one that deals solely with describing the nature of the motion and one that goes beyond description to explain the causes of the motion. These two subdivisions are known, respectively, as *kinematics* and *kinetics*.

Each of these subdivisions can themselves be broken down further according to whether the motion under consideration is linear or angular. Thus, for example, the branch of kinematics that deals with the description of linear motion is known as *linear kinematics* and the branch that deals with the description of angular motion is known as *angular kinematics*.

This chapter contains a discussion of quantities used in linear kinematics—that is, to describe linear motions.

## DISTANCE AND DISPLACEMENT

The Engadin ski marathon is one of the world's most famous cross-country skiing events. Conducted in the Engadin Valley, Switzerland, the race starts in Maloja and finishes near Zuoz (Fig. 77). Assuming only minor deviations from the measured route, the length of the path followed by each of the 10,000–12,000 skiers who completes the race each year (that is, the *distance* they cover) is 42 km. At the end of this arduous event, the skiers are only 33.7 km from the point at which they started. In other words, they have traveled only 33.7 km "as the crow flies." In mechanics, the length of the straight line joining the start and the finish—together with some indication of the direction involved—is known as the *displacement*. Thus, those who finish the Engadin ski marathon cover a distance of 42 km and undergo a displacement of 33.7 km in a northeasterly direction.

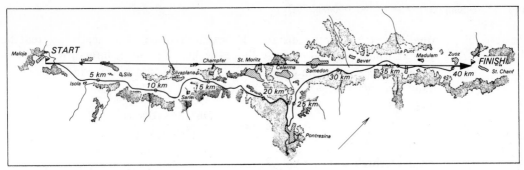

**Fig. 77.** *Distance and displacement in the Engadin ski-marathon.*

## SPEED AND VELOCITY

Table 5 lists the world records (as of January 1, 1980) for selected track events, together with the equivalent average speeds and average velocities. Such a tabulation is bound to raise some questions. For example, "Why is it necessary to list both speeds and velocities—aren't they the same thing?"; "Why *average* speed and *average* velocity—what other kind is there?"; and "Why are the average velocities such ridiculous values—for instance, why are some of them zero and the others so very slow?"

**TABLE 5   Average Speeds and Velocities for Selected World Records in the Track Events**

| Distance (Sex)* | Record | Average Speed (m/s) | Average Velocity† (m/s) |
|---|---|---|---|
| 200 m (W) | 21.71 | 9.21 | 4.68 |
| 200 m (M) | 19.72 | 10.14 | 5.16 |
| 400 m (W) | 48.60 | 8.23 | 0 |
| 400 m (M) | 43.86 | 9.12 | 0 |
| 800 m (W) | 1:54.9 | 6.96 | 0 |
| 800 m (M) | 1:42.4 | 7.81 | 0 |
| 1,500 m (W) | 3:56.0 | 6.36 | 0.27 |
| 1,500 m (M) | 3:32.1 | 7.07 | 0.30 |
| 3,000 m (W) | 8:27.2 | 5.92 | 0.20 |
| 5,000 m (M) | 13:08.4 | 6.34 | 0.13 |
| 10,000 m (M) | 27:22.5 | 6.09 | 0 |

*M and W refer to men and women, respectively.
†A velocity, like a displacement, has both a magnitude and a direction. However, since the directions are of no interest in the present case, only magnitudes are presented here.

While the words speed and velocity are often used interchangeably in general conversation, in the language of mechanics they have distinctly different meanings. The *speed* of a body is found by dividing the distance it covers by the time it takes to cover that distance. If the time involved is sufficiently long that the rate at which the body is traveling can change—and in human motion this generally means anything more than a small fraction of a second—the speed value obtained in this way is an *average speed*. Thus,

$$\text{Average speed} = \frac{\text{Distance covered}}{\text{Time taken}}$$

or, in algebraic terms,

$$\bar{s} = \frac{l}{t} \tag{1}$$

where $\bar{s}$ = "*s* bar" = the average speed (in this book, average values for speed and other quantities are indicated by an overbar), $l$ = the length of the path followed, in other words, the distance, and $t$ = the time taken. If the time involved is very brief, the value obtained is an *instantaneous speed*.

The velocity of a body, on the other hand, is found by dividing the displacement it undergoes by the time taken for it to undergo that displacement. For relatively long time intervals, the value obtained in this way is an *average velocity*:

$$\text{Average velocity} = \frac{\text{Displacement}}{\text{Time taken}}$$

or, in algebraic terms,

$$\bar{v} = \frac{d}{t} \tag{2}$$

where $v$ = "*v* bar" = average velocity, $d$ = the displacement, and $t$ = the time taken, as before.[1] For very brief intervals, the value obtained by dividing the displacement by the time taken is an *instantaneous velocity*.

It should now be obvious why the average velocity values in Table 5 make little sense. In those events like the 400 m and 800 m where the race starts and finishes in the same place—that is, ignoring minor differences due to the staggered start—the displacements that runners experience are zero and their average velocities are also zero, no matter how well they run. In those events where the start and the finish are not in the same place—the 200 m and 1500 m, for example—the average velocity is computed using the length of an imaginary line joining these two points in the numerator. Under such circumstances, it is no wonder that the

---

[1]The equations for instantaneous speed and velocity are identical to Eq. 1 and 2 except that the overbars are deleted.

value obtained is of little interest. Indeed, it would be fair to say that average velocities are rarely of much interest. The values that generally convey the most useful information are the average speed, as Table 5 suggests, and the instantaneous velocity. The importance of this latter quantity should become more evident as further mechanical concepts are introduced.

## ACCELERATION

The velocity of a human body often changes during the course of its motion. For example, even when a runner appears to be running at an even pace, a detailed analysis of his motion inevitably reveals a decrease in his forward velocity as his foot strikes the ground followed by an increase as he forcefully extends his driving leg (Fig. 78(a)). If the gain in forward velocity at the end of the supporting phase is equal to the loss at the beginning, the runner leaves the ground with the same forward velocity he had when he made contact with it. This contributes to the illusion that he is moving at a constant forward velocity. Variations in forward velocity also occur in most other forms of human locomotion. In breaststroke swimming, for example, the slowing of the forward velocity (Fig. 78(b)) is sometimes so pronounced that the performer momentarily comes to a stop or, worse yet, actually acquires a backward velocity! Needless to say, such backward motion is not a characteristic of good technique in the event.

The rate at which the velocity of a body changes with respect to time is known as its *acceleration*. The average acceleration of a body is given by the following equation:

$$\text{Average acceleration} = \frac{\text{Change in velocity}}{\text{Time taken}}$$

$$= \frac{\text{Final velocity} - \text{initial velocity}}{\text{Time taken}}$$

or, finally, in algebraic form

$$\bar{a} = \frac{v_f - v_i}{t} \tag{3}$$

where $\bar{a}$ = "*a* bar" = the average acceleration, $v_f$ and $v_i$ = the final and initial velocities, respectively, and $t$ = the time taken. (Like velocity from which it derives, the value obtained is an average acceleration if the time was long enough to permit the acceleration to have changed and an *instantaneous acceleration* when it was not long enough for this to have occurred.)

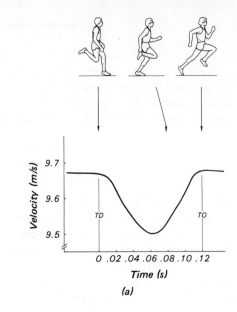

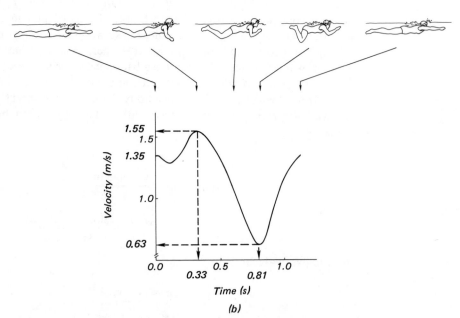

**Fig. 78.** *Horizontal velocity-time curves for (a) the support phase of a running stride, and (b) one cycle of breaststroke swimming. (Data for the running stride courtesy Ralph W. Mann, University of Kentucky, Lexington, Kentucky. Data for the swimming cycle based on paper by Keith McElroy and Brian Blanksby "Intra-cycle Velocity Fluctuations of Highly Skilled Breaststroke Swimmers,"* Australian Journal for Health, Physical Education and Recreation *(March 1976, pp. 25–34.)*

The average acceleration of the swimmer in Fig. 78(b) during the first 0.33 s of the stroke cycle can be found using Eq. 3:

$$\bar{a} = \frac{v_f - v_i}{t}$$

$$= \frac{(1.55 - 1.35) \text{ m/s}}{0.33 \text{ s}}$$

$$= 0.61 \text{ m/s/s } or \text{ m/s}^2$$

(The units used to describe accelerations are often the source of difficulties. This stems from the fact that the unit of time must be stated twice, once as part of the composite unit used to describe the change in velocity and once as the unit to describe the length of time involved.) The average acceleration of the swimmer during the next 0.48 s of the cycle can be found in similar fashion:

$$\bar{a} = \frac{v_f - v_i}{t}$$

$$= \frac{0.63 - 1.55) \text{ m/s}}{0.48 \text{ s}}$$

$$= -1.92 \text{ m/s}^2$$

It is common practice to refer to a body being accelerated when the magnitude of its velocity is increasing and decelerated when it is decreasing. While, strictly speaking, the body is being accelerated in both instances—deceleration is simply a particular form of acceleration—there is no real harm in such usage. Problems arise, however, when the terms *positive acceleration* and *negative acceleration* are used for the same purpose—that is, in reference to bodies that are being "speeded up" and "slowed down," respectively. While the results obtained to this point—a positive acceleration when the swimmer "speeded up" and a negative acceleration when he "slowed down"—give no indication of difficulty, consider what happens when the swimmer completes the first length of a race and turns for the second length. If the swimmer gets a good push-off from the end of the pool, the magnitude of the velocity at the start of the second length is almost certainly greater than it was at the end of the first. Suppose that the two magnitudes in question were 2.0 m/s and 1.5 m/s, respectively, that the change from one to the other required 1.0 s, and that the direction of motion during the first length was chosen as the positive direction. (In mechanics, it is customary to indicate directions by the use of positive and negative signs.) The average acceleration of the swimmer during this interval is thus

$$\bar{a} = \frac{v_f - v_i}{t}$$

$$= \frac{-2.0 - 1.5}{1.0} \text{ m/s}^2$$

(Because the swimmer is moving in the negative direction at the start of the second length, the final velocity is negative.)

$$\bar{a} = -3.5 \text{ m/s}^2$$

Thus, although the swimmer "speeded up" as a result of the turn, the acceleration experienced was negative. Because complications like this arise whenever the body can reverse its direction of motion, it is advisable to confine the use of the terms positive and negative acceleration to indicate "speeding up" and "slowing down" to those cases in which the body moves in a straight line in one direction.

### Acceleration Due to Gravity

All bodies on or near the surface of the earth are attracted toward it by *gravity*. If these bodies are airborne, they are accelerated toward the earth's surface at a constant acceleration of approximately 9.81 m/s$^2$—the exact value varies slightly from place to place (p. 158). The acceleration due to gravity is of such importance to the study of mechanics that it is given a special designation, the letter *g*. This use of the letter *g* to indicate the acceleration due to gravity has one unfortunate side effect—the letter is often wrongly assumed to represent gravity rather than the acceleration due to gravity, a quite different quantity.

## VECTORS AND SCALARS

Some quantities, like distance and speed, are completely defined when their magnitude is stated. Such quantities are called *scalars*. Others, like displacement, velocity, and acceleration, are not completely defined until both their magnitude and the direction in which they act are stated. Such quantities are called *vectors*.

Like a vector, an arrow can also be said to have a magnitude (its length) and a direction (indicated by the arrowhead). For this reason, vectors are often represented by arrows that have been drawn so that the length of the arrow represents the magnitude of the vector to some previously selected scale—for example, 1 cm = 5 m/s$^2$—and its direction represents the direction in which the vector acts.

## Resultant Vector

Two arrows representing like quantities—for example, two displacements, two velocities, or two accelerations—can be added to determine the combined effect of the two quantities. Consider, for example, the case of a ball carrier being tackled in the course of a football game—Fig. 79(a). If the ball carrier is running down the field at 8 m/s and is hit from the side by a defensive player who imparts a velocity of 6 m/s to him in a direction at right angles to his original motion, the combined effect of these two velocities can be determined using a simple geometric construction known as the *parallelogram of vectors*. This construction involves the following series of steps (Fig. 79(b)):

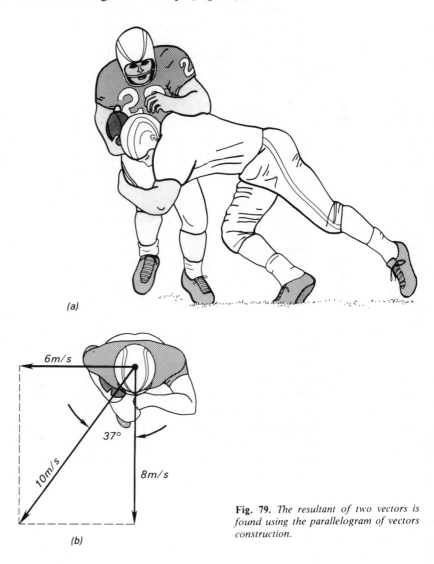

(a)

6m/s

10m/s

37°

8m/s

(b)

Fig. 79. *The resultant of two vectors is found using the parallelogram of vectors construction.*

1. Make a rough sketch of the body involved—in this case, the ballcarrier—and place a dot near the center of the body. This is the point through which the vectors will be assumed to act. (This sketch of the body is not an essential part of the procedure but, rather, an aid to visualizing the motion and avoiding errors.)

2. Select a scale so that the arrows representing the vector quantities involved—in this case, the velocities—will be as large as space permits. As a general rule, the larger the scale, the less the error that is likely to occur during the construction. In the present case (Fig. 79(b))—the scale chosen was 1 cm = 1.9 m/s.

3. Start with the dot (step 1) and draw an arrow, pointing away from the body, to represent the first of the two vectors.

4. Place the baseline of a protractor on the vector so that its center coincides with the dot and mark the direction in which the second vector acts relative to the first—that is, the angle between the lines of action of the two vectors. In the present case, the second vector acts at an angle of 90° to the first.

5. Start with the dot and draw an arrow, pointing away from the body, to represent the second vector.

6. Construct the remaining two sides of the parallelogram of which the arrows drawn in steps 3 and 5 are adjacent sides.

7. Construct the diagonal from the dot to the opposite corner of the parallelogram and place an arrowhead on the end of the diagonal farther from the dot.

8. Measure the length of this arrow and convert this measurement to the value it represents using the scale selected in step 2. Measure also the angle that this arrow makes with either of the other two arrows. The vector represented by this third arrow is known as the *resultant* and is equal to the sum of the original two vectors. In the present case, the resultant velocity of the ball carrier is 10 m/s in a direction 37° to the right of his original path. In other words, as a result of being hit by the defensive player, the runner's velocity is momentarily increased in magnitude to 10 m/s and changed in direction.

It is well to note that the accuracy of any graphical procedure depends on the care that is taken with each of the several steps leading up to the final result. For this reason, every effort should be made to ensure that all lengths and angles are measured correctly and precisely.

When the angle between two vectors is 90°, the magnitude of their resultant can readily be computed using the theorem of Pythagoras (p. 420). For example, the magnitude of the ball carrier's resultant velocity is ($\sqrt{8^2 + 6^2} = \sqrt{64 + 36} = \sqrt{100} =$) 10 m/s. Computational methods may also be used to determine the magnitude of the resultant when the angle between the two vectors is not 90° and to determine the direction in which the resultant acts regardless of the angle between the two vectors. Information concerning these methods, which requires a knowledge of trigonometry and is thus beyond the scope of this book, is readily available.[2,3]

## Vector Components

While it is often necessary to determine the resultant of two vectors, there are also many occasions when it is desirable to reverse this process—that is, to take a vector and resolve it into *components* that act in given directions. For example, when analyzing the motion of a shot—or of any other body projected into the air—it is usually necessary to resolve the velocity at which the implement is released into its horizontal and vertical components. To illustrate, consider the case of a shot released with a velocity of 14 m/s at an angle of 40° above the horizontal. The horizontal and vertical components of the velocity of release can be determined as follows (Fig. 80):

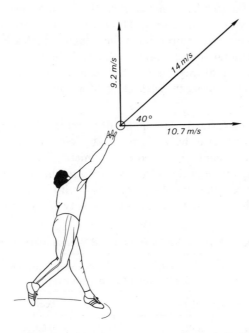

**Fig. 80.** *The parallelogram of vectors' construction can also be used to find the components of a vector.*

[2]James G. Hay, *The Biomechanics of Sports Techniques* (Englewood Cliffs, N.J.: Prentice-Hall, 1978), pp. 25–26.

[3]Katharine F. Wells and Kathryn Luttgens, *Kinesiology: Scientific Basis of Human Motion* (Philadelphia: W. B. Saunders, 1976), pp. 261–65.

1. Make a rough sketch of the body involved (the shot) and place a dot near the center of the body.
2. With a ruler and a protractor, draw lines through this dot to represent the horizontal and vertical directions.
3. Select a scale so that arrows drawn to represent the velocity of release and its horizontal and vertical components can be as large as space permits.
4. Place the baseline of a protractor on the line representing the horizontal direction so that its center coincides with the dot, and mark the direction in which the shot is released relative to the horizontal.
5. Start with the dot and draw an arrow, pointing away from the body, to represent the velocity of release. (Since the shot is released at an angle of 40° above the horizontal, the angle between the arrow representing the velocity of release and the horizontal must, of course, be 40°.)
6. Through the point of this arrow draw lines that correspond to the horizontal and vertical directions.
7. Where they meet the lines drawn through the other end of the arrow, place arrowheads on the two lines that start at the dot and serve to indicate the horizontal and vertical directions. The arrows thus formed represent the horizontal and vertical components of the velocity of release or, a little more simply, the horizontal and vertical velocities of release.
8. Measure the lengths of these arrows and convert the measurements to the values they represent using the scale selected in step 3.

The horizontal and vertical components of a vector, like the resultant vector discussed earlier, can also be determined by computational means. The procedure involved, a thinly disguised trigonometrical procedure, consists of multiplying the magnitude of the release velocity by the appropriate factors in Table 6. For example, in the case of a shot released with a velocity of 14 m/s at an angle of 40° to the horizontal, the horizontal velocity of release (HV) is given by

**TABLE 6  Multiplication Factors to Determine the Horizontal and Vertical Velocities of Release**

| Angle of Release* | Multiply Resultant Velocity by Factors Indicated to Get Required Component Velocity | |
|:---:|:---:|:---:|
| | Horizontal Velocity | Vertical Velocity |
| 0° | 1.000 | 0.000 |
| 10° | 0.985 | 0.174 |
| 20° | 0.940 | 0.342 |
| 30° | 0.866 | 0.500 |
| 40° | 0.766 | 0.643 |
| 50° | 0.643 | 0.766 |
| 60° | 0.500 | 0.866 |
| 70° | 0.342 | 0.940 |
| 80° | 0.174 | 0.985 |
| 90° | 0.000 | 1.000 |

*The *angle of release* is defined as the angle that the velocity of release makes with the horizontal.

$$HV = 14 \times 0.766$$
$$= 10.724 \text{ m/s}$$

and the vertical velocity of release (VV) by

$$VV = 14 \times 0.643$$
$$= 9.002 \text{ m/s}$$

## PROJECTILE MOTION

Projectile motion—that is, the motion of bodies that have been projected into the air—occurs with considerable frequency in physical education and athletics. Shot putters, golfers, and baseball, basketball, and volleyball players project a shot or a ball; discus and javelin throwers, archers, and badminton players project variously shaped objects; and jumpers, divers, and gymnasts project themselves into the air. In all these cases, the outcome depends largely on the performer's ability to control or (if control is impossible) to predict, the result of the projectile motion involved. It is therefore important that physical education teachers and coaches have some knowledge of the mechanical factors that govern such motion.

When a body is projected into the air (and thus becomes a *projectile*), two things act to change its motion—gravity, which pulls it toward the earth with a constant acceleration of 9.81 m/s$^2$, and air resistance, which serves to retard its forward motion and, under certain circumstances, to cause it to be lifted (p. 242). For the purposes of the following discussion, the effects of air resistance are assumed to be negligible—as indeed they are for many of the projectile motions encountered in physical education and athletics. The effects of air resistance will be discussed in Chapter 13.

### *Trajectory*

If gravity did not act to change its motion, a projectile would continue to travel indefinitely with the same velocity it had at the moment of release—that is, at the moment it became a projectile. Consider, for example, the case of the shot referred to earlier (p. 127). In this case, the shot was released with a velocity of 14 m/s at an angle of 40° to the horizontal or, to state it another way, with a horizontal velocity of 10.7 m/s and a vertical velocity of 9.0 m/s. In the absence of external influences, the shot would thus travel 10.7 m horizontally and 9.0 m vertically during every second of its flight (Fig. 81).

While shot putters might well wish that they could throw the shot out of sight in this fashion, in reality the result described is only half correct. Because gravity

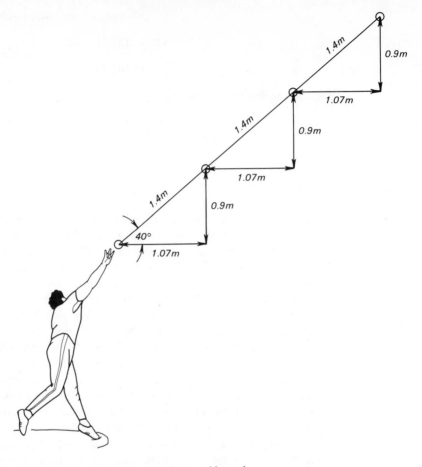

**Fig. 81.** *If gravity could be ignored, a shot would travel equal horizontal and vertical distances in equal periods of time. Thus, a shot released at 14 m/s at an angle of 40° to the horizontal would travel 1.07 m horizontally and 0.9 m vertically, for each 0.1 s for its flight.*

acts in a vertical direction and thus has no influence on the horizontal motion of the shot, nothing acts to alter this horizontal motion. As a result, the horizontal velocity of the shot remains constant throughout its flight, and it travels equal horizontal distances in equal intervals of time.

However, because the shot is subjected by gravity to a constant vertical acceleration of $9.81 \text{ m/s}^2$, its vertical velocity changes from some positive value at release—that is, assuming that the upward direction is positive—to zero at the peak of the flight, to some negative value immediately before landing. One effect of these changes in velocity is a progressive increase in the distance that the shot falls below the path it would follow in the absence of gravity. This distance, *l*, which can be computed using the equation

$$l = \frac{1}{2} gt^2 \tag{4}$$

is given in Table 7 for selected times of flight. The path actually followed by the shot can be obtained by modifying the straight-line path shown in Fig. 81 to allow for the influence of gravity (Fig. 82). The resulting flight path or *trajectory* is a smooth symmetrical curve. Similar curves—members of a family known in geometry as *parabolas*—are obtained when the flight paths of other projectiles

**TABLE 7   Distances a Projectile Falls Due to Gravity in Selected Time Intervals***

| Time (s) | 0.1 | 0.2 | 0.3 | 0.4 | 0.5 | 1.0 | 1.5 | 2.0 | 2.5 | 3.0 |
|---|---|---|---|---|---|---|---|---|---|---|
| Distance (m) | 0.05 | 0.20 | 0.44 | 0.78 | 1.23 | 4.91 | 11.04 | 19.62 | 30.66 | 44.15 |

*These are the distances that a projectile falls *below the path it would follow in the absence of gravity.*

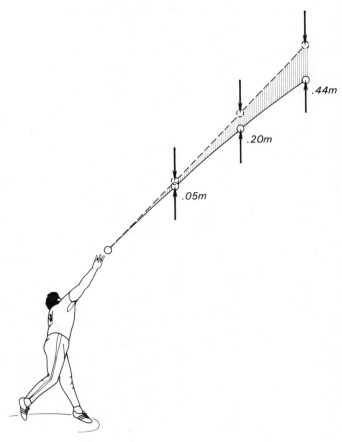

.44m

.20m

.05m

**Fig. 82.** *Gravity causes a projectile to deviate from the path it would follow in the absence of gravity.*

are plotted in similar fashion. (The term trajectory is sometimes confused with the term projectile, which looks and sounds rather similar. A projectile is a body that is projected into the air; a trajectory is the flight path followed by such a body.)

## Range

The horizontal distance that a projectile travels during its flight is of interest not only to shot putters but to tennis players, football punters, golfers, and many others.

The horizontal distance traveled by a body in motion is determined simply by multiplying its average horizontal velocity by the time involved. Thus, a cyclist riding with an average horizontal velocity of 40 km/h, will travel 40 km in 1 h, 80 km in 2 h, and so on. Similarly, a shot released with a horizontal velocity of 9 m/s will travel 9 m horizontally in 1 s of flight, 18 m in 2 s and so on. (The use of an instantaneous velocity in this last computation is justified because, in projectile motion, average and instantaneous horizontal velocities are equal.)

The total distance that a cyclist travels obviously depends on the average horizontal velocity and on the length of time spent riding. The horizontal distance that the shot travels—the *range* of the throw—similarly depends on the horizontal velocity of the shot and the time it spends traveling at that velocity—that is, on the *time of flight*. Thus, the performance of a shot putter can only be improved by increasing the horizontal velocity of the shot at release and/or the time of flight.

Such a conclusion raises the obvious question—"What determines the time of flight?" As indicated previously (p. 129), all projectiles experience a constant downward vertical acceleration of 9.81 m/s$^2$ due to the influence of gravity. Thus, ignoring the effects of air resistance, the magnitude of the vertical velocity of a projectile is reduced by 9.81 m/s for every second of its upward flight and is increased by the same amount for every second of its downward flight. For example, an arrow released with an upward vertical velocity of 24.53 m/s in an attempt on a distance-shooting record will have a vertical velocity of (24.53 − 9.81 =) 14.72 m/s at the end of the first second of flight, (14.72 − 9.81 =) 4.91 m/s at the end of the second second, and so on. This reduction in the arrow's vertical velocity will continue until it reaches zero at the peak of its flight. At this point—reached (24.53/9.81 =) 2.5 s after release—gravity begins to increase the magnitude of the arrow's vertical velocity so that after 1 s it will be traveling downward at (0 + 9.81 =) 9.81 m/s, after 2 s at (9.81 + 9.81 =) 19.62 m/s, and so on. Because the vertical velocity of the arrow is now increasing at the same rate as it was decreasing earlier, it will take exactly the same time for it to return from the peak to the level from which it was released as it did to travel from release to peak—that is, 2.5 s. Thus, the vertical velocity at release directly determines the time of the upward flight and, since the two times are equal, indirectly determines the time of its downward flight from the peak to the level at which it was released. If the point at which the arrow landed is lower than the point of release, the time required to complete its downward flight will be increased by a corresponding amount. In this case, the time taken for the arrow to reach the peak of its flight is governed by its vertical velocity at release, and the time taken for it to travel from the peak to its point of landing is governed, additionally, by the vertical distance between the points of release and landing. (This distance—the height of release relative to the height of landing—is referred to throughout this text as the *relative height of release*.) The total time of flight is determined, therefore, by just two factors—the vertical velocity and the relative height of release. Thus, an archer can only increase the time of flight of an arrow, and therefore the distance that it travels, by increasing one or both of these factors.

The conclusions reached in the previous two paragraphs have very important practical implications for, considered together, they indicate the only way in which the horizontal range of a projectile can be modified—that is, by a change in one or more of the factors that determine that range. These factors are the horizontal and vertical velocities at release and the relative height of release. Addition of the first two of these, using the parallelogram of vectors (p. 125), yields the magnitude and direction of the velocity of release. These latter are frequently referred to as the *speed of release* and the *angle of release*. The range of a projectile may thus be said to be determined by its speed, angle, and relative height of release.

### Other Characteristics

The process used to deduce that the range of a projectile is determined by the speed, angle, and relative height of release might also be used to identify the factors that determine other characteristics of a projectile motion. If this is done, the same result is obtained in every case, for the speed, angle, and relative height of release determine not only the range of a projectile but every other characteristic of its motion as well. They govern its location at any given time after release and thus, of course, the form of its trajectory. They govern the time it takes to reach a given point on its path. Finally, they govern the velocity it has at any given instant or at any given location on its path.

### Angle of Release

The angle at which a body is projected into the air is frequently discussed in physical education and athletics. This angle—usually referred to as the angle of release in throwing and striking activities and the angle of takeoff in jumping activities—is the acute angle between the velocity vector and the horizontal at the instant of release or takeoff (Fig. 83). The size of the angle used varies considerably according to the purpose of the activity. For example, activities like the shot put, long jump, and ski jump, in which the performer seeks a large horizontal range, generally involve the use of smaller angles than do those like the high jump and most trampoline stunts, in which height is the objective.

**Fig. 83.** *The angle of release is the acute angle between the horizontal and the velocity vector at the instant of release.*

The angles of release and takeoff reported in the research literature for various activities (Table 8) often differ quite markedly from what is generally thought. For example, ski jumpers do *not* take off in an upward direction, long jumpers do *not* take off at anything like 45°, and high jumpers rarely take off at angles of more than 60°.

The relatively small angles of release reported for some of the activities in which the performer attempts to obtain a large horizontal range raises the obvious question, "What is the optimum angle of release or takeoff in such cases?" The answer to this question has three parts. When the projectile lands at the same level as that from which it was released, the optimum angle is 45°. When the projectile lands at a level below that at which it was released—as in the shot put and long jump—the optimum angle is always less than 45°. The exact

**TABLE 8    Angles of Takeoff and Release Used in Selected Activities**

| Activity | Angle* (degrees) | Source |
|---|:---:|---|
| Ski jumping | −4 | Komi et al.[a] |
| Tennis serve | −3–15 | Owens and Lee[b] |
| Racing dive | 5–22 | Heusner[c] |
| Triple jump (step) | 11–13 | Nett[d] |
| Volleyball float serve | 13–20 | McElroy[e] |
| Triple jump (hop) | 16–18 | Nett[d] |
| Triple jump (jump) | 16–21 | Nett[d] |
| Long jump | 17–22.5 | Nigg[f] |
| Javelin throw | 25–34 | Nigg et al.[g] |
| Discus throw | 35–38.5 | Terauds[h] |
| Shot put | 33.7–40 | Kuhlow and Heger[i] |
| Hammer throw | 40 | Kuzneyetsov[j] |
| High jump (Fosbury flop) | 40–48 | Dapena[k] |
| High jump (straddle) | 42–53 | Dapena[l] |
| Handspring vault (board) | 52 | Fetz and Opavsky[m] |
| Front somersault (tuck) | 53 | Knight et al.[n] |
| Round-off, back somersault (layout) | 58 | Fetz and Opavsky[o] |
| Round-off, back somersault (tuck) | 63.5 | Fetz and Opavsky[p] |
| Standing back somersault (tuck) | 75 | Fetz and Opavsky[q] |

*Angles above the horizontal are considered to be positive and angles below the horizontal to be negative. Where one number is given, this is the mean of all the values obtained. Two numbers indicate the range of the recorded values.

[a] Paavo Komi, Richard C. Nelson, and Matti Pulli, *Biomechanics of Skijumping* (Jyväskylä: University of Jyväskylä, 1974), p. 25.

[b] Mary S. Owens and Hong Y. Lee, "A Determination of Velocities and Angles of Projection for the Tennis Serve," *Research Quarterly*, 40 (December 1969), 750–54.

[c] William W. Heusner, "Theoretical Specifications for the Racing Dive: Optimum Angle of Take-off," *Research Quarterly*, 30 (March 1959), 25–37.

[d] Toni Nett, *Die Technik beim Hürdenlauf und Sprung* (Berlin: Verlag Bartels & Wernitz, 1961), p. 114.

[e] G. Keith McElroy, "Understanding the Volleyball Float Serve," *Sports Coach*, 3 (Winter 1979), 35–37.

[f] Benno M. Nigg, *Sprung Springen Sprünge* (Zürich: Juris Verlag, 1974), p. 65.

[g] B. Nigg, K. Roethlin and J. Wartenweiler, "Biomechanische Messungen beim Speerwerfen," *Jugend und Sport*, 31 (1974), 172–74.

[h] Juris Terauds, "Some Release Characteristics of International Discus Throwing," *Track and Field Quarterly Review*, 75 (March 1975), 54–57.

[i] Angela Kuhlow and Werner Heger, "Die Technik des Kugelstossens der Männer bei den Olympischen Spielen 1972 in München," *Leistüngssport* (Supplement), 2 (March 1975), 49.

[j] W. Kuzneyetsov, "Speed and Path of the Hammer," *Track Technique*, 25 (September 1966), 796–98.

[k] Jesus Dapena, "Mechanics of Translation in the Fosbury Flop," *Medicine and Science in Sports and Exercise*, 12 (Spring 1980), 37–44.

[l] Jesus Dapena, Personal Communication, January 1980.

[m] Friedrich Fetz and Pavle Opavsky, *Biomechanik des Turnens* (Frankfurt: Limpert Verlag, 1968), p. 246.

[n] Stephen Knight, Barry D. Wilson, and James G. Hay, "Biomechanical Determinants of Success in Performing a Front Somersault," *International Gymnast*, 20 (March 1978), 54–56.

[o] Fetz and Opavsky, *Biomechanik des Turnens*, p. 183.

[p] Fetz and Opavsky, *Ibid.*, p. 181.

[q] Fetz and Opavsky, *Ibid.*, p. 180.

magnitude of the optimum angle in such cases depends on the velocity and relative height of release. For example, the optimum angle for the able-bodied shot putter in Fig. 84(a) is 42° while that for his paraplegic counterpart in Fig. 84(b) is only 39°. Finally, when the projectile lands at a level above that from which it was released—a rare occurrence in sport—the optimum angle is greater than 45° and again depends on its velocity and relative height of release.

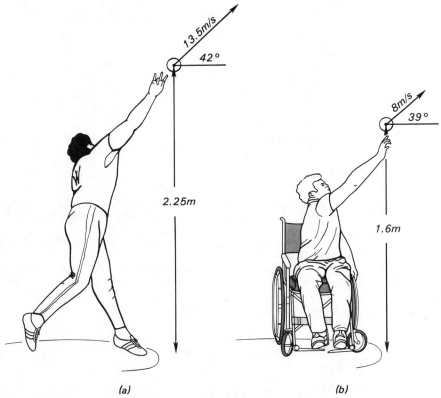

(a)                          (b)

**Fig. 84.** *Optimum angles of release for (a) an able-bodied shot putter, and (b) a paraplegic shot putter. (Note: When a projectile lands below its point of release, the optimum angle of release is a function of its velocity and relative height of release).*

There are several ways in which these results may be tested or demonstrated. A garden hose may be directed so that the water leaves the nozzle at a variety of angles and the distance the water travels noted. If the hose is held at, or near, ground level, it will be readily seen that the maximum distance is obtained when the water leaves the nozzle at approximately 45°. Another method[4] uses the

---

[4]Ronald F. Kirby, "A Projection Board Illustrator," *Journal of Health, Physical Education and Recreation*, 41 (May 1972), 75.

principle employed in Fig. 82. In this method, thin strips of wood are cut in proportion to the distances that a projectile falls due to gravity (Table 7) and are then fastened, in order, to a long piece of wood. When the resulting device is held against a vertical surface so that the long piece of wood indicates the direction in which the projectile is released, the ends of the thin strips indicate points on the projectile's flight path (Fig. 85). If a ground line is drawn on the vertical surface, the angle of release indicated by the long piece of wood can be varied systematically to determine the optimum for a given case.

**Fig. 85.** *Model used to demonstrate basic characteristics of projectile motion.*

## Relative Importance of Factors

In a practical teaching or coaching situation, it is generally not sufficient to know that the range of a projectile is governed by just three factors. To give each appropriate emphasis, one must also know something about their relative importance in determining the outcome. One must have some idea about the answer to such questions as, "Will an increase in the relative height of release produce a greater improvement in the performance than a comparable increase in the speed or angle of release?"

It is possible to obtain some indication of the relative importance of the three factors by taking a particular case and increasing each of the factors, in turn, by a comparable amount. Consider, for example, the case of a long jumper who had a speed at takeoff of 8.90 m/s, an angle of takeoff of 20°, and a relative height of takeoff of 0.45 m and who jumped a distance of 7.00 m. The horizontal distance that the jumper traveled during the airborne phase of the jump can be computed using the known values for the speed, angle, and relative height of takeoff. This distance was 6.23 m. (The known values for the jump are shown in Table 9, column 1.) Now, if each of the three factors governing the range of a projectile is increased by 5 percent while the other two remain the same and the horizontal range of the airborne phase is recomputed, it is possible to obtain an indication of the influence that each exerts on the final result (Table 9, columns 2–4). Such an analysis reveals that an increase in the speed of release yields an increase in the distance of the jump that is substantially greater than that produced by corresponding increases in the other two factors. In short, the speed of release appears to be much more important than the angle and relative height of release. While this finding is consistent with practical experience, it should be pointed out that it is virtually impossible in practice to change one factor without bringing about changes in the other two. Further, for a given case, it may be much easier to make a change in one factor than it is to make a comparable change in others. Thus, although the speed of release is generally the most important of the three takeoff factors, it is possible that one of the others may be the most important under certain circumstances.

**TABLE 9**  **Effect of 5% Increases in Speed, Angle, and Relative Height of Takeoff for a Long Jump**

| Variable | Values for Actual Jump (1) | Values for Hypothetical Jumps Under Different Conditions | | |
|---|---|---|---|---|
| | | Speed of Takeoff Increased 5% (2) | Angle of Takeoff Increased 5% (3) | Relative Height of Takeoff Increased 5% (4) |
| Speed of takeoff | 8.90 m/s | 9.35 m/s | 8.90 m/s | 8.90 m/s |
| Angle of takeoff | 20° | 20° | 21° | 20° |
| Relative height of takeoff | 0.45 m | 0.45 m | 0.45 m | 0.47 m |
| Horizontal range | 6.23 m | 6.77 m | 6.39 m | 6.27 m |
| Change in horizontal range | — | 0.54 m | 0.16 m | 0.04 m |
| Distance of jump | 7.00 m | 7.54 m | 7.16 m | 7.04 m |

*Summary*

The major conclusions reached in the foregoing anaylsis of projectile motion (air resistance neglected) may be summarized as follows:

1. The horizontal velocity of a projectile is constant. The projectile thus travels equal distances in a horizontal direction in equal intervals of time.
2. The vertical velocity of a projectile is constantly changing due to the influence of gravity. The projectile thus travels unequal distances in a vertical direction in equal intervals of time.
3. The constant horizontal velocity and constantly changing vertical velocity of a projectile cause it to follow a parabolic flight path (or trajectory).
4. The horizontal range of a projectile depends on its horizontal velocity and its time of flight.
5. The time of flight of a projectile depends on its vertical velocity at release (or takeoff) and the relative height of release—that is, the difference between the height at which it was released and the height at which it landed.
6. The horizontal range of a projectile—and all other characteristics of its motion—are determined by its speed, angle, and relative height of release (or takeoff).
7. For maximum range, the optimum angle of release is 45° if release and landing are at the same height, less than 45° if release is at a greater height than landing, and more than 45° if release is at a lesser height than landing.
8. In general, the speed of release exerts a greater influence on the range obtained than do either the angle or the relative height of release.

**EXERCISES**

Give answers of magnitude less than 1.0—that is, $> -1.0$ and $< 1.0$—correct to two decimal places and all other answers correct to one decimal place.

1. The Comrades Marathon is one of the world's most famous ultra-long-distance running events. Run in South Africa, over a distance that is more than twice the length of a normal marathon, the race attracts over 4,000 runners each year. Fig. 86 shows two views of the course—(a) a plan, or overhead, view and (b) an elevation, or side, view—together with linear scales and, in (a), an arrow indicating North. Using this information, a rule, and a protractor, work out the horizontal and vertical displacements that a runner experiences in going from the start in Pietermaritzburg to the finish in Durban. Using the relevant information, a rule, and a piece of nylon cord—or something else that is thin, flexible, and inextensible—work out

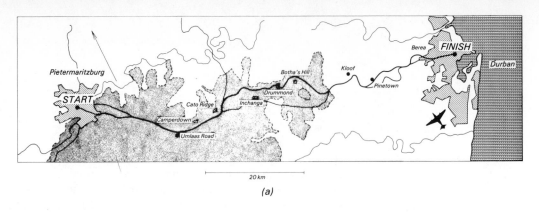

*(a)*

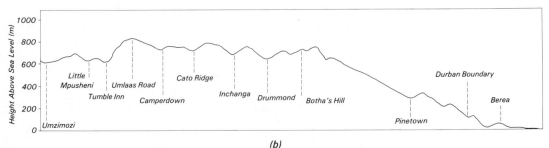

*(b)*

the length of the course as shown in Fig. 86(a). Measurement errors aside, is this the actual length of the course, more than the actual length, or less than the actual length? Finally, assume that a runner records a time of 6 h 30 min and compute the average speed and average velocity of the runner over the course. (The accuracy of the measurements obtained in this exercise will likely be improved if the maps are enlarged with the aid of an opaque projector, a pantograph, or some other device before the measurements are taken.)

2. A swimmer covers the first 50 m length of a 100 m backstroke race at an average speed of 1.5 m/s and the second length at an average speed of 1.2 m/s. (a) What were the times for the first and second lengths, respectively? (b) What was the average speed for the whole race? (c) What was the average velocity for the whole race?

3. If a runner has the same forward velocity at takeoff as at touchdown—that is, at the end and the beginning of the support phase, respectively—how does the runner's average forward velocity during this phase compare with the average forward velocity during the next airborne phase?

4. In the course of an 800 m race, a swimmer's times for two consecutive laps are identical. A physicist watching the race comments that the swimmer is moving with a constant velocity. The observer's companion, also a physicist, disagrees. (a) Can they both be right? (b) If not, which of the two is wrong and why?

5.  In August 1975, John Walker (New Zealand) became the first man to run a mile (1.6 km) in under 3 min 50 s when he set a world record of 3 min 49.4 s in Goteborg, Sweden. Four years later, Sebastian Coe (Great Britain) broke Walker's record with a 3 min 49.0 s performance in Oslo, Norway. The lap times for these two athletes in their respective record-breaking runs were

| | Walker | Coe |
|---|---|---|
| | 55.8 s | 57.5 s |
| | 1 min 55.1 s | 1 min 54.5 s |
| | 2 min 53.0 s | 2 min 52.0 s |
| | 3 min 49.4 s | 3 min 49.0 s |

Using these times, compute the average speeds for each runner over the full distance, over the first and second half miles, and over each of the four quarter-mile laps. Enter these values in the following table and note how the differences between the two athletes become more clearly apparent as the distance over which the average speed is computed becomes shorter.

| | Average Speed | |
|---|---|---|
| | Walker | Coe |
| Mile (1609 m) | | |
| 1st 880 yd (805 m) | | |
| 2nd 880 yd | | |
| 1st 440 yd (402 m) | | |
| 2nd 440 yd | | |
| 3rd 440 yd | | |
| 4th 440 yd | | |

6.  A softball player sprints from first base to second base. If her instantaneous horizontal velocity is 9.3 m/s when she is 8 m from second base and 7.5 m/s when she is 5 m from second base 0.4 s later, what is her average acceleration over that 3 m interval? (Assume that she travels in the positive direction when moving from first to second.)

7. Ebert[5] reported that a tennis ball served at 150 mph (241 km/h) crosses the net some 39 ft (11.89 m) from the service line at 108 mph (174 km/h). (a) Can you compute the average acceleration of the ball from the moment it leaves the racket until it crosses the net? (b) If so, make this computation. If not, what additional information would you need before you could make the computation?

8. A basketball player does a standing upward jump to gather in a rebound but misses the ball completely. If the upward direction is regarded as positive, which of the following statements best describes the acceleration that the player experiences due to gravity during the flight?
   (a) It is negative throughout.
   (b) It is negative on the way up, zero at the peak of the flight, and positive on the way down.
   (c) It is positive on the way up, zero at the peak of the flight, and negative on the way down.
   (d) It is positive throughout.

9. A cross-country skier travels 5 km due North, makes a right-angled turn, and travels a further 12 km due East. (a) If the distance is covered in 51 min, what is the skier's average speed in km/h? (b) What is the magnitude of the skier's average velocity in km/h?

10. A halfback in football takes a handoff from the quarterback in the middle of the field at the opponent's 23 yd-line and runs in a straight line toward the near right corner of the end zone. After traveling 10 yd, the back is running at 8.5 m/s. Make a scale drawing of the relevant part of the field. (A regulation-sized field is $53\frac{1}{3}$ yd wide.) Indicate the position of the player with a dot and the player's velocity with an arrow drawn to some appropriate scale (1 cm = 0.5 m/s, for example). Use the graphical method for resolving a vector into components and determine (a) the downfield component and (b) the crossfield component of the player's velocity.

11. (a) At what point does a pole vaulter become a projectile during the course of a successful vault? (b) Is it possible for the vaulter to become a projectile more than once after leaving the ground and still have a successful vault? If so, how?

12. At what point in the course of a jump does a water-ski jumper become a projectile?

13. (a) Fig. 87 is a stroboscopic photograph of a golf ball bouncing on an artificial surface. Using the black-and-white scale included in the photograph—and ignoring the fact that it is a little behind the plane in which the ball moved—determine the horizontal distances between consecutive posi-

[5]Lynn J. Ebert, "Materials in Sports: A New Undergraduate Course in Engineering," *Machine Design*, 48 (August 12, 1976), 19.

tions of the ball during each airborne phase. (Each interval of the scale is 0.5 m long.) What could account for any differences in these horizontal distances found within a given airborne phase?

(b) If the time between consecutive exposures of the film—and, thus, between the consecutive positions recorded on the film—was .04 s, what was the horizontal velocity of the ball during each airborne phase? Is this an average or an instantaneous value for the horizontal velocity?

14. The famed cliff divers of Acapulco, Mexico take off from a rocky cliff face 26.5 m above the surface of the water. In the course of their flight—of which all but the last 1–2 m is over rock—they travel 8 m horizontally. (a) If the time taken from takeoff to entry is 2.3 s, what is the diver's horizontal velocity at takeoff? (Ignore air resistance.) (b) For approximately how long is the diver over the water before entry?

15. A shot putter releases the shot at 13.5 m/s at an angle of 40° to the horizontal and from a height of 2.30 m above the ground. (a) Take a sheet of graph paper and rule vertical and horizontal axes starting near the bottom, left-hand corner of the sheet. (b) Label these axes in meters using the same scale for both axes. The vertical axis should range from 0 to 25 m and the horizontal axis from 0 to 25 m. (c) From the point on the vertical axis representing a height of 2.30 m (the height of release), rule a line at an angle of 40° (the angle of release) to the horizontal direction. (d) Compute the distance that the shot would have traveled along this line in consecutive 0.1 s intervals following release in the absence of gravity and plot the corresponding points on the line. (e) From Table 7, determine the distance the shot would have fallen below the line in the 0.1 s, 0.2 s, 0.3 s, . . . , and so on immediately following release and plot the corresponding points on the graph. (f) Join these points with a smooth curve and, from the point at which it crosses the horizontal axis of the graph, determine the range of the throw. (g) Determine how high the shot rises above its point of release and estimate how long it takes to reach its peak height. (h) Estimate the time of flight of the shot. (i) Determine what the horizontal range of the shot would have been if it had been released with the same velocity from ground level. (It would, of course, be a most unusual shot-putting contest in which such a thing could actually occur!) (j) Estimate what its time of flight would have been in this latter case.

16. A gymnast leaves the ground with a vertical velocity of 3 m/s and a horizontal velocity of 2.5 m/s and lands, after completing a front somersault, at the same height as takeoff. (a) What was the gymnast's vertical velocity at the peak of flight? (b) What was the gymnast's vertical velocity at the instant of touchdown? (c) What was the gymnast's horizontal velocity at the peak of flight? (d) What was the gymnast's horizontal velocity at the instant of touchdown? (e) If the gymnast took 0.3 s to reach the peak of flight, how long did it take for the gymnast to descend from that point to the point of regaining contact with the ground? (f) If the total time of flight was 0.3 s plus the answer to the previous question, how far did the gymnast travel horizontally during the airborne phase of the motion?

17. Angles of takeoff and release for various athletic activities are listed in Table 8. Consult the research literature on the biomechanics of sport and compile a list of ten additional examples of such angles. Cite the source of your information in each case. [*Hint: Research Quarterly* and *Medicine and Science in Sports and Exercise* are two journals that frequently include articles on the biomechanics of sport. *Biomechanics I-VII*, the proceedings of the International Congresses on Biomechanics, are other likely sources of such information.]

## RECOMMENDED READINGS

Dyson, Geoffrey H. G., *The Mechanics of Athletics*. New York: Holmes & Meier, 1977, pp. 14–27 (Motion).

Rackham, George, *Diving Complete*. London: Faber and Faber, 1975, pp. 151–164 (Time).

Rogers, Eric M., *Physics for the Inquiring Mind*. Princeton, N.J.: Princeton University Press, 1977, pp. 36–52 (Projectiles: Geometical Addition: Vectors).

Tricker, R. A. R., and B. J. K., *The Science of Movement*. London: Mills and Boon, 1966, pp. 1–12 (The Problems of Movement).

# Chapter 10

# Describing Angular Motion (Angular Kinematics)

The quantities used to describe angular motions are very similar (indeed, directly equivalent) to those used to describe linear motions.

## ANGULAR DISTANCE AND ANGULAR DISPLACEMENT

The *angular distance* through which a rotating body moves is the angle between its initial and final positions measured following the path followed by the body. Thus, the gymnast who swings her legs forward through 120° and then backward through 150° so that she can move from a sitting to a standing position on the beam (Fig. 88) moves her legs through an angular distance of (120° + 150° =) 270°. The *angular displacement* through which a rotating body moves is equal in magnitude to the smaller of the two angles between its initial and final positions. The direction of the angular displacement—which must also be stated—is usually indicated by the word clockwise or counterclockwise or, even more commonly, by the use of a plus or a minus sign. In the latter case, the counterclockwise direction is normally considered to be positive, and the clockwise direction negative. The two angles between the initial and final positions of the gymnast's legs in Fig. 88 are 30° and 330°. The magnitude of their angular displacement is thus 30°. The direction in which her legs must be considered to move if they are to rotate through 30° from their initial position to their final position is the clockwise or negative direction. The angular displacement involved is, thus, −30°.

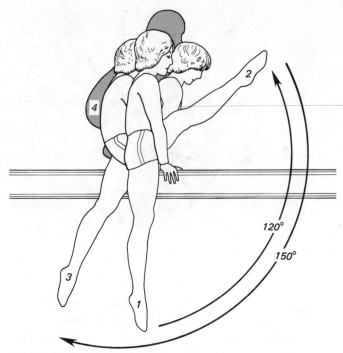

**Fig. 88.** *Angular distance and displacement in the execution of an exercise on the balance beam.*

## ANGULAR SPEED AND ANGULAR VELOCITY

The average angular speed of a rotating body is found by dividing the angular distance through which it has moved by the time taken:

$$\text{Average angular speed} = \frac{\text{Angular distance}}{\text{Time taken}}$$

or, in algebraic terms,

$$\bar{\sigma} = \frac{\phi}{t} \tag{5}$$

where $\bar{\sigma}$ = "sigma bar" = the average angular speed, $\phi$ = "phi" = the angular distance, and $t$ = the time taken. The average angular velocity is found in similar fashion:

$$\text{Average angular velocity} = \frac{\text{Angular displacement}}{\text{Time taken}}$$

or, in algebraic terms,

$$\bar{\omega} = \frac{\theta}{t} \tag{6}$$

where $\bar{\omega}$ = "omega bar" = the average angular velocity, $\theta$ = "theta" = the angular displacement, and $t$ = the time taken. Thus, if the girl in Fig. 88 takes 1.5 s to swing her legs forward (from their initial position) and then backward (to their final position), their average angular speed is (270° / 1.5 s =) 180° / s and their average angular velocity is (−30° / 1.5 s =) −20° / s.

## ANGULAR ACCELERATION

The average angular acceleration of a body is found by dividing the change in its angular velocity by the time taken:

$$\text{Average angular acceleration} = \frac{\text{Change in angular velocity}}{\text{Time taken}}$$

$$= \frac{\text{Final angular velocity} - \text{initial angular velocity}}{\text{Time taken}}$$

or, in algebraic form,

$$\bar{\alpha} = \frac{\omega_f - \omega_i}{t} \tag{7}$$

where $\bar{\alpha}$ = "alpha bar" = the average angular acceleration, $\omega_f$ and $\omega_i$ = the final and initial angular velocities, respectively, and $t$ = the time taken. Thus, if the legs of the girl in Fig. 88 have an angular velocity of −210° / s after completing 150° of their backward swing and an angular velocity of 0° / s, 0.7 s earlier at the limit of their forward swing, their average angular acceleration is [(−210° / s − 0° / s)/0.7 s =] −300° / s².

(As with the corresponding quantities used to describe linear motion (pp. 119–124), an angular speed, velocity, or acceleration may be either an average or an instantaneous quantity.)

## ANGULAR MOTION VECTORS

To find the resultant of two vectors or to resolve a vector into its components requires that each vector first be represented by an arrow. This presents some difficulties when the motion of interest and the vectors that describe or explain it are angular or rotary. While the representation of linear motion vectors—that is, vectors used to describe or explain linear motion—with straight arrows and angular motion vectors by curved arrows might have some initial appeal, the addition and resolution of vectors represented by curved arrows is, at least, very difficult and probably impossible. A simple convention—the *right-hand thumb rule*—is used to overcome this problem. According to this convention, an angular motion vector can be represented by an arrow that is drawn so that, when the fingers of the right hand are curled in the direction of rotation, the direction of the arrow coincides with the direction indicated by the extended thumb (Fig. 89). The length of the arrow is determined using an appropriate scale. The right-hand thumb rule can be used, for example, to represent the angular velocity of a diver performing a front $1\frac{1}{2}$ somersault dive (Fig. 90(a)), a reverse $1\frac{1}{2}$ somersault dive (Fig. 90(b)), and even a double-twisting, front $1\frac{1}{2}$ somersault dive

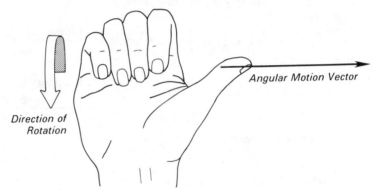

*Direction of Rotation*

*Angular Motion Vector*

**Fig. 89.** *Angular motion vectors are represented by arrows with the aid of the right-hand thumb rule.*

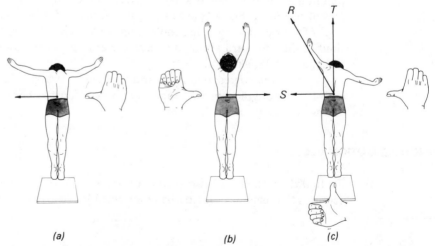

(a)                     (b)                     (c)

**Fig. 90.** *The right hand thumb rule applied to forward and backward somersaulting dives—(a) and (b), respectively—and to a forward somersaulting and twisting dive (c).*

(Fig. 90(c)). In this last case, the diver leaves the board with both a somersaulting angular velocity and a twisting angular velocity. The resultant angular velocity is the vector sum of these two and is found using the parallelogram of vectors in the usual way.

## VELOCITY AND ANGULAR VELOCITY

Punters, batters, boxers, and a score of others often seek to have some part of their bodies, or some piece of equipment they are using, moving at the maximum possible velocity at the moment it makes contact with a second body. The velocity of the part that makes contact with the second body is invariably the result of either a simple angular motion or a complex general motion that involves a combination of several angular motions. Consider the case of the young volleyball player executing an underarm serve in Fig. 91. In this case, the velocity imparted to the ball—and thus its chances of clearing the net—depends, among other things, on the velocity of the player's hand at the instant of contact. This, in turn, depends on the angular velocity of his arm at the same instant. The relationship between the velocity, $v$, and the angular velocity, $\omega$, is summarized in the equation:

$$v = \omega r \tag{8}$$

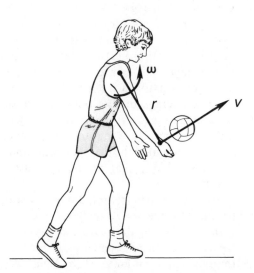

**Fig. 91.** *The velocity of the hand is equal to the product of the angular velocity of the arm and the distance from the hand to the axis of rotation.*

where $r$ = the radius, that is, the distance between the axis of rotation and the point of contact. This relationship suggests that there are just two ways in which the contact velocity of a rotating body can be increased—by increasing its angular velocity and/or by increasing the radius. Thus, if the boy in Fig. 91 is having trouble getting the ball over the net, he should probably concentrate on swinging his arm hard (to maximize $\omega$) and keeping it straight (to maximize $r$).

While the relationship summarized in Eq. 8 may be readily used in making simple mechanical analyses of human motion, special care must be exercised when using it to compute actual numerical values. This is because the angular velocity must be expressed—using a rather odd unit of angular measurement called a *radian*—in rad/s. However, provided it is not overlooked, this condition poses no real problems, because angular velocities expressed in °/s can readily be converted to radians per second by dividing by 57.3 (1 rad = 57.3°).

## EXERCISES

Give answers of magnitude less than 1.0—that is, $>-1.0$ and $<1.0$—correct to two decimal places and all other answers correct to one decimal place.

1. A gymnast does $2\frac{1}{2}$ consecutive giant swings on a horizontal bar—that is, $2\frac{1}{2}$ circles of the bar in an essentially straight or extended position. (a)Through what angular distances does the gymnast's body pass in the process? (State your answer in degrees and in radians.) (b) What angular displacement does the gymnast's body experience? (Assume that the gymnast moves in a clockwise direction and that this has arbitrarily been designated as the negative direction.) (c) If it takes 2.2, 2.0, and 0.9 s to complete the first, second, and half swings, respectively, what were the gymnast's average angular speeds for each of these three parts of the total performance? (d) What were the average angular velocities for each of the three parts? (e) What was the average angular speed for the total performance? (f) What was the average angular velocity for the total performance? (g) If the angular velocity was 7 rad/s at the end of the first half swing and 4.5 rad/s at the end of the next half swing 1.2 s later, what was the gymnast's average angular acceleration over that period?

2. A diver performed a front, double-twisting, $1\frac{1}{2}$ somersault dive from the 3 m board. (a) If the diver's body was inclined forward at an angle of 30° to the vertical on leaving the board and was vertical on entering the water 1.3 s later, what angular distance and displacement did the diver experience as a result of the somersaulting motion? (b) What were the average speed and average velocity of somersaulting? (Assume that the diver's somersaulting motion was in a clockwise, or negative, direction.) (c) What angular distance and displace-

ment were experienced as a result of the diver's twisting motion? (d) What were the average speed and average velocity of twisting? (Assume that the twisting motion was to the diver's left and that this was the positive direction for twisting.)

3. Imagine that you are sitting in the end zone at a football game watching a ball carrier running directly toward you. A defensive player, moving in the opposite direction hits the ball carrier at knee level and causes him to fall forward. At the moment this happens, you take a photograph of the action. (a) If you were to draw an arrow on this photograph to represent the angular velocity of the ball carrier, in what direction would you draw it? (b) If the ball carrier spun to his left in an attempt to avoid the tackler, in what direction would you draw the arrow representing the angular velocity at which he performed this spinning motion? (c) If the ball carrier was spinning to his left and falling forward at the same time, in what direction would you draw the arrow representing his resultant angular velocity? (These questions are probably best answered by drawing a sketch of the ball carrier. (Use Figure 79 (a), p. 125, as a guide and superimpose the appropriate arrows.)

4. There are many putting techniques used in golf. Two of the more popular are the wrist technique (in which the putter is rotated about an axis through the golfer's wrists while the rest of the body is kept stationary) and the arm technique (in which the golfer's arms and the putter are rotated as a unit about an axis through the golfer's chest approximately midway between the shoulders). The distance from the contact point on the clubhead to the axis for a given golfer is 0.86 m if the wrist technique is used and 1.32 m if the arm technique is used. If the clubhead must be moving at 3.5 m/s at contact to sink the putt, at what angular velocity must the club be moving at this time in each case?

## RECOMMENDED READING

Hay, James G., *The Biomechanics of Sports Techniques* (2nd ed.). Englewood Cliffs, N.J.: Prentice-Hall, 1978, pp. 45–55 (Angular Kinematics).

# Chapter 11

# Explaining Linear Motion (Linear Kinetics)

Kinetics is the branch of mechanics that deals with what causes a body to move in the way that it does. This chapter contains a discussion of quantities used in linear kinetics—that is, to explain linear motions.

## INERTIA

To perform a leg press on a weight machine (Fig. 92), the performer must move a load from a state of rest to one of motion and then, at the end of the exercise,

Fig. 92. *All bodies exhibit a characteristic reluctance to change what they are doing.*

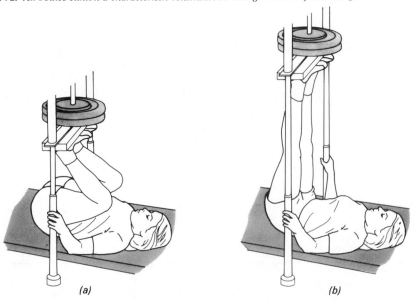

(a)                                                          (b)

reverse the process to bring it to rest in preparation for the next repetition. In each instance, the load seems to be reluctant to change its state. It seems, when at rest, to prefer to remain at rest and, when in motion, to remain in motion. In mechanics, this property of a body, this reluctance to change what it is doing, is known as its *inertia*.

## MASS

The quantity of matter in a body—its *mass*—is a measure of the body's inertia. For example, if another disc were added to the load in Fig. 92, the mass to be lifted—measured in kilograms (kg)—would be increased accordingly. The resistance of the load to efforts to get it moving and, once moving, to change its motion—that is, its inertia—would also be increased.

## FORCE

In attempting to change the initial state of the load, the woman in Fig. 92 (a) pushes vertically upward on the undersurface of the carriage supporting it. If she pushes hard enough, the load moves. If she does not push hard enough to move the load, she nonetheless causes it to more closely approach the point at which it will begin to move than it did at the outset. In other words, she tends to set it in motion.

A push (or pull) that alters, or tends to alter, the state of motion of a body is called a *force*. If the body is at rest, a force exerted by another body will set it in motion or at least will tend to do so. Similarly, if the body is moving in a straight line, a force exerted by another body will change, or tend to change, the velocity at which it is moving.

Forces are vector quantities—that is, they have a magnitude and a direction and can be added or resolved using the parallelogram of vectors—and are measured in Newtons (N).

In analyses of human motion, it is customary to regard the human body as a system of bones, muscles, ligaments, and other tissues and the forces exerted by one on another—as, for example, when a muscle contracts and exerts forces on the bones to which it is attached—as *internal forces*. Conversely, forces exerted on the body from outside the system—by gravity or by contact with some other body—are regarded as *external forces*.

## MOMENTUM

The term "momentum" is widely used in athletics to describe dominance at some stage in a team game and as an alternative to the term "speed." In mechanics, momentum has a narrow and precisely defined meaning. Momentum is the quantity of motion of a body and is equal to the product of the body's mass and velocity. That is,

$$\text{Momentum} = \text{Mass} \times \text{velocity}$$

or, algebraically,

$$M = mv \tag{9}$$

where $M$ = the momentum, $m$ = the mass, and $v$ = the velocity. Thus, a downhill skier with a mass of 70 kg and a velocity of 30 m/s has a momentum of 2100 kg-m/s.

Momentum is particularly important in impact situations because the result of the impact depends very largely on the momentum possessed by each of the bodies involved. The concept of momentum is often used, for example, as the basis for the argument that a small, fast man can be just as effective in contact situations in football as a larger, slower man. Unfortunately for small fast men, this is true only within very narrow limits. For example, a 100 kg man who runs the 100 m in 13 s has an average momentum of:

$$M = 100 \text{ kg} \times \frac{100 \text{ m}}{13 \text{ s}}$$

$$= 769 \text{ kg-m/s}$$

To match this, a 90 kg man must run the 100 m in 11.7 s, an 80 kg man must do it in a world-class time of 10.4 s; and a 70 kg man must completely shatter the world record by running the distance in 9.1 s! Clearly, then, a small, fast man can only be as effective as a larger, slower one if the difference in mass is small and the difference in velocity is large.

## NEWTON'S LAWS OF MOTION

By far the greatest single contributor to present knowledge in the field of mechanics was an Englishman, Isaac Newton (1642–1727). Among Newton's many far-reaching achievements was his formulation of the three laws of motion that, to this day, bear his name. These laws summarize and develop the concepts discussed in the previous sections.

### Newton's First Law

Newton's first law—the *law of inertia*—states that:

*Every body at rest, or moving with constant velocity in a straight line, will continue in that state unless compelled to change by an external force exerted upon it.*

Thus, when a curler pushes off from the hack (the starting block fixed on the ice) to deliver a stone, he and the stone would continue sliding forward across the ice (Fig. 93) were it not for the external forces that act to eventually bring them to a halt.

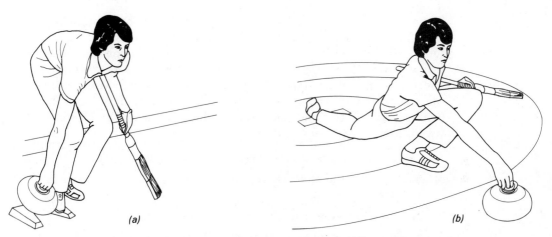

(a)                    (b)

**Fig. 93.** *External forces eventually bring the curler and the stone to rest.*

### Newton's Second Law

For bodies whose mass remains constant throughout the motion (and this includes most bodies of interest in physical education, athletics, and rehabilitation) Newton's second law—*the law of acceleration*—states that:

*The acceleration of a body is proportional to the force causing it and takes place in the direction in which the force acts.*

With an appropriate choice of units and the use of algebraic notation, this statement can be reduced to the well-known form

$$F = ma \tag{10}$$

where $F$ = the force exerted, $m$ = the mass, and $a$ = the acceleration. Thus, when a lineman in a football game charges an opponent, the acceleration he imparts to that man is directly proportional to the force he exerts and indirectly proportional to the man's mass. In other words, the harder he pushes and the smaller his opponent, the more likely he is to be able to drive him back.

Equation 10 provides a basis for defining the units commonly used in the measurement of force. Thus, 1 Newton (N) is that force that will produce an acceleration of 1 m/s$^2$ in a body of 1 kg mass:

$$1 \text{ N} = 1 \text{ kg} \times 1 \text{ m/s}^2$$

## Newton's Third Law

When a swimmer thrusts against the starting block at the beginning of a race, he exerts force in a downward and backward direction. As a result of this thrusting action, his body is driven in an upward and forward direction (Fig. 94(a)). The swimmer is thus accelerated in one direction when he exerts force in the opposite direction. At first glance, this appears to be in conflict with Newton's second law, which states that the acceleration of a body is in the same direction as the force that produces it. This apparent contradiction is explained by Newton's third law—the *law of reaction*—which states that

*For every action, there is an equal and opposite reaction.*

or, in less general and more useful terms,

*For every force that is exerted by one body on another, there is an equal and opposite force exerted by the second body on the first.*

Thus, when the swimmer exerts a force of, say, 1000 N in a downward and backward direction *against the block*, the block exerts a force of 1000 N in an upward and forward direction *against the swimmer* (Fig. 94(b)). It is customary to speak of one of these forces as the *action* and the other as the *reaction*, but which is which is purely arbitrary. (One of the most common causes of error in the application of Newton's third law is a failure to recognize that the action and reaction do not both act on the same body. It must be emphasized, therefore, that the action is a force exerted on one body and that the reaction is a contrary force exerted on the other body.)

It is sometimes difficult to accept that the action and reaction are equal in magnitude when the observed effects in a particular case are so markedly different. For example, the force exerted on the swimmer in Fig. 94 causes him to be driven forward into his dive, whereas that exerted against the block in reaction seems to have no effect at all. This disparity is due to the difference between the masses of the bodies involved. If the swimmer has a mass of 75 kg, a force of 1000 N will cause him to be accelerated at 13.3 m/s$^2$ (Newton's second law), an acceleration sufficiently large that its effects can readily be discerned by an observer. The block, however, is fixed firmly to the end of the pool, which, in turn, is imbedded firmly in the ground. Therefore, the swimmer is interacting, in effect, with a body whose mass is equal to that of the block plus the pool plus the earth, a body of enormous mass. The acceleration that a force of 1000 N produces in such a body is, thus, very small—so small, in fact, that it is imperceptible to the human eye.

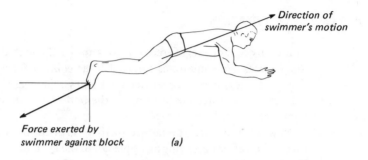

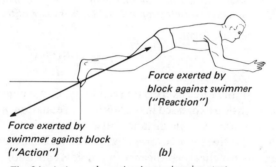

**Fig. 94.** *Action and reaction in a swimming start.*

Volleyball players are sometimes instructed to stamp their feet hard on the floor as they land, preparatory to going up for a spike or a block. Basketball players, high jumpers, and long jumpers are sometimes given similar instructions. These instructions are based on the idea that the larger the vertical force exerted by the performer against the takeoff surface, the larger the vertical force available in reaction to drive the performer into the air. While it is, of course, true that large downward forces evoke large upward forces in reaction, these latter forces serve no useful purpose unless they are exerted at the appropriate time. In the cases mentioned, large vertical forces are exerted against the floor or ground at the very beginning of the takeoff, as the performer lands at the end of the preceding step. The reaction to these forces occurs at the same time and *not*, when it would actually be useful, sometime later in the takeoff. Thus, because an action and the reaction it evokes occur at precisely the same time, the practice of instructing people to stamp their feet so that they can jump higher cannot be justified.

## NEWTON'S LAW OF GRAVITATION

Bodies in contact exert forces on one another through their point(s) of contact. Such forces are sometimes referred to as *contact forces*. In addition to these contact forces, other forces act on bodies regardless of whether they are in contact. These forces tend to make the bodies gravitate, or move, toward one another.

Newton described the nature of these gravitational forces in a law that he formulated after observing an apple fall to the ground—or, as legend would have it, after being hit on the head by a falling apple. This law—the *law of gravitation*—states that

> All bodies attract one another with a force proportional to the product of their masses and inversely proportional to the square of the distance between them.

Or, in algebraic form,

$$F = G\left(\frac{m_1 m_2}{d^2}\right) \qquad (11)$$

where $F =$ the force exerted on each body, $G =$ a constant needed to transform the relationship expressed in the law into an equation, $m_1$ and $m_2 =$ the masses of the bodies, and $d =$ the distance between them.

The attractive forces exerted by one body on another are generally so small as to be of no practical significance in physical education, athletics, and rehabilitation. There is, however, one exception to this—the force exerted by the earth, *gravity*.

As indicated in Eq. 11, the force exerted on a body by the earth varies with the mass of the body and with its distance from the center of the earth. This latter distance varies from one place to another on the earth's surface. It is greater in places that are well above sea level than it is in places at or below sea level; and, due to the earth's being somewhat flattened at the poles, it is greater near the equator than it is near the poles. As a result, a body will experience a slightly lesser attraction due to gravity in, say, Mexico City, which is near the equator and at high altitude (2278 m), than it would in Fairbanks, Alaska, which is near the North Pole and close to sea level (137 m). Differences like these have a small, but perceptible, effect on athletic performances. For example, Heiskanen[1] reported in 1955 that

[1] Weikko A. Heiskanen, "The Earth's Gravity," *Scientific American*, 193 (September 1955), 164–65.

*Other things being equal, we should expect jumpers to jump higher and javelin throwers to hurl the javelin farther in the 1956 Olympic Games in Melbourne, Australia, than they did in the 1952 Games at Helsinki. Making corrections for the smaller force of gravity at Melbourne because of its closer approach to the equator, we calculate that the world record for the javelin throw is equivalent to 15.75 centimeters (six inches) farther at Melbourne than at Helsinki; for the hammer throw, 11.54 centimeters (five inches) farther; for the broad jump, 3.63 centimeters (1½ inches) longer.*

## WEIGHT

The gravitational force exerted on a body by the earth is called the *weight* of the body and, like the acceleration produced by this force (*g*), is given a special designation, *W*.

The difference between the mass and the weight of a body is often the source of some confusion. This confusion can be dispelled by considering how variations in location—and, thus, in the attraction of gravity—affect the two quantities. Because the quantity of matter in a body does not change as the body is moved from one place on the earth to another, the mass of the body is obviously constant. This is not true, however, of the weight that, like the distance to the center of the earth, varies slightly from place to place. Since one quantity is constant in value and the other varies with location, these two are obviously not exactly the same or directly equivalent. They are, however, very closely related. Consider the case of a pole vaulter at the peak of his upward flight. At this moment, as at all other moments during his flight, he is acted on by a downward force—his weight, *W*—which causes him to accelerate toward the pit with an acceleration, *g*. The basic relationship between his mass and his weight can be found by taking the equation representing Newton's second law:

$$F = ma$$

And, substituting the values appropriate to this case,

$$W = mg \tag{12}$$

Thus, if $g = 9.81 \text{ m/s}^2$, a vaulter of 80 kg mass weighs ($80 \times 9.81 =$) 785 N. A simple rearrangement of Eq. 12:

$$m = \frac{W}{g}$$

makes it easy to determine the mass when the weight is known. For example, an 883 N vaulter has a mass of ($883/9.81 =$) 90 kg.

Joggers who wish to buy a new pair of road shoes, football coaches who must order new shoes for their teams, and skiers who need new pairs of skis all have one thing in common. Whether they seek a shoe that provides maximum traction or a ski that permits maximum slippage, they have a bewildering array of designs from which to choose. Because traction and slippage are very important considerations in these and other sports, the variety of available designs raises an obvious and important question—"What are the factors that govern whether a body will slide and, if it does, how easily it will slide?"

Whenever one body moves, or tends to move, across the surface of another, force is created. This force, which acts tangential to the point(s) of contact of the two bodies and opposes the motion, or impending motion, is called *friction*. The magnitude of the friction varies in a rather unusual way. Consider the simple test shown in Fig. 95(a).[2] In this test, a shoe sole is glued to a block of wood that is then placed, sole downward, on a table. The block is connected by a string, over a freely rotating pulley, to a weight tray. A weight—say, 1 N—is then placed in the tray. Due to the presence of the pulley, which redirects the force, the addition of this weight causes a horizontal force of 1 N to be exerted against the block of wood. If the block does not move, the friction that opposes the impending motion must be equal in magnitude and opposite in direction to the applied force (Fig. 95(b)). (This conclusion is justified because there are only two horizontal forces acting upon the block—that due to the applied load and that due to friction.) At this point a second 1 N weight is placed in the tray. If the block still does not move, the friction must have increased in magnitude so that it again matches the applied load (Fig. 95(c)). If this process is continued, a point is eventually reached at which the block is still at rest, but the friction has reached the limit beyond which it is incapable of increasing further. This value is called the *limiting friction* and is equal in magnitude to the applied load at that time (5 N in Fig. 95(d)). If a further weight is added, the block begins to slide, and the opposing friction becomes less than its limiting value (Fig. 95(e)).

This simple testing procedure could be used to determine how the soles of different materials and construction behave when used on different surfaces. To do this, different soles could be glued to identical blocks of wood—or to the different faces of the same block—and tested on a tabletop to which different surfaces could be attached. As expected, the results of such a test reveal marked

[2]Peter R. Francis, "Biomechanics of Human Movement: Course Guide and Lab Manual," Mimeographed notes (Iowa State University), pp. 20–24.

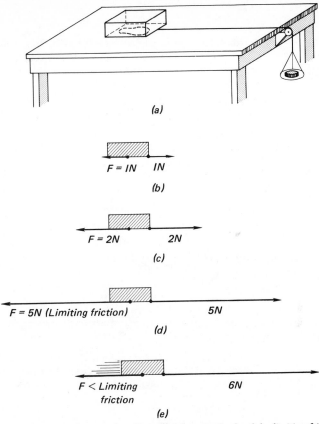

(a)

$F = IN$   $IN$

(b)

$F = 2N$   $2N$

(c)

$F = 5N$ (Limiting friction)   $5N$

(d)

$F <$ Limiting friction   $6N$

(e)

**Fig. 95.** *Procedure used to determine the magnitude of the limiting friction.*

differences in the limiting friction values recorded (Table 10). They also serve to illustrate an obvious, yet important, point—that limiting friction depends on the nature of the surfaces involved.

Now it might be argued that the results obtained in a test of this kind are not relevant to the practical situation because the weight of the body being supported by the sole—the block of wood—is much less than is that of the body for whom the shoe was designed. Although the use of very heavy weights creates problems in the conduct of the experiment, the effects of adding relatively light weights can readily be studied by systematically loading the upper surface of the block of wood. (The use of barbell discs and a piece of broomstick inserted in the upper surface of the block simplifies the process.) As expected, the results of such an experiment reveal that the limiting friction depends on the weight of the body involved or, put another way, on the force holding the two surfaces together (Table 11, columns 1 and 2).

#### TABLE 10  Limiting Friction for Different Shoe Soles and Different Surfaces*
#### (Weight of block plus sole = 105 N)

| Sole | Limiting Friction (N) |
|---|---|
| Different soles on cement sidewalk | |
|     Ripple | 102 |
|     Rubber | 102 |
|     Composition cleats | 98 |
|     Composition | 89 |
|     Nylon cleats | 71 |
|     Leather | 53 |
| Composition sole on different surfaces | |
|     Cement | 89 |
|     Asphalt | 71 |
|     Rubber mat | 71 |
|     "Tartan" (smooth) | 62 |
|     Wood | 58 |
|     Tile | 53 |
|     Carpet | 48 |

*Data courtesy Peter R. Francis, Iowa State University, Ames, Iowa.

#### TABLE 11  Limiting Friction and Coefficient of Friction for a Composition Sole
#### on "Astroturf" Surface Under Different Loads*

| Weight of Block and Additional Barbell Discs(N) (1) | Limiting Friction (N) (2) | Coefficient of Friction (3) |
|---|---|---|
| 40.1 | 11.1 | 0.3 |
| 51.2 | 17.8 | 0.3 |
| 62.3 | 17.8 | 0.3 |
| 73.4 | 24.5 | 0.3 |
| 84.6 | 28.9 | 0.3 |
| 95.7 | 33.4 | 0.3 |

*Force data have been converted from nonmetric to metric units. Differences in coefficients of friction when these are expressed correct to more than one decimal place are almost certainly due to lack of precision in force measurement (±2 N), to small changes in bearing surfaces from trial to trial, and to effects of rounding during conversion of units. (Data courtesy Peter R. Francis, Iowa State University, Ames, Iowa.)

The basic results obtained in these experiments—that limiting friction depends on the nature of the surfaces involved and on the force holding them together—are summarized in what is sometimes called the *law of friction*:

*For two dry surfaces, the limiting friction is equal to the product of the force holding the surfaces together—the so-called normal reaction—and a constant that depends on the nature of the surfaces.*

Or, in algebraic terms,

$$F = \mu R \qquad (13)$$

where $F$ = the limiting friction, $\mu$ = "mu" = the constant known as the *coefficient of limiting friction*, and $R$ = the normal reaction. The law of friction applies only to dry surfaces. If the surfaces are wet, limiting friction is markedly reduced. For example, when the composition blocks used in the rim brakes of a bicycle are wet, the limiting friction drops to one tenth of its dry value.[3]

A simple rearrangement of Eq. 13:

$$\mu = \frac{F}{R}$$

is used to obtain the value of the coefficient of limiting friction in a given case. Thus, for the ripple sole and cement surface (Table 10), the coefficient of limiting friction was

$$\mu = \frac{102 \text{ N}}{105 \text{ N}}$$

$$= .97$$

Similar computations have been made using the data in Table 11 to demonstrate that the coefficient of limiting friction is a constant that depends only on the nature of the bearing surfaces involved—Table 11, column 3.

Many examples could be cited to demonstrate the application of the law of friction in physical education, athletics, and rehabilitation. Track athletes; road runners; football, basketball, and baseball players; golfers; and many others wear shoes that are specially designed to maximize $\mu$—and, thus, the limiting friction—while retaining their other essential characteristics. The manufacturers of golf clubs place grooves on the face of the club to minimize slippage between club and ball. Gymnasts, weight lifters, and others use powders, resin, and other materials to alter the nature of the bearing surfaces and thereby increase the magnitude of the limiting friction. Bowlers wear one shoe with a sole designed to permit sliding and the other with a sole designed to prevent it. Skiers use

[3]Brian D. Hanson, "Wet-Weather-Effective Bicycle Rim Brake: An Exercise in Product-Development," M.S. thesis (Cambridge, Mass.: M.I.T., 1971).

techniques known as "up-unweighting" and "down-unweighting" to reduce the friction between skis and snow and thus facilitate turning. In all these cases, and in others like them, people seek to modify limiting friction in the only ways open to them—by changing the coefficient of limiting friction and/or the normal reaction.

***Sliding friction.***   As a body begins to slide, the friction opposing its motion decreases. Thus,—as is well known—it is easier to keep a body sliding than it is to start it sliding. The factors governing the magnitude of the sliding friction encountered in a given case are the normal reaction and a coefficient—the *coefficient of sliding friction*, $\mu_s$—that varies with the surfaces involved and with the velocity at which one moves over the other.

## IMPULSE

A force platform—a flat, rectangular device designed to sense the forces exerted against its upper surface—and a strip chart recorder are often used to determine the forces exerted against the ground during walking, running, and other activities. When a sprinting athlete places his foot on a force platform set flush in the track, he exerts forces on it in vertical, lateral (or sideward), and frontal (forward–backward) directions. A typical record of the forces exerted in the frontal direction (Fig. 96(a)) shows that the runner experiences negative forces (causing him to slow down) during the first part of his contact with the ground and positive forces (causing him to speed up) during the second part. Whether he slows down, maintains the same horizontal velocity, or speeds up as a result of this overall process depends, of course, on how his final gain in horizontal velocity compares with his initial loss.

The average force exerted on a body in a given direction can be found by dividing the area under the corresponding force–time curve, expressed in units of force and time, by the time involved. Thus, for the first part of the curve shown in Fig. 96(a);

$$\bar{F} = \frac{\text{Area}}{\text{Time}}$$

$$= \frac{-24 \text{ N} \cdot \text{s}}{.09 \text{ s}}$$

$$= -267 \text{ N}$$

The quantity represented by the area under the curve—or, more precisely, by the area between the curve and the time axis—is called the *impulse*. Thus, if $J$ = the impulse

$$\overline{F} = \frac{J}{t}$$

and

$$J = \overline{F}t$$

The average acceleration experienced by a body is a function of the average force exerted (Newton's second law):

$$\overline{F} = m\overline{a}$$

Substituting $J/t$ for $\overline{F}$ and $(V_f - V_i)/t$ for $\overline{a}$ (Eq. 10) yields

$$\frac{J}{t} = m\left(\frac{V_f - V_i}{t}\right)$$

or

$$J = mV_f - mV_i \tag{14}$$

or, in other words, the change in momentum of a body—the right-hand side of the equation—is equal to the impulse that produces it. This is the so-called *impulse-momentum relationship*.

The question of whether the horizontal velocity of a sprinter is increased, decreased, or unaltered as the result of a given contact with the ground can be resolved by comparing the areas under the positive and negative parts of the frontal force–time curve—that is, by comparing the positive and negative impulses. If the initial negative impulse is less than the final positive impulse—as it is, for example, at the start of a race—the athlete's horizontal velocity is increased (Fig. 96(a)). If the two are equal—as, for example, when the athlete is running at top speed—there is no change in the athlete's horizontal velocity (Fig. 96(b)). Finally, if the negative impulse is greater than the positive impulse—as it is when the athlete overstrides after passing the finish line, the runner's horizontal velocity is decreased (Fig. 96(c)).

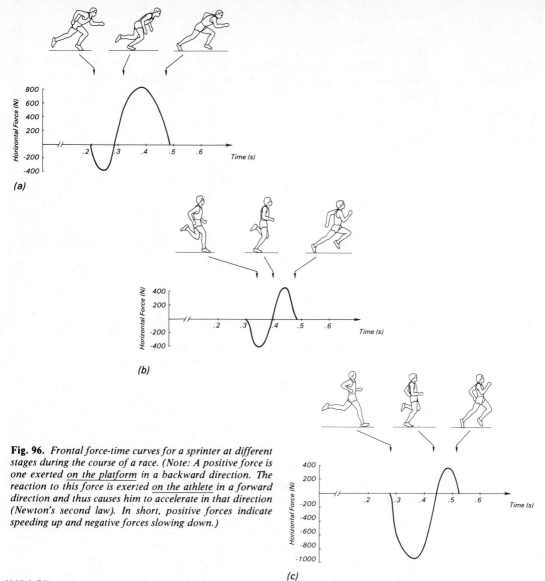

**Fig. 96.** *Frontal force-time curves for a sprinter at different stages during the course of a race. (Note: A positive force is one exerted <u>on the platform</u> in a backward direction. The reaction to this force is exerted <u>on the athlete</u> in a forward direction and thus causes him to accelerate in that direction (Newton's second law). In short, positive forces indicate speeding up and negative forces slowing down.)*

## IMPACT

Because of the very large number of physical education and athletic activities that involve the impact of one body with another, it is highly desirable that physical education teachers and coaches have a good understanding of the factors that govern the results of such impacts.

When two bodies hit, they either remain in contact or they bounce apart. In the first case, the bodies tend to become one body—as, for example, when a shot imbeds itself in a soggy field—and the impact is said to be *inelastic*. In the second case, both the bodies involved in the impact are first deformed and then, due to the springy nature of the materials of which they are made, are restored to their original shape. This sequence of deformation and restoration is dramatically demonstrated by the impact of a soft rubber ball and a hard surface shown in Fig. 97. Such impacts are said to be *elastic*. The property of a body that causes it to return to its original shape after being deformed in an impact—or, indeed, in any other way—is called its *elasticity*.

**Fig. 97.** *During an elastic impact between a soft rubber ball and a hard surface (a) the ball approaches the surface, (b) deformation begins, (c) the ball is completely flattened, (d) restoration begins on the upper side, (e) it extends to the bottom side, and (f) the still-distorted ball continues upward on the rebound. The jet on which the ball appears to be rising is from the chalk-covered impact surface. (Photos courtesy Dunlop Sports Company Limited.)*

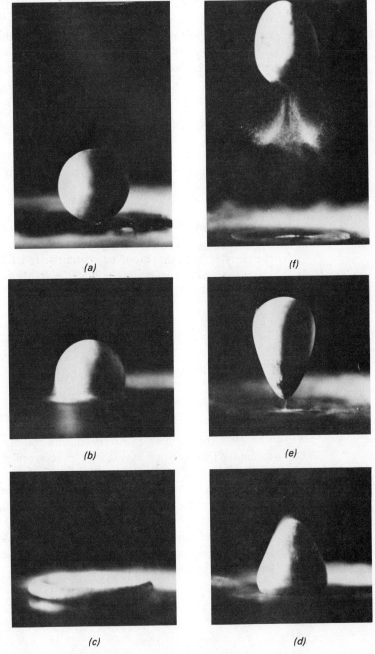

(a)

(f)

(b)

(e)

(c)

(d)

The material of which a body is made and the manner in which it is made influence its elasticity. Thus, an air-filled ball made of a relatively soft rubber (a paddle ball or racket ball, for example) is much less elastic than is a solid ball made of a similar but highly compressed material (a golf ball or a superball, for instance).

The velocity at which two bodies separate after an impact depends on the velocity at which they came together initially and on their elasticity—the slower the velocity at which they come together and the less their elasticity, the slower they are likely to separate. The velocity at which they come together—the *velocity of impact*—is equal to the difference in the velocities of the two bodies immediately before impact. That is, in algebraic terms,

$$u_1 - u_2$$

where $u_1$ and $u_2$ are the velocities of the bodies 1 and 2 immediately before impact. The velocity at which the two bodies move apart—*the velocity of separation*—is similarly equal to the difference in the velocities of the two bodies involved, immediately after they break contact with each other. That is,

$$v_1 - v_2$$

where $v_1$ and $v_2$ are the velocities of bodies 1 and 2 immediately after impact.

The relationship among the velocities of impact and separation and the elasticity of the bodies involved is summarized in *Newton's law of impact*, which states that, when two bodies collide, the velocity of separation bears a constant relationship to the velocity of impact:

$$\frac{\text{Velocity of separation}}{\text{Velocity of impact}} = -e$$

or, algebraically,

$$\frac{v_1 - v_2}{u_1 - u_2} = -e \tag{15}$$

where $e =$ the *coefficient of restitution*, a constant determined by the nature of the bodies involved in the impact. (This law applies to the velocities of bodies that impact directly, or "head on," and to the components of the velocities perpendicular to the surface of contact of bodies that impact obliquely. The application of the law in the analysis of oblique impacts will be discussed later in this chapter.)

Some care must be taken to ensure that the velocities entered into Eq. 15 are correctly designated with respect to direction. Consider, for example, the very simple case of a basketball that is dropped from a height of 2 m onto the floor

(Fig. 98). As the ball strikes the floor, it has a downward vertical velocity of 6.3 m/s and, when it leaves the floor, an upward vertical velocity of 5.0 m/s. If the upward direction is designated as positive, the coefficient of restitution for the impact is computed as follows:

$$e = -\frac{v_1 - v_2}{u_1 - u_2}$$

$$= -\frac{5.0 - 0}{-6.3 - 0}$$

$$= -\frac{5.0}{-6.3}$$

$$= 0.8$$

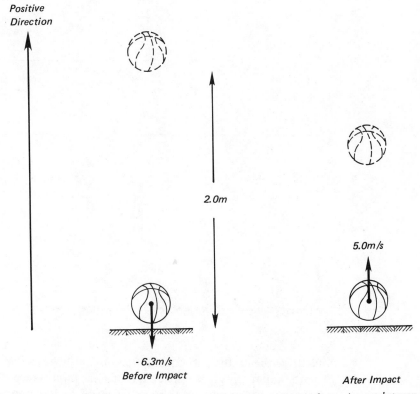

Positive Direction

2.0m

5.0m/s

- 6.3m/s
Before Impact

After Impact

**Fig. 98.** *Computation to determine the coefficient of restitution for an impact between basketball and floor.*

(The second body involved in the impact—the floor—is stationary throughout and, thus, $u_2 = v_2 = 0$.) Still more care must be exercised when both bodies are moving before and after the impact. Consider the case in which a tennis ball traveling at 25 m/s meets a racket traveling in the opposite direction at 35 m/s and then, after the impact, the racket and ball travel in the same direction with velocities of 10 m/s and 30 m/s, respectively (Fig. 99). If the direction in which the ball was moving before impact is arbitrarily designated as positive, the coefficient of restitution is given by

$$e = -\frac{v_1 - v_2}{u_1 - u_2}$$

$$= -\frac{-30 - (-10)}{25 - (-35)}$$

$$= -\frac{-20}{60}$$

$$= 0.33$$

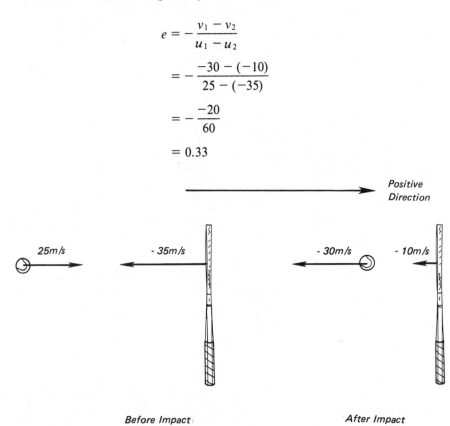

**Fig. 99.** *Care must be exercised in computing the coefficient of restitution when both bodies are free to move.*

Because the coefficient of restitution obviously has a considerable influence on the result of any given impact, it is of some importance that the factors that influence the magnitude of this coefficient be understood. These factors can, perhaps, best be considered by taking a special impact situation, varying the conditions of the impact, and observing the results. The most convenient impact situation for such purposes is one in which a body is dropped, under the influence of gravity, onto a fixed horizontal surface—as, for example, when a basketball official drops the ball from a prescribed height onto the floor to test whether it is suitable for use. Because one of the bodies involved in an impact of this type is at rest, both before and after impact, two of the terms in Eq. 15 are eliminated. Further, because the other body moves only under the influence of gravity, the remaining terms on the left-hand side of Eq. 15 can be replaced by the equivalent vertical displacements. This process of simplifying and modifying Eq. 15 thus yields the equation

$$e = \sqrt{\frac{h_b}{h_d}} \tag{16}$$

where $h_b$ = the height to which the body bounced and $h_d$ = the height from which it was dropped. (It is important to emphasize here that Eq. 15 applies only to the impact situation described. It does *not* apply to situations in which both bodies are free to move or to those in which the moving body is impelled by influences other than gravity.)

The use of Eq. 16 reveals that the NCAA rule requiring that a basketball dropped from a height of 1.83 m (measured to the bottom of the ball) bounce to a height of 1.24–1.37 m (measured to the top of the ball) is equivalent to one requiring that the coefficient of restitution lie within the range from

$$e = \sqrt{\frac{1.24 - 0.25}{1.83}} \qquad \text{to} \qquad e = \sqrt{\frac{1.37 - 0.25}{1.83}}$$

$$= .74 \qquad\qquad\qquad = .78$$

(The diameter of the ball has been assumed equal to 0.25 m, a value close to the midrange of permissible diameters.)

As already suggested, the simple testing procedure used by basketball officials and others can also be used to examine the factors that influence the magnitude of the coefficient of restitution. One might, for example, collect a wide variety of balls and note the height to which they bounce when dropped from a fixed height—say, 2 m—onto a fixed surface. One might also select one ball and drop it from a fixed height onto a variety of different surfaces. As no doubt expected, such simple tests would reveal that the magnitude of the coefficient of restitution depends on the nature of both of the bodies involved in an impact and not just on one of them. Since the coefficient of restitution is thus a measure of the behavior of both of the bodies involved in an impact, it is incorrect to speak of "the coefficient of restitution of a body."

Another factor that influences the magnitude of the coefficient of restitution—and whose influence might also be examined using the simple test already described—is the temperature of the ball. When a ball is heated or cooled for several hours, its "nature" changes. If the ball is solid, like a golf ball, changes in its temperature produce changes in the elastic properties of the materials from which it is made. These, in turn, produce changes in the behavior of the ball, in particular, an increased height of bounce with heating and a decreased one with cooling. If the ball is an air-filled one like a squash or tennis ball, changes in its temperature also produce changes in internal pressure and thus, too, in the way in which it behaves in an impact situation. When it is heated, the internal pressure increases and the ball has a relatively high bounce. Conversely, when it is cooled, the internal pressure decreases and it has a relatively low bounce.

When a ball is deformed in the course of an impact, heat is generated and the temperature of the ball is increased. In golf, for example, the temperature of the ball rises about $0.5°$ C as it is driven off the tee.[4] While this rise in temperature has little practical significance when only a single impact is involved, or when the ball may cool off again in the long intervals between impacts, it is of considerable importance in games like squash and tennis where repeated impacts produce a prolonged elevation in the temperature of the ball. Indeed, squash players deliberately warm up the ball by rallying for some minutes before the start of a game and whenever a new ball is introduced to replace one that is no longer useable.

While somewhat difficult to demonstrate convincingly using a simple "ball drop" test, the coefficient of restitution is also influenced to some extent by the velocity of impact. Large velocities of impact tend to produce somewhat smaller *e* values than do small velocities of impact.

Impacts that occur in physical education and athletic activities may be divided into two categories—those in which a moving body impacts with a fixed surface and those in which a moving body impacts with a second body that is also free to move. The first category includes impacts between the ball and the floor, wall, or ceiling in court games like handball, racketball, or squash; the backboard in basketball; and the cushion in billiards. The second includes virtually all striking activities—bowling; kicking in football; hitting in golf, softball, baseball, and the racket sports; heading in soccer; and so on.

### Impact with a Fixed Surface

What happens when a basketball strikes the floor in the course of a bounce pass (or when any other body strikes a fixed surface) can, perhaps, best be described in terms of the changes that occur in its velocity. As the ball approaches the floor, it

[4]C. B. Daish, *The Physics of Ball Games* (London: English Universities Press, 1972), p. 16.

is traveling both horizontally and vertically—that is, both parallel and perpendicular to the floor surface. When it makes contact with the floor, it tends to continue traveling in the same manner (Newton's first law) and, as a result, exerts both horizontal and vertical forces against the floor. The floor, in reaction, exerts equal and opposite forces on the ball (Newton's third law). It is these that change the magnitude and direction of the ball's velocity.

The change in the vertical velocity depends on the elasticity of both the ball and the floor and can be determined using Eq. 15 which reduces to

$$\frac{v_1}{u_1} = -e \qquad \text{or} \qquad v_1 = -eu_1$$

when one of the bodies is stationary throughout. Thus, if the vertical velocity of the ball is, say, 5 m/s immediately prior to impact and the coefficient of restitution 0.8, the vertical velocity of the ball after impact is

$$v_1 = -(0.8 \times 5)$$

$$= -4.0 \text{ m/s}$$

The change in the horizontal velocity depends on the horizontal force—the friction—acting on the ball. If the friction were negligible, as it would be in the unlikely event that the floor were made of ice or something equally slick, the horizontal velocity would be unaltered by the impact. The ball would simply maintain its preimpact horizontal velocity by sliding during the period of contact. Under more realistic circumstances, the horizontal velocity of the ball is noticeably reduced as a result of the horizontal forces exerted upon it during the impact.

The magnitude—and, in some cases, the direction—of the horizontal forces exerted on the ball are influenced by the amount of spin imparted to it. Consider, once again, the simple case of a basketball that is dropped onto a wooden floor. If no spin is imparted to the ball, it will rebound vertically to some fraction of the height from which it was dropped (Fig. 100(a)). When spin is imparted to the ball, that part of it that makes contact with the floor is traveling horizontally at the instant of impact and, as a result, evokes an opposing horizontal reaction from the floor. This horizontal reaction (friction) accelerates the ball in the direction in which it acts (Newton's second law) and thus causes the ball to travel horizontally (as well as vertically) during its subsequent flight (Figs. 100(b) and (c)). (The vertical motion of the ball is unaffected by the horizontal forces evoked as a result of the spin imparted to the ball. The ball thus rebounds to the same height—a height determined by the height from which it was dropped and the coefficient of restitution, (Eq. 16)—in all three cases.)

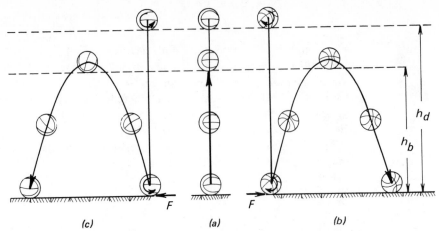

**Fig. 100.** *The effects of (b) topspin and (c) backspin on the path followed by a basketball after impact with the floor. The no-spin case is shown in (a) as a basis for comparison.*

A similar but somewhat more complicated situation arises when a ball is traveling horizontally—as well as vertically—prior to contact. Consider the case of a tennis ball that bounces on the court on its way to the receiver. If no spin has been imparted to it, all parts of it will have the same horizontal velocity at the instant it lands on the court (Fig. 101(a)). Under such circumstances, the opposing friction will depend on the velocity of the ball and the nature of the surfaces in contact. If spin has been imparted to the ball, the horizontal velocity of each point on the ball is, in effect, the sum of two contributing, or component, velocities—a *translational component* due to the horizontal motion of the center of the ball and a *rotational component* due to the rotation of the ball about an axis through its center (Figs. 101(b) and (c)).

When topspin is applied to the ball, the component velocities of that part of it that makes contact with the court tend to offset, or cancel, each other (Fig. 101(b)). As a result, the forward velocity with which this part strikes the court, the evoked frictional reaction, and the decrease in the ball's forward velocity are all less than when no spin is imparted to the ball. (If the rotational component of the horizontal velocity is greater than the translational component, the frictional reaction is directed forward and serves to increase the horizontal velocity of the ball.) The net effect (no pun intended!) is that the application of topspin causes the ball to come off the court faster and at a lower angle than it would in the absence of spin. It is for these reasons that topspin is widely used in tennis, table tennis, and other games as an offensive weapon.

If backspin is applied to the ball, the results obtained are essentially the reverse of those produced when topspin is applied (Fig. 101(c)). The translational and rotational components complement each other, the evoked frictional reaction is increased (compared with what it would be in the absence of spin), and the postimpact horizontal velocity of the ball is markedly less than in either the

174

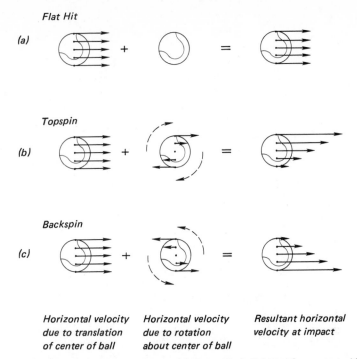

*Flat Hit*

(a)

*Topspin*

(b)

*Backspin*

(c)

Horizontal velocity
due to translation
of center of ball

Horizontal velocity
due to rotation
about center of ball

Resultant horizontal
velocity at impact

**Fig. 101.** *The horizontal velocity of that part of a ball that makes contact with the ground is equal to the sum of its translational and rotational components.*

topspin or no-spin cases. (If the ball has a small horizontal velocity and a very large amount of backspin—as sometimes occurs in the execution of a drop shot—the friction encountered on landing may actually be sufficient to reverse the direction of the ball's horizontal motion. This effect is observed, on occasion, when a golfer plays a short, lofted shot to the green.) Because the application of backspin causes the ball to come off the court more slowly and at a higher angle than it would if no spin were applied, backspin is often used in defensive situations.

The use of topspin and backspin for offensive and defensive purposes, respectively, is probably exemplified nowhere more dramatically than in the game of table tennis. Here, one often sees players, standing close to the table, hitting topspin drives, while their opponents, standing several meters beyond the other end of the table, return them with delicately controlled backspin shots.

The effect that spin has on the horizontal and resultant velocities of a tennis ball after impact with the court, and on the subsequent flight path of the ball, are summarized in Fig. 102. Several points should be noted. First, the horizontal velocity, the resultant velocity, and the angle to the vertical at which the ball leaves the court are least for the backspin and greatest for the topspin case. Second, the vertical velocity at which the ball leaves the court and, thus, the

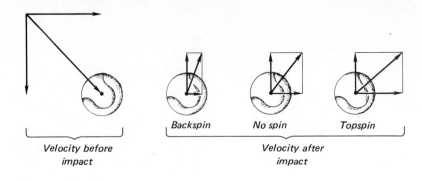

Backspin    No spin    Topspin

Velocity before          Velocity after
impact                       impact

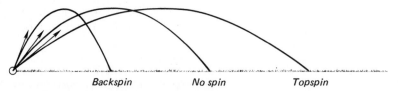

Backspin          No spin          Topspin

**Fig. 102.** *Trajectories followed by a tennis ball with (a) no spin, (b) topspin, and (c) backspin at the instant it strikes the court. (Note: For purposes of illustration, the vertical scale has been made much larger than the horizontal scale.)*

maximum height to which it rises, is the same in all three cases. Finally, Fig. 102 is intended to demonstrate the effects that spin has on the horizontal, vertical, and resultant velocities after impact and on the subsequent trajectory of the ball. For this purpose, it is assumed that the velocity of the ball immediately prior to impact is the same in all three cases. This is rarely the case in practice, however, for topspin shots are normally hit hard and backspin shots rather delicately. As a result, topspin shots usually bounce much higher than do backspin ones. It must be recognized, however, that this difference in the height of the bounce is due *not* to the spin but rather to the velocity imparted to the ball.

The relationship between the direction in which the ball moves after impact and that in which it moved before impact is of interest in many sports. In basketball, a player executing a bounce pass needs to know how the ball should be directed toward the floor so that it will bounce up into the hands of the intended receiver. The player also needs to know how to direct the ball onto the backboard so that it will rebound into the basket in a lay-up shot. Athletes in a score of other sports and games have similar needs.

The angles between the velocity vector and the vertical immediately before impact and immediately after impact are called the *angle of incidence* and the *angle of reflection*, respectively (Fig. 103). The relative magnitude of the angles of

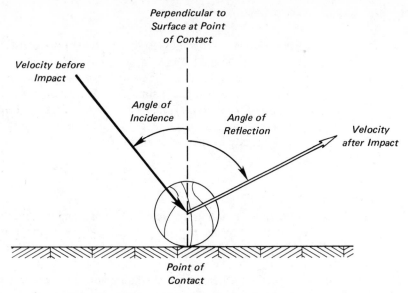

Perpendicular to
Surface at Point
of Contact

Velocity before
Impact

Angle of
Incidence

Angle of
Reflection

Velocity
after Impact

Point of
Contact

**Fig. 103.** *Angles of incidence and reflection.*

incidence and reflection depends on how the change in vertical velocity (due to elastic effects) compares with the change in horizontal velocity (due to frictional effects). If the effects of friction are less than those of elasticity, the angle of reflection is greater than the angle of incidence. If they are equal, the angles are also equal. Finally, if the effects of friction are greater than those of elasticity, the angle of reflection is less than the angle of incidence.

## Impact Between Moving Bodies

An elastic impact between two bodies that are moving (or are free to move) is somewhat more difficult to analyze than is an impact between a moving body and a fixed surface. Consider the case of a bowling ball when it strikes a pin. In this case, the ball has a certain amount of horizontal momentum prior to impact and, due to the forces that the pin exerts on it, a lesser amount after impact. The pin, on the other hand, has zero horizontal momentum prior to impact and, due to the forces that the ball exerts upon it, a substantial amount after impact.

Now, if one considers the ball and the pin to be parts of a ball-plus-pin system, the forces that they exert on each other are internal forces and have no effect on the total momentum of the system (Newton's first law). Thus, ignoring the presumably trivial effects of friction and air resistance, the total momentum of the system remains the same because there are no external forces that act to alter it. The momentum of the system prior to impact is thus equal to the momentum of the system after impact—or, to express it algebraically:

$$m_1u_1 + m_2u_2 = m_1v_1 + m_2v_2 \qquad (17)$$

where $m_1$ and $m_2$ = the masses of the bodies involved, $u_1$ and $u_2$ = their velocities prior to impact, and $v_1$ and $v_2$ = their velocities after impact.

What happens as a result of the elastic impacts like that between a bowling ball and pin can be determined using Eq. 17 (also known as the *principle of conservation of momentum*) and Eq. 15 (Newton's law of impact). For example, Eqs. 15 and 17 can be combined to yield the following expression for the horizontal velocity of the pin after impact:

$$v_P = \frac{m_B u_B (1 + e)}{m_B + m_P} \tag{18}$$

where the subscripts $B$ and $P$ refer to the ball and pin, respectively.[5]

Several conclusions can be reached by examining this equation closely. Specifically, the equation indicates that, if all the other factors remain the same,

1. The greater the velocity of the ball prior to impact, the greater the velocity of the pin after impact.
2. The greater the coefficient of restitution, the greater the velocity of the pin after impact.
3. The greater the mass of the pin, the smaller the velocity of the pin after impact.

The effect that changes in the mass of the ball have on the velocity of the pin are a little less easy to determine by inspection than are the effects produced by changes in the other factors because the mass of the ball appears in both the numerator and denominator of the fraction on the right-hand side of the equation. A simple way to overcome this problem is to take a series of realistic values and compute the velocity of the pin for different values of the mass of the ball. For example, if

$$m_B = 5 \text{ kg}$$

$$m_P = 1.5 \text{ kg}$$

$$u_B = 6 \text{ m/s}$$

$$e = 0.3$$

$$v_P = \frac{5 \times 6(1 + 0.3)}{5 + 1.5}$$

$$= 6.00 \text{ m/s}$$

[5] For the sake of simplicity, the impact is treated here as a planar or two-dimensional problem. The "horizontal velocity of the pin after impact" is thus the velocity it possesses in the forward direction defined by the motion of the ball.

Now, if $m_B$ is increased first to 6 kg

$$v_P = \frac{6 \times 6(1 + 0.3)}{6 + 1.5}$$

$$= 6.24 \text{ m/s}$$

and then to 7 kg

$$v_P = \frac{7 \times 6(1 + 0.3)}{7 + 1.5}$$

$$= 6.42 \text{ m/s}$$

it becomes clear, as might have been expected, that the greater the mass of the ball, the greater the velocity of the pin.

On the basis of these several conclusions, it is apparent that the velocity of the pin after impact, and thus its ability to knock over other pins, depends on two factors over which the bowler has some degree of control—the mass of the ball and its velocity prior to impact. However, unless the ball is being released at a velocity much less than the bowler is capable of imparting to it, the bowler cannot expect to obtain more "pin action"—that is, a higher velocity of the pin after impact—by simultaneously increasing both the mass of the ball and its velocity prior to impact. Because these two factors are interrelated—the greater the mass of the ball, the less the velocity that can be imparted to it—the bowler must try instead to find that combination of the two that yields the best results.

In summary, it should be noted that there are just five factors included in Eqs. 15 and 17—the masses and preimpact velocities of the bodies involved and their mutual coefficient of restitution. The outcome of an elastic impact between two bodies that are moving (or are free to move) is governed by the magnitudes (and, in the case of the velocities, by the directions) of these five factors.

## PRESSURE

When a gymnast performs a one-handed handstand, she supports her weight on the palm of her hand (Fig. 104). If she weighs 500 N and the area of her hand in contact with the floor is 100 cm$^2$, the average load supported per unit of area—that is, the average *pressure*— is (500 N/100 cm$^2$ =) 5 N/cm$^2$. If the same gymnast performs a regular handstand, and the area of her hands in contact with the floor is 200 cm$^2$, the average pressure is decreased to (500 N/200 cm$^2$ =) 2.5 N/cm$^2$, because the area of support has been increased. The weight that must be supported—her body weight—is, of course, unaltered. Finally, if the gymnast performs a headstand and the area of her head and hands in contact with the floor is 250 cm$^2$, the average pressure is reduced further to (500 N/250 cm$^2$ =) 2N/cm$^2$. The load that must be supported is once again unaltered.

**Fig. 104.** *The average pressure is large when the area of support is small.*

The pressure to which any given part of the human body can be exposed without it suffering damage is naturally limited. It is therefore essential, from a safety standpoint, that physical education and sports activities do not involve the application of greater pressures than the body is capable of withstanding. (Although the words "load" or "force" are often substituted for the word "pressure" in discussions of injuries, it is usually the last of these that is the most appropriate. Very large loads or forces can be either harmful or harmless depending on the area over which they are distributed. Very large pressures are rarely harmless.)

Many sports involve the use of protective equipment that has been designed to minimize the pressure to which the body is exposed and thus to reduce the risk of injury. Football, baseball, and ice hockey helmets are designed to spread the force from its point of application over as wide an area as is needed to reduce pressure to tolerable levels. Facemasks, gloves, and other items of protective equipment serve the same function.

Some sports also involve the use of special techniques designed to spread forces—especially impact forces—over a large area. Among these are the martial arts, like judo, which incorporate the use of breakfalls to reduce the pressures exerted on the body during the landing following a throw.

# WORK

Several terms used in everyday language have special, restricted meanings when used in science. Among the more confusing of these is the term *work*. In mechanics, work is defined as the product of the force applied to a body and the distance through which the body moves *in the direction in which the force acts*.

This definition of work often leads to confusion because it is at variance with the everyday concept of work. Consider, for example, an athlete who holds a barbell at full arm's length following the completion of a bench press (Fig. 105(b)). To do this, he must exert an upward force on the barbell equal to the downward force exerted on it by gravity—that is, equal to its weight. However, according to the definition of the term, he does not have to perform work. Because the body upon which the force is exerted is stationary, and the distance it moves in the direction of the force is thus zero, the product of force and distance—the work done—is also zero. To highlight the distinction that must therefore be made between muscular effort—which is often equated with work in everyday language—and work, as defined in mechanics, the latter quantity is often referred to as *mechanical work*.

The mechanical work done in a given situation can be positive, zero, or negative. If the athlete in Fig. 105(a) exerts a constant upward force of 1000 N as he moves the barbell through the middle 0.3 m of its upward motion, the work he does is (1000 N $\times$ 0.3 m =) 300 N $\cdot$ m or 300 joules (1 joule (J) = 1 N $\cdot$ m). This work is said to be positive because the direction in which the body moves is the same as that in which the force acts. If the athlete exerts an upward force of 1000 N to control the downward motion of the bar as he lowers it through the same distance (Fig. 105(c)), the work he does is again 300 J. However, because the direction in which the body moves is opposite to that in which the force is applied, this work is said to be negative. It is useful to note that positive work is done when the muscles producing the force—in this case, the elbow extensors and shoulder flexors—are shortening and negative work when they are lengthening. Thus, from a muscular standpoint, positive and negative work can be equated with concentric and eccentric contractions, respectively.

During the execution of a bench press, the barbell is acted upon by two vertical forces—an upward force exerted by the performer and a downward force exerted by gravity. The total work done on the barbell is equal to the sum of the work done by these two forces or, to state it in a different way, to the work done by their resultant. If the barbell in Fig. 105 weights 800 N, the total work done on it during the middle 0.3 m of its upward motion is equal to the sum of the work done by the athlete and by gravity:

$$(1000 \text{ N} \times 0.3 \text{ m}) - (800 \text{ N} \times 0.3 \text{ m}) = 60 \text{ J}$$

or to the work done by their resultant

$$(1000 \text{ N} - 800 \text{ N}) \times 0.3 \text{ m} = 60 \text{ J}$$

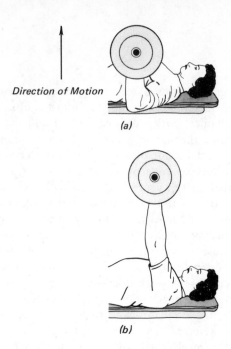

Direction of Motion

*(a)*

*(b)*

Direction of Motion

*(c)*

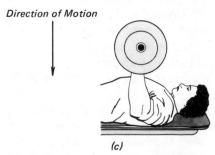

**Fig. 105.** *(a) Positive work is done as the barbell is lifted, (b) zero work is done as it is held aloft, and (c) negative work is done as it is lowered.*

## POWER

The rate at which mechanical work is performed is termed the *power* and is determined by simply dividing the work done by the time taken. Thus, the power developed by a weight lifter who does 600 J of work in raising a barbell overhead and who completes this task in 1.5 s is (600 J / 1.5 s =) 400 J/s or 400 watts (1 watt (W) = 1 J/s).

## ENERGY

Unlike work, with which it is closely related, energy has essentially the same meaning whether used in everyday language or in a scientific context. Energy is the capacity to do work.

Energy can exist in many forms—including mechanical, chemical, electrical, heat, and sound—and, given the appropriate circumstances, can be transformed from one to another.

Mechanical energy—the only form of energy to be considered here—is the energy that a body has because of its motion, because of its position relative to the surface of the earth, or because it has been pushed or pulled out of its normal shape.

### Kinetic Energy

The energy that a body possesses because it is in motion is called its *kinetic energy*. When the body—or some point taken to represent the body—experiences linear motion, the energy that it possesses as a result of that motion is referred to as the *kinetic energy of translation*. When the body experiences angular motion, the energy that it possesses by virtue of that motion is called the *kinetic energy of rotation*. Only the kinetic energy of translation—shortened, for the sake of brevity, to kinetic energy—will be considered here.

The energy that a body possesses as a result of its linear motion is given by the equation

$$\text{K.E.} = \tfrac{1}{2}mv^2 \tag{19}$$

where K.E. = the kinetic energy, $m$ = the mass of the body, and $v$ = the velocity at which it is moving. Thus, a 100 kg back running at 9 m/s has a kinetic energy of $(1/2 \times 100 \times 9^2 =)$ 4050 kg $\cdot$ (m/s)$^2$ or 4050 joules (1J = 1 kg $\times$ (m/s$^2$)). In short, he has a very large capacity to do work on anyone who happens to get in his way!

### Potential Energy

The energy that a body possesses because it is some distance above the level of the earth's surface and can thus do work in returning to that level is called its *potential energy*. The potential energy of a body is given by the equation

$$\text{P.E.} = Wh \tag{20}$$

Where P.E. = the potential energy, $W$ = the weight of the body, and $h$ = the height of the body above the earth's surface. Thus, a 600 N diver at the peak of flight some 3 m above the water has a potential energy of (600 N × 3 m =) 1800 J.

When a springboard diver leaves the board he has a certain amount of both kinetic and potential energies. As he rises to the peak of his flight, his kinetic energy decreases and his potential energy increases. Then, as he begins his descent to the water, this process is reversed—his kinetic energy increases and his potential energy decreases. The *law of conservation of mechanical energy* states that the sum of the kinetic and potential energies of an airborne body is constant. In such cases, the gains in one must be matched exactly by the losses in the other. Thus, when the diver gains potential energy during his ascent, he simultaneously loses exactly the same amount of kinetic energy. Similarly, his losses in potential energy during the descent are exactly matched by his gains in kinetic energy.

The law of conservation of mechanical energy applies to all bodies that experience airborne motion. It also applies, for practical purposes, in some other cases. Consider, for example, a procedure that is sometimes used to determine the velocities of bullets, arrows, and similar missiles. The apparatus used is called a ballistic pendulum and consists of a block of wood or some other suitable material suspended on long, vertical strings (Fig. 106). The missile is fired horizontally into the end of the block, which swings backward and upward as a result of the impact. The height to which the block and the imbedded missile are lifted is noted. The velocity of the missile is then computed by equating the mechanical energy of the system immediately before impact with that at the end of the backward and upward swing of the pendulum. (Changes in the total mechanical energy of the system due to the work done by the supporting strings are generally ignored.) For example, if an arrow of mass .035 kg were fired into a block of mass 10 kg and the pendulum were lifted 0.5 m, the velocity of the arrow could be determined as follows:

$$\frac{\text{Mechanical energy}}{\text{before impact}} = \frac{\text{Mechancial energy}}{\text{at end of swing}}$$

$$\frac{\text{Kinetic energy of arrow}}{\text{before impact}} = \frac{\text{Potential energy or arrow plus}}{\text{block at end of swing}}$$

$$\tfrac{1}{2}m_a v_a^2 = (m_a + m_b)gh$$

$$v_a = \sqrt{\frac{2(m_a + m_b)gh}{m_a}}$$

$$= \sqrt{\frac{2(.035 + 10)(9.81)(0.5)}{.035}}$$

$$= 53.0 \ \text{m/s}$$

where $m_a$ and $m_b$ = the masses of the arrow and block, respectively, $v_a$ = the velocity of the arrow, and $h$ = the height the pendulum was lifted above its initial position.

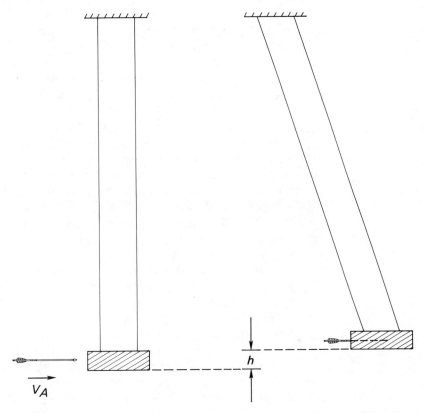

**Fig. 106.** *The velocity of an arrow can be determined using a ballistic pendulum. Two positions are shown: (a) immediately before impact, and (b) at the limit of the upward swing of the pendulum. The velocity of the arrow is designated by $V_A$ and the change in height of the pendulum by h.*

## Work–Energy Relationship

On entry into the water, a diver's velocity is gradually reduced to zero as a result of the forces exerted upon him or her by the water. In this process, the diver gives up all kinetic energy and some potential energy to do work on the water. (The level that is considered to be the "surface of the earth" and at which the potential energy of a body is zero is defined somewhat arbitrarily. For the purposes of the present example, the bottom of the pool, rather than the surface of the water, should be considered the baseline or ground level.) Since work is equal to the product of the force and the distance through which the body moves in the direction of the force (p. 181), the average force to which the diver is subjected in this process depends to a large extent on the depth to which he or she penetrates the water—or, in other words, the distance over which the water exerts retarding forces. If the diver enters the water in orthodox fashion and descends, as is quite common, to depths in the order of 3–5 m, the forces to which he or she is subjected are usually relatively small. However, if, by accident or design, the diver enters the water with a horizontal rather than a vertical orientation, the distance over which the motion is arrested is very small and the forces to which he or she is exposed are very large. Incidentally, some idea of the magnitude of the forces involved in such cases was obtained in a study of a seventy-year-old stuntman who specialized in high diving into shallow pools.[6] This man regularly gave public performances in which he dived from a height of 10.5 m to land, in a back-arched, prone position, in a pool containing 33 cm of water. When the stuntman dove from the more modest heights of 1.5–4.6 m used in this study, the instruments attached to his body revealed that the forces to which he was subjected produced accelerations of up to 224 times the acceleration due to gravity as he landed in the pool. Extrapolating from their experimental results, the investigators estimated that the diver was subjected to accelerations of up to 499 times g when he dived from his customary 10.5 m.

The relationship between the distance over which a body is brought to rest and the magnitude of the forces involved in the process is well known to participants in many other activities. Gymnasts reduce the forces to which they are subjected in the landing following a dismount by flexing at hip, knee, and ankle joints and thus increasing the distance over which their bodies are brought to rest. Baseball and cricket players reduce the forces exerted on their hands by allowing their arms—and, in some cases, their bodies—to "give" as they catch the ball. This action increases the distance over which the ball is brought to rest, thus reducing the chance of a dropped catch or, worse yet, an injury to the hands.

### Strain Energy

The capacity to do work that a body has as a result of being pushed or pulled out of its normal shape is called its *strain energy*. (Some authorities lump potential energy and strain energy together and refer to them both as potential energy.[7,8] However, since this tends to obscure some essential differences between them, these two forms of mechanical energy are considered separately here.)

[6]David C. Viano, Richard M. Schreck, and John D. States, "Dive Impact Tests and Medical Aspects of a 70-Year-Old Stunt Diver," Research publication (Detroit: General Motors Corporation, 1975).

[7]G. H. G. Dyson, *The Mechanics of Athletics* (London: University London Press, 1976), p. 322.

[8]Katherine F. Wells and Kathryn Luttgens, *Kinesiology: Scientific Basis of Human Motion* (Philadelphia: W. B. Saunders, 1976), p. 322.

Strain energy is of greatest importance in those physical education and athletic activities in which mechanical energy is stored at one stage in the activity and released again at some other stage. A pole vaulter, for example, develops kinetic energy during the approach run and then at takeoff stores some of this energy in the pole in the form of strain energy, for use later in the vault. A springboard diver depresses the board as he or she lands from the hurdle step immediately preceding the takeoff and in so doing stores strain energy that will be returned later in the form of kinetic and potential energies.

In both these activities and in others like weight lifting, trampolining, and gymnastics—especially the bar events—the timing of those movements associated with the development of strain energy and the subsequent release of that energy are critical to the success of the performance. For example, if a pole vaulter fails to time his movements so that he puts a deep bend in the pole early in the vault, he is unlikely to be able to swing forward far enough to land in the pit, let alone clear the bar on the way. Even if he does store a large amount of energy in the pole during the early phases of the vault, the pole vaulter is by no means assured success. To be successful, he must also be in the correct position and moving appropriately when the pole gives up its strain energy to perform work on him in the later phases of the vault.

## EXERCISES

Give answers of magnitude less than 1.0 (that is, $>-1.0$ and $<1.0$) correct to two decimal places and all other answers correct to one decimal place.

1. A bowler releases a ball of 7.2 kg mass at 3.3 m/s. At what velocity would a 4.5 kg ball have to be released so that it has the same momentum as the more massive ball?
2. A road cyclist riding at 25 km/h on a level surface encounters air resistance and rolling (or frictional) resistance of 9.3 N and 6.5 N, respectively. If the forces exerted on the pedals result in the tires exerting a total backward force of 24.2 N on the ground, (a) what is the resultant horizontal force acting on the cyclist-plus-machine system? (b) If the mass of the system is 80 kg, what is its horizontal acceleration? (c) What is the resultant vertical force acting on the system? (d) What is its vertical acceleration?
3. A weight lifter exerts an upward force of 2000 N on a 150 kg barbell. What vertical acceleration does it experience?
4. A bowling ball of 7.2 kg mass strikes a pin of 1.5 kg mass and exerts a force of 100 N on it. (a) To what force is the ball subjected? (b) Ignoring air resistance and friction, what accelerations do the two bodies experience?

5. The heaviest player on the Pittsburgh Steelers football team, which won the Super Bowl in January 1980, was offensive tackle Jon Kolb, who weighed 1166 N (262 lb). What was his mass?
6. The acceleration due to gravity is much less on the moon ($1.57 \, \text{m/s}^2$) than it is on Earth ($9.81 \, \text{m/s}^2$). How much would Kolb weigh on the moon?
7. The following table shows the results obtained when several vaulting horses were tested, in a manner similar to that depicted in Fig. 95 (p. 161), to determine how much horizontal force could be exerted against them before they began to slide.

| Make | Weight (N) | Force to Initiate Sliding (N) |
|------|-----------|-------------------------------|
| American | 1530 | 805 |
| Gym Master | 1588 | 779 |
| Medart | 2011 | 716 |
| Mitufa | 756 | 187 |
| Nissen | 1190 | 587 |

(a) Is the force needed to initiate sliding the same as the limiting friction? (b) For which horse was the highest coefficient of friction recorded? (c) For which horse was the lowest coefficient recorded? (d) Would the limiting friction be the same, higher, or lower when a gymnast was performing a vault over the horse? (e) Would the coefficient of friction be the same, higher, or lower in this case?

8. Figure 107 (p. 189) is a record of the horizontal forces exerted against the starting block by a swimmer of 72.5 kg mass during the start of a race. Using this record, determine the horizontal impulse exerted on the swimmer in reaction and the horizontal velocity of the swimmer at the instant the swimmer left the block. The steps listed should be followed: (a) Enlarge the record with the aid of an opaque projector, pantograph, or similar device and transfer it to a sheet of graph paper. (b) Determine the area between the curve and the time axis. This can be done by carefully counting the squares or by using a planimeter designed for the purpose. In general, the area obtained in this manner will not be in the required N-s units. The next step is to determine the necessary conversion factor. (c) Mark on the graph paper a square or rectangle with a base (or horizontal side) equivalent to a known time and a height (or vertical side) equivalent to a known force. The area of this figure is thus the equivalent of the product of a force and a time or, in other words, an impulse. (d) Determine the area of the figure and compute the required

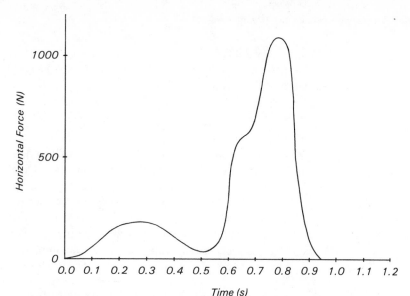

Horizontal Force (N)

Time (s)

**Fig. 107.** *Horizontal force-time curve for a swimming start.*

conversion factor (C.F.):

$$C.F. = \frac{\text{Impulse represented by figure}}{\text{Area of figure}}$$

(e) Multiply the area obtained in (b) by C.F. to determine the horizontal impulse exerted on the swimmer. (f) Use the impulse–momentum relationship (p. 165) to compute the swimmer's horizontal velocity at takeoff.

9. The rules of the United States Squash Rackets Association require that, at a temperature of 70–74° F (21.1–23.3° C), the ball must rebound off a steel plate 24–26 in. (61.0–66.0 cm) when dropped from a height of 100 in. (254.0 cm). If these requirements were restated in terms of the coefficients of restitution involved, within what range must the coefficient lie?

The rules state further that, at a temperature of 83–84° F (28.3–28.9° C), the ball must rebound to a height of 27–30 in. (68.6–76.2 cm). What is the corresponding range within which the coefficient of restitution must lie? (It is estimated that, after 10 minutes of play, the ball's temperature is increased by just over 10° F.)

10. Which of the following series of values is most likely to represent, respectively, the coefficients of restitution for a superball, a tennis ball, and a softball dropped from a height of 2 m onto a concrete floor? Explain the basis for your choice.

    a.  0.92, 0.75, 0.94

    b.  5.86, 3.21, 1.65

    c.  0.31, 0.67, 0.89

    d.  0.92, 0.75, 0.36

    e.  0.45. 3.21. 5.86

11.  a.  Measure the peak height attained by the golf ball during each of the airborne phases shown in Fig. 87 (p. 142) and, using Eq. 16 (p. 171) compute the coefficient of restitution for each of the elastic impacts involved. (To what part of the ball should the measurement of peak height be made, for this purpose?) How would you account for any differences that you found among the coefficients of restitution? (Theoretically, there are at least four possible reasons why differences might occur.)

      b.  Draw tangents to the trajectory of the ball immediately prior to and after each impact and, with a protractor, measure the corresponding angles of incidence and reflection. For each impact, compare these angles and determine whether the influence of elasticity on the vertical velocity of the ball was the same, more, or less than the influence of friction on the horizontal velocity.

12.  A tennis ball bounces on a grass court and the coefficient of restitution associated with the impact is 0.6. (a) If the horizontal velocity of the ball is reduced by 10 percent as a result of the impact, how does the angle of reflection compare with the angle of incidence? (b) How does it compare if the reduction in horizontal velocity is 70 percent?

13.  Explain with the aid of carefully drawn diagrams how a tabletennis ball might be made to pass over the net, land on the far side, and bounce back over the net without being hit by the player at that far end. (Incidentally, do the rules of the game make any provision for such an event?)

14.  Cochrane and Stobbs[9] have demonstrated that when the head of a golf club is free to move as if it were not connected to the shaft of the club, the ball will travel as far as it does when a normal club is used. The mass and velocity of the clubhead can thus be used with Eq. 18 (p. 178) to compute what happens under given circumstances. If the masses of the clubhead and ball are 0.21 kg and 0.05 kg, respectively, and their mutual coefficient of restitution is 0.67, how fast must the clubhead be traveling at impact if the ball is to leave the club at 61.5 m/s—a velocity that will produce a drive on the order of 205 m under normal conditions.

15.  An Indian fakir who weighs 500 N lies on a bed of nails so that he is supported by a total of 3,000 nails. If each nail is blunted so that its upper surface—the surface in contact with the fakir—is 0.75 mm$^2$, what is the

[9]Alastair Cochran and John Stobbs, *The Search for the Perfect Swing* (London: Heinemann Educational Books, 1968), pp. 145–47.

average pressure at the points of contact between fakir and nail? (State your answer in $N/mm^2$.)

16. If the average frictional force exerted by the snow on the skis of a competitor in a 2000 m downhill event is 14 N, how much work will be done on the skier by friction during the course of the event?

17. When a Swedish ski jumper crashed during the 90 m jump at the XIII Winter Olympics in Lake Placid, a television commentator covering the event said, "It's just as well that the snow is packed hard because if it were soft, he'd just sink right in and be much more likely to get seriously injured. As it is, he just skids and bounces on down the hill." This statement is at variance with the widely held view that soft materials make for safer landings than do hard ones and may thus come as a surprise to many. Can you explain why it is correct, in mechanical terms? [*Hint*: Use the work–energy relationship, discussed on p. 185, as a basis for comparing what happens to the athlete when the snow is soft and when it is packed hard.]

18. The sequence of Fig. 108 shows the final daring trick in a trampoline show. In this trick, the gymnast went high in the air to perform a back somersaulting stunt only to find that the spotters who had been standing around the trampoline had picked it up and walked off with it. Apparently unable to do anything else, the gymnast crashed to the floor where he lay until the spotters returned to carry his seemingly lifeless body away. The gymnast survived this very dangerous trick by landing in an "all fours" position, which enabled him to decelerate his body over a relatively long distance. (a) If the gymnast weighed 650 N and had a downward velocity of 9 m/s at the moment he first made contact with the mat placed beneath the trampoline, what would his kinetic energy have been at that moment? (Ignore his horizontal motion.) (b) How much of this kinetic energy did he lose in the process of halting his downward motion? (c) If his center of gravity was 0.65 m above the mat at the moment he first made contact with the mat and 0.05 m above it when he finally came to rest, how much potential energy did he lose in the process? (d) How much work was done on his body in bringing it to rest? [*Hint*: See work–energy relationship, p. 185.] (e) What was the average vertical force to which he was subjected during the landing? (*Warning*: The sequence of Fig. 108 was taken from a film of a highly skilled gymnast who performed the trick in a few shows several years ago. Shortly after the film was taken, he gave up performing it because he considered it much too dangerous. Readers are urged to accept this judgment. *The trick is inherently very dangerous and should not be attempted again under any circumstances.*)

19. A weight lifter performing a dead lift—a lift in which the arms are kept straight throughout and the barbell is raised from the floor to a position resting against the front of the lifter's thighs—lifts a 200 kg barbell a distance of 0.62 m in 2.8 s. What was the average power developed in the process?

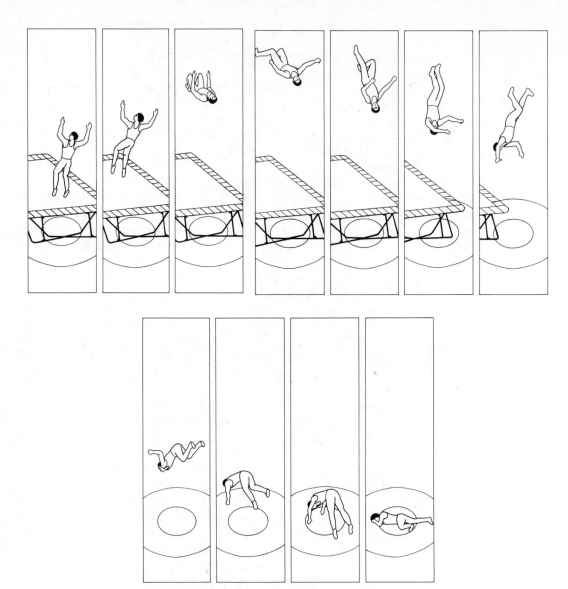

**Fig. 108.** *Decreasing the force of the impact in a dangerous trampoline stunt.*

## RECOMMENDED READINGS

Daish, C. B., *The Physics of Ball Games*. London: English Universities Press, 1972, pp. 4–12 (Impact), 13–21 (More about Impact), 35–40 (Power), 91–98 (Bouncing).

Dyson, Geoffrey, H. G., *The Mechanics of Athletics*. New York: Holmes &

Meier, 1977, pp. 28–33 (Forces (1): Newton's Laws).

Ecker, Tom, *Track and Field Dynamics* (2nd ed.). Los Altos, Calif.: Tafnews Press, 1974.

Faria, Irvin E., and Peter R. Cavanagh, *The Physiology and Biomechanics of Cycling*. New York: John Wiley, 1978, pp. 51–56 (How Powerful Am I?) 57–70 (What is Slowing Me Down?, Hills and Tires).

Hay, James G., *The Biomechanics of Sports Techniques* (2nd ed.). Englewood Cliffs, N.J.: Prentice-Hall, 1978, pp. 56–104 (Linear Kinetics).

Kelley, David L., *Kinesiology Fundamentals of Motion Description*. Englewood Cliffs, N.J.: Prentice-Hall, 1971, pp. 87–97 (An Introduction to Forces).

Rogers, Eric M., *Physics for the Inquiring Mind*. Princeton, N.J.: Princeton University Press, 1977, pp. 53–60 (Forces as Vectors), pp. 106–134 (Force and Motion), 135–153 (Crashes and Collisons: Momentum).

Tricker, R. A. R., and B. J. K., *The Science of Movement*. London: Mills and Boon, 1966, pp. 13–27 (The Role of Friction in Movement), 83–99 (The Conservation of Momentum), 100–106 (Force and Mass).

# Chapter 12

# Explaining Angular Motion (Angular Kinetics)

The quantities used to explain angular motion are very similar to those used to explain linear motion.

## ECCENTRIC FORCE

If an exponent of karate punches a bag suspended between two springs so that the line of action of the force he exerts passes through the center of the bag, the bag is translated in the direction of the blow (Fig. 109(a)). If he strikes the upper half of the bag, the bag is simultaneously translated in the direction of the blow and rotated clockwise—that is, as viewed in Fig. 109(b). Finally, if he strikes the lower half of the bag, the bag is simultaneously translated in the direction of the blow and rotated counterclockwise (Fig. 109(c)).

A force whose line of action does not pass through the center of the body on which it acts is called an *eccentric force*. Such forces cause, or tend to cause, the body to translate and rotate simultaneously—that is, to undergo general motion.

**Fig. 109.** *A force through the center of a body causes it to translate. A force that is not directed through the center of the body—an eccentric force—causes it to translate and rotate simultaneously.*

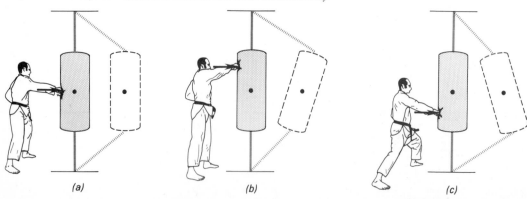

|       |       |       |
|-------|-------|-------|
| *(a)* | *(b)* | *(c)* |

## COUPLE

When a tightrope walker adjusts the position of the balancing pole that he carries, he generally exerts an upward force via one hand and a downward force of equal magnitude via the other (Fig. 110). Acting alone, the upward force would tend to translate the pole upward and, in the case in Fig. 110 to rotate it clockwise. Similarly, the downward force acting alone would tend to translate the pole downward and rotate it clockwise. When the two forces are exerted at the same time, their translatory effects cancel each other and their rotational effects combine to cause the pole to rotate, and, correctly executed, to correct any imbalance in the performer's motion. A combination like this of two forces of equal magnitude acting in opposite directions is called a *couple*.

The contents of the last two sections provide a simple, but important, explanation of the basic causes of the three forms of motion discussed earlier (pp. 113–117):

1. A force that acts through the center of a body—occasionally referred to as a centric force—causes translation.
2. Two equal and opposite forces—a couple—cause rotation.
3. A force that does not act through the center of a body—an eccentric force—causes both translation and rotation or, in other words, a general motion.

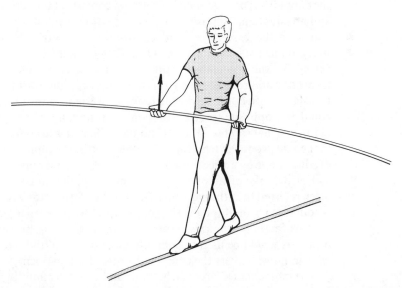

**Fig. 110.** *Equal and opposite vertical forces exerted by a tight-rope walker constitute a couple that serves to rotate his balancing pole and thereby enable him to adjust his balance.*

## MOMENT

The effect produced when a couple is exerted on a body depends, among other things, on what is termed the *moment of the couple*—that is, the product of the magnitude of the forces involved and the distance between their lines of action. In algebraic terms, the moment of the couple—also known as the *torque*—is given by

$$M = F \times d_F \qquad (21)$$

where $M$ = the moment or torque, $F$ = the magnitude of one of the forces making up the couple, and $d_F$ = the shortest, or perpendicular, distance between the lines of action of the forces. This distance is known as the *moment arm* of the couple.

Thus, the tightrope walker in Fig. 110 can make a given adjustment in his balance by having his hands close together on the pole and exerting relatively large forces or by having them well separated and exerting correspondingly smaller forces. Because tightrope walking is an activity that requires a very delicate control of the forces exerted, and because it is more likely that such control can be exercised if the forces exerted are small rather than large, tightrope walkers invariably hold the pole with their hands well separated.

When a body is fixed at some point and is free only to rotate about that point, the couple that produces such a rotation consists of the force applied to the body and an equal and opposite force exerted at that point where the body is fixed. Consider the case of a child exerting an eccentric, horizontal force on a playground roundabout (Fig. 111). If there were no other horizontal forces acting on the roundabout, this eccentric force would cause it to be both translated and rotated. Since this does not occur, it must be concluded that the translatory effect of the eccentric force exerted by the child is matched by an equal and opposite force exerted on the roundabout at the fixed central point about which it rotates—the only point at which it is in contact with another body that could produce such a force. In other words, the rotation of the roundabout is produced not by an eccentric force, as might at first appear, but by a couple.

For the sake of simplicity, the moment applied to the body in such cases is often referred to as the *moment of the applied force*. However, since this moment is computed by multiplying the applied force by the perpendicular distance from its line of action to the axis of rotation—in other words, by using exactly the same means as would be used to compute the moment of the couple—the use of this term in no way alters the computed value of the moment.

The moment of a force can be computed *either* by multiplying the force by the perpendicular distance from the line of action of the force to the axis of rotation

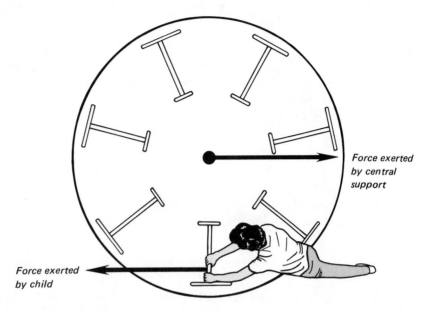

*Force exerted by central support*

*Force exerted by child*

**Fig. 111.** *An eccentric force that appears to cause rotation without translation is, in reality, one of two forces constituting a couple.*

*or* by multiplying the component of the force at right angles to a line joining the line of action and the axis of rotation, with the length of that line. For example, the moment of the force provided by the weight stack of the seated leg press machine in Fig. 112 is

$$1500 \text{ N} \times 0.95 \text{ m} = 1425 \text{ N} \cdot \text{m}$$

when computed using the first method and

$$1425 \text{ N} \times 1.0 \text{ m} = 1425 \text{ N} \cdot \text{m}$$

when computed using the second method.

In many cases, a body is acted upon not by one couple but by several simultaneously, and what happens to the body as a result depends on the magnitudes and directions of the moments exerted by these several couples. Consider the woman using the seated leg press machine in Fig. 112. In this case, the body acted upon is the open L-shaped member, which is pivoted at a central point and on which forces are exerted by the feet of the woman at one end and by the weight stack at the other. The moment that tends to turn this member in the clockwise direction is that provided by the weight stack—1425 N · m. The moment that tends to turn it in the opposite direction is that exerted by the woman. (This assumes that the plates comprising the top 1500 N of the weight

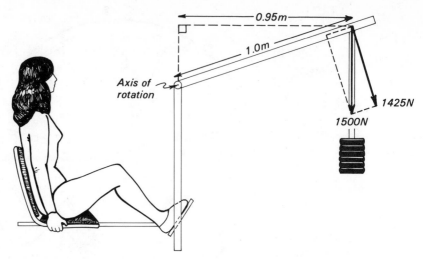

**Fig. 112.** *The moment of a force may be computed in two equivalent ways.*

stack are not resting on the plates below. If they were, the upward vertical force exerted by these plates would contribute an additional counterclockwise moment.) What happens under such circumstances depends on the magnitudes of these two contrary moments. If the moment exerted by the woman exceeds that of the weights, the member moves in a counterclockwise direction and the weights are lifted. If it does not, the member rotates in the opposite direction and the weights are lowered.

The sum of two or more moments, computed with due regard to differences in direction, is called the *resultant moment*. Thus, if the woman in Fig. 112 exerted a moment of 1500 N · m and if the counterclockwise direction were taken as positive as it usually is, the resultant moment would be

$$1500 \text{ N} \cdot \text{m} - 1425 \text{ N} \cdot \text{m} = 75 \text{ N} \cdot \text{m}$$

One additional feature of such weight-training devices is relevant to the discussion of resultant moments and thus deserves mention. Because the length of the moment arm of the weight stack exceeds that of the forces applied by the user of the device, the amount of weight necessary to provide a given resistance is actually less than that resistance. For example, if the moment arm of the weight were twice the length of the moment arm of the applied force, a weight of 500 N would be sufficient to provide a resistance of 1000 N. Thus, the weights marked on the plates do not indicate the weight of the plates themselves but rather the resistance they provide.

## EQUILIBRIUM

When the sum of the forces and the sum of moments acting upon a body are both equal to zero, the body is said to be in *equilibrium*. This situation most frequently arises when the body is stationary, in which case it is said to be in a state of *static equilibrium*.

A gymnast performing a back lever on the rings is acted upon by three external forces—his weight and the forces exerted on his hands by the rings (Fig. 113). Since the gymnast is not being accelerated horizontally, the horizontal components of the forces exerted on him by the rings must sum to zero. Similarly, since he is not being accelerated vertically, the vertical components of these forces and his weight—the only other vertical force to which he is subjected—must also sum to zero. Finally, since he is not being angularly accelerated, the moments of these various forces about any arbitrary point in space must sum to zero. The gymnast is, thus, in a state of static equilibrium.

**Fig. 113.** *A gymnast performing a back lever on the still rings is in a state of static equilibrium.*

When a body is moving with constant linear and angular velocities—which can only occur if the sum of the forces and sum of the torques acting on it is zero—it is said to be in a state of *dynamic equilibrium*. A sky diver who leaps from an airplane is subjected to two vertical forces—the downward force exerted by gravity, which causes him to accelerate toward the earth, and the upward force exerted by the air. This latter force increases as the velocity of the sky diver increases until eventually—at what is known as the *terminal velocity*—it reaches a magnitude equal to that of his weight. If the weight and air resistance forces act along the same line at this time, the sum of the forces and the sum of the moments acting on the sky diver are both equal to zero. The sky diver is, thus, in a state of dynamic equilibrium.

As the somewhat contrived nature of the preceding example might suggest, examples of dynamic equilibrium are difficult to find in physical education and athletic activities. This is because gravity, air resistance, and the other external forces that act on moving bodies generally cause them to experience linear and/or angular acceleration.

The magnitude and direction of the unknown forces that act on a body in equilibrium can often be determined by equating the sum of the forces and the sum of the moments that act on the body to zero. Suppose, for example, that one wished to determine the forces that a pole vaulter needs to exert to hold his pole in a horizontal position at the start of his approach run. Suppose, too, that the weight of the pole, the point at which it may be considered to act, the positions of the vaulter's hands, and the directions of the forces exerted via the hands were as indicated in Fig. 114(a). The force exerted via the top hand ($F_{TH}$) can be found by taking moments about A, the point at which the force exerted via the bottom hand is applied to the pole:

$$\Sigma M_A = 0$$

where $\Sigma$ is used to mean "the sum of"

$$(F_{TH} \times 0.5 \text{ m}) - (20 \text{ N} \times 2.0 \text{ m}) = 0$$

$$(F_{TH} \times 0.5 \text{ m}) = 40 \text{ N} \cdot \text{m}$$

$$F_{TH} = \frac{40 \text{ N} \cdot \text{m}}{0.5 \text{ m}}$$

$$= 80 \text{ N}$$

With the force exerted via the top hand known, the force exerted via the bottom hand ($F_{BH}$) can now be found by simply equating the sum of all the forces acting in the vertical direction to zero:

$$\Sigma F_V = 0$$

$$F_{BH} - F_{TH} - W = 0$$

$$F_{BH} - 80 \text{ N} - 20 \text{ N} = 0$$

$$F_{BH} = 100 \text{ N}$$

Results of practical interest can often be obtained with calculations of this kind. For example, in the case of the pole vaulter, it can be demonstrated that

1. The higher the grip of the top hand on the pole, the greater the force that must be exerted via this hand to hold the pole in a horizontal position—assuming a constant distance between the hands.
2. For any given position of the top hand, the greater the handspread—that is, the distance between the hands—the less the force required to hold the pole in a horizontal position. For example, if the vaulter in Fig. 114(a) increases his handspread from 0.5 m to 1.0 m, the forces that must be exerted via top and bottom hands decreases to 30 N and 50 N, respectively. It is for this reason that vaulters who place the top hand very close to the upper end of the pole often use very wide handspreads.
3. If the pole is inclined above the horizontal—as is usually the case (Fig. 114(b))—the ratio of the moment arms of the weight and the vertical force

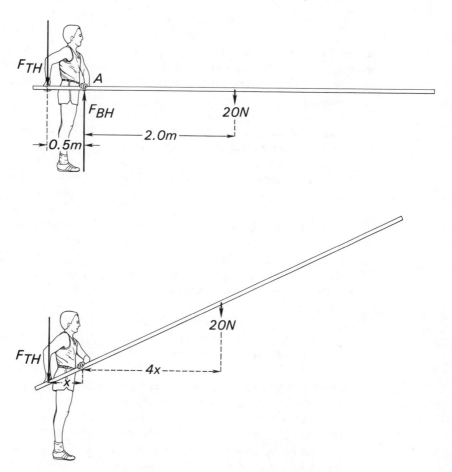

**Fig. 114.** *The basic conditions that must be satisfied if a body is in equilibrium may be used to determine the magnitude and direction of an unknown force.*

exerted via the top hand is the same as when the pole is held horizontally, that is, 4 to 1 in the case of Fig. 114. There are at least two possible reasons why vaulters generally find it easier to hold the pole in the inclined position than in the horizontal position—it permits a more completely extended and less fatiguing position of the top arm, and, if force is exerted via the top hand *at right angles to the line of the pole*, it permits a decrease in the magnitude of the force required to hold the pole in position.

## CENTER OF GRAVITY

Every animate or inanimate body may be considered to consist of an infinite number of tiny particles. Each of the particles is acted upon by gravity—that is, it is subjected to a downward vertical force due to its proximity to the earth (Newton's law of gravitation, p. 158). In any given case, the resultant of all these tiny weight forces is the weight of the whole body, a force that also acts vertically downward.

Since the line of action of this force—the so-called *line of gravity* or *gravity line*—is often very important in analyses of human motion, various methods have been developed to determine its location. Suppose—as is often the case when evaluating bats, clubs, and rackets—that one wants to locate the line of gravity of a tennis racket. One method devised for such purposes involves laying the racket across a sharp edge and moving it back and forth until it balances horizontally (Fig. 115(a)). To achieve this balanced state, the two forces acting on the racket—its weight and the vertical reaction provided by the support—must be equal in magnitude and must lie in the same vertical plane. (If they are not, the racket will be accelerated in the direction of the larger force and/or rotated about the support due to the moment of its weight. The fact that neither of these things occurs can only mean, therefore, that the weight and the vertical reaction are equal and in the same vertical plane.) The procedure used to this point has resulted in the identification of a vertical plane that contains the line of gravity. The exact location of the line of gravity within that plane has yet to be found. There are at least two ways in which this might now be done. If the racket is neither warped nor of some unusual asymetric construction, it is reasonable to assume that the line of gravity lies on the plane of symmetry that divides the racket into front and back halves. The line of intersection of this vertical plane and that located by balancing the racket will then coincide with the line of gravity (Fig. 115(b)). If the assumption of symmetry is not warranted, the vertical plane located by balancing the racket can be marked on the racket and the latter balanced a second time with its longitudinal axis at a different angle to the supporting edge. The line of intersection of the two vertical planes obtained by balancing the racket indicates the line of gravity (Fig. 115(c)).

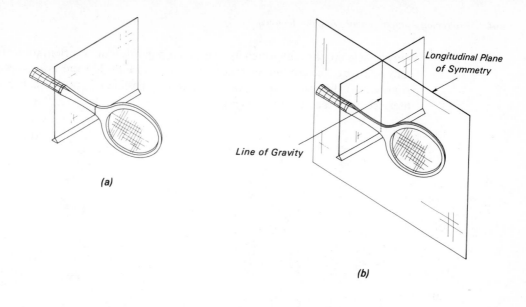

*(a)*

Longitudinal Plane
of Symmetry

Line of Gravity

*(b)*

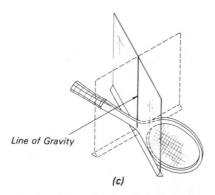

Line of Gravity

*(c)*

**Fig. 115.** *Balance method to locate the line
of gravity with racket horizontal.*

This procedure for determining the location of the line of gravity can be repeated for any number of orientations of the body. For example, if the tennis racket were balanced on its butt end as shown in Fig. 116, a line of gravity that coincided with the longitudinal axis of the racket would be found—again assuming that the racket is symmetrical.

In the course of such repeated determinations of the line of gravity, it may be observed that there is one point that always lies on the line of gravity, regardless of the orientation of the racket relative to the ground—that is, regardless of whether it is balanced horizontally or vertically. This point, known as the *center of gravity* of the body, is located at the intersection of the lines of gravity determined for two different orientations of the body. The center of gravity of a body is the point through which its weight acts regardless of the orientation of the body relative to the ground. For the tennis racket of Figs. 115 and 116, the center of gravity is that point in the racket at which the line of gravity in Fig. 115 intersects with that in Fig. 116 (Fig. 117).

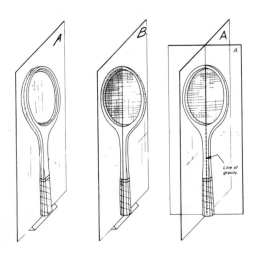

**Fig. 116.** *Balance method to locate the line of gravity with racket vertical.*

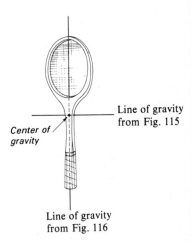

**Fig. 117.** *The location of the center of gravity of the racket is obtained by combining the two previous results.*

Although the balance method for locating the center of gravity of a body is fine in theory, the difficulties involved in obtaining the required balance positions—especially with bodies of irregular shape—limits its use in practice. To overcome this problem, an alternative method—the suspension method—is often used. The theory underlying this method of locating the center of gravity is identical to that for the balance method. In this method, the body—a field hockey stick in Fig. 118—is suspended from a fixed point and set oscillating about that point. When the body finally comes to rest, the direction of the vertical line through the point of support is marked on the body (Fig. 118(a)). This is usually accomplished with the aid of a plumb bob suspended from the same fixed point. The body is then suspended from another point, not on the line already determined, and the process repeated (Fig. 118(b)). Finally, if the asymmetry of the body requires it, the body is suspended from a third point and the process repeated (Fig. 118(c)). The point at which the determined lines intersect is the center of gravity of the body.

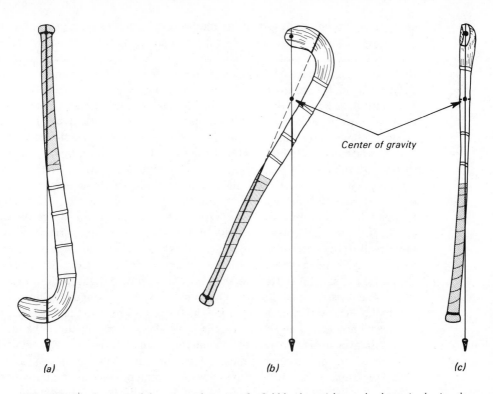

Center of gravity

(a)                    (b)                    (c)

**Fig. 118.** *The location of the center of gravity of a field hockey stick may be determined using the suspension method.*

The location of the center of gravity of the field hockey stick in Fig. 118 may well seem to be in error for it lies outside the physical substance of the stick rather than within it. This, however, is *not* an error but rather an example of a perhaps unexpected characteristic of centers of gravity—they need not lie within the substance of the bodies with which they are associated. The centers of gravity of many bodies used in physical education and athletics lie beyond the limits of the materials of which they are made. For example, the centers of gravity of hoops, tires; rings used in gymnastics; air-filled balls used in tennis, soccer, basketball, and other games; and helmets used in football, baseball, and hockey all lie in the air space enclosed, or partially enclosed, by these bodies and not within the substance of the bodies themselves. The center of gravity of the human body can also lie outside the body, especially if it is in a position of marked hip and trunk flexion or extension.

Another characteristic of centers of gravity worthy of note is that they move relative to the body as the parts of the body are themselves moved. An unusual example of this was provided some years ago by a rather unscrupulous athlete who thought that he could throw the javelin farther if its center of gravity were some distance behind the grip rather than within the limits of the grip as the rules demanded. To exploit this idea, the thrower placed small clips inside the tip and tail of his javelin and inserted a ball bearing in the former. He then reassembled the javelin and submitted it for inspection prior to a major competition. With the ball bearing in its forward position, the center of gravity of the javelin was found to lie within the limits of the grip. Having thus passed the initial test, the thrower's next task was to adjust the location of the javelin's center of gravity. To do this, he simply banged the tail of the javelin on the ground, which dislodged the ball bearing and caused it to fall into the clip at the tail. The thrower then executed his throw with the now illegally balanced javelin. When the javelin landed—not at the expected great distance, it might be added—the sudden impact once again dislodged the ball bearing and caused it to return to its original position in the forward end of the implement. The javelin was thus, once again, outwardly legal.

The fact that the center of gravity of a body moves as the parts of the body move relative to one another can be used to explain how some basketball players can actually "hang in the air" during the execution of a shot. Consider the case of the player in Fig. 119(a). Nearing the peak of his upward flight (position 1), he is in a characteristic jump-shooting position—body fully extended, elbows high, and ball overhead. From this position, he flexes his knees so that his lower legs are nearly horizontal as he reaches the peak of his jump (position 2) and then lowers them again as he begins to descend (position 3). If these movements of the lower legs are appropriately timed, the amount the center of gravity is raised *within the trunk* by the lifting of the legs exactly matches the amount it is raised as it completes its predetermined upward motion. Similarly, the amount the center of gravity is lowered *within the trunk* by the lowering of the legs is exactly equal to the amount it is lowered as it starts downward on its predetermined path. Under such circumstances, the player's head, neck, and trunk stay in exactly the same position throughout the period involved. In short, the player literally "hangs in the air." If the player adjusts the positions of his arms and the ball in addition to the position of his lower legs, as in Fig. 119(b), the distance the center of gravity is raised or lowered within the trunk is increased, and the length of time for which he "hangs in the air" is extended.[1]

Because of the nonrigid nature of the human body, the process of locating its center of gravity is often more difficult than is that of locating the center of gravity of an essentially rigid body like a tennis racket or field hockey stick. It is difficult, for instance, to determine the location of the center of gravity of a human body by balancing it on a sharp edge. Although it has been done,[2] it is also rather difficult to locate the center of gravity by the suspension method.

[1] Robert D. Bishop and James G. Hay, "Basketball: The Mechanics of Hanging in the Air." *Medicine and Science in Sports*, 11 (Fall 1979), 274–77.
[2] R. A. R. and B. J. K. Tricker, *The Science of Movement* (London: Mills and Boon, 1966), pp. 232–33.

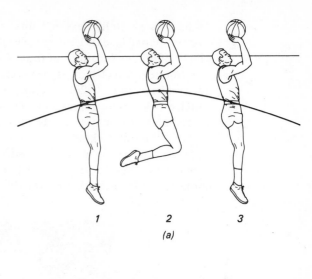

1     2     3

(a)

Fig. 119. *Techniques that might enable a player to "hang in the air." (The position of the player's center of gravity is indicated by a dot, and the path followed by that point is indicated by a curved line. Consecutive positions of the player, which would otherwise be superimposed, have been separated in the interest of clarity.)*

1     2     3

(b)

One method used to try and overcome this problem is to construct a mannikin of the human body and to determine the location of the center of gravity of this mannikin using either the balance or suspension methods. While the very simplicity of this approach has obvious appeal, the differences that exist in the distribution of the weight in such a mannikin and in the human body often result in very large errors.[3] Some attempts have been made to make due allowance for these differences. One of these[4]—a modification and simplification of a method proposed over seventy years ago by Fischer[5]—consists of a mannikin to which is

[3] James G. Hay, *The Biomechanics of Sports Techniques* (Englewood Cliffs, N.J.: Prentice-Hall, 1978), p. 131.

[4] E. Hoerler, "Ein Schwerpunktsmodell," *Jugend und Sport*, 29, no. 2 (1971).

[5] O. Fischer, *Der Gang des Menschen, II Teil* (Leipzig: Teubner, 1899).

attached a skillfully designed linkage system (Fig. 120). The design of this system—based on data gathered in cadaver experiments in which the weights and center of gravity locations of the individual segments of the human body were determined—is such that, as the various parts of the body are moved, one joint in the system (G in Fig. 120) trace the path that would be followed by the center of gravity of the person represented by the mannikin (Fig. 121). Full constructional details for this mannikin can be found in Appendix D.

Mannikins of the kind shown in Fig. 120 can only be used to locate the center of gravity of a person in a body position that the mannikin is capable of adopting. Their use is thus limited to a relatively small range of positions. Fortunately, more versatile methods are available for locating the center of gravity of the

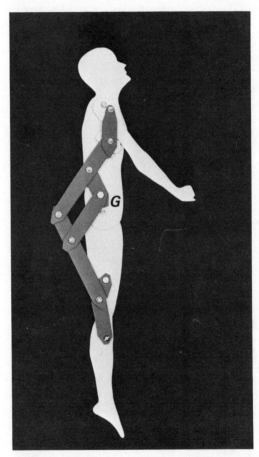

**Fig. 120.** *Mannikin used to estimate the location of the center of gravity. (The joint in the linkage system that represents the center of gravity is indicated by the letter G.)*

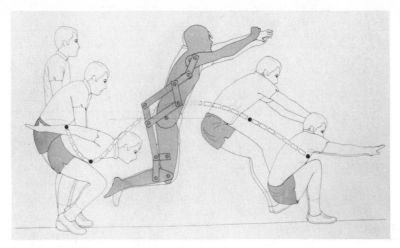

**Fig. 121.** *A center of gravity mannikin may be used to determine the path followed by the center of gravity throughout the course of a performance.*

human body. One of these involves the use of a *reaction board* (Fig. 122). This is a large board (usually about 2.5 m × 1.0 m) with two sharp edges fixed to its undersurface (Fig. 122(a)). These edges, set a known distance apart, rest one on the platform of a set of scales and the other on a block of wood of a height equal to that of the platform. The location of the center of gravity of a person in some prescribed position is determined in four simple steps.

1. The weight indicated on the scale is recorded (Fig. 122(b)). (This is not the weight of the reaction board, as is sometimes erroneously stated, but rather that part of it supported by the scale. The remainder is, of course, supported by the block of wood.)
2. The subject assumes the prescribed position on the reaction board (Fig. 122(c)).
3. The weight indicated on the scale is recorded.
4. The location of the subject's center of gravity relative to the long axis of the board—or, more specifically, the distance of the center of gravity from the block-supported edge—is computed.

When the reaction board is placed in position, it is in a state of equilibrium under the influence of three forces—its weight ($W_b$) and the vertical reactions exerted on it by the block of wood and the platform of the scale ($R_A$ and $R_1$, respectively). Since the board is in equilibrium, the sum of the moments of these forces about any point is equal to zero. Taking moments about A, the point at which the board is supported on the block (Fig. 122(b)), yields

$$(R_1 \times d) - (W_b \times X_b) = 0$$

or

$$(R_1 \times d) = (W_b \times X_b)$$

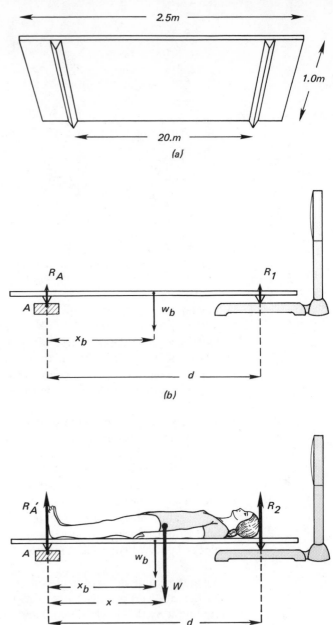

**Fig. 122.** *Reaction board method for locating the center of gravity: (a) oblique inferior view of reaction board showing typical dimensions and arrangement of supporting edges, (b) locating the center of gravity of the board, and (c) locating the center of gravity of the subject.*

where $d$ = the known distance between the supporting edges and $X_b$ = the unknown horizontal distance from A to the line of action of $W_b$. (In accord with the established convention, counterclockwise moments are considered to be positive and clockwise moments to be negative.)

When the subject has taken up her position on the board (Fig. 122(c)), the combined subject-plus-board system is in equilibrium under the influence of four forces—the weights of the subject (W) and the board, and the vertical reactions exerted on it at the two edges, ($R_A'$ and $R_2$). Since the system is in equilibrium, the sum of the moments of these forces about any point is equal to zero. This fact is now used to determine the location of the center of gravity of the subject or, more precisely, the distance of her center of gravity from the supporting edge at A. Taking moments about A,

$$(R_2 \times d) - (W \times X) - (W_b \times X_b) = 0$$

Substituting for ($W_b \times X_b$),

$$(R_2 \times d) - (W \times X) - (R_1 \times d) = 0$$

and rearranging yields

$$X = \frac{(R_2 - R_1)d}{W} \qquad (22)$$

It should be noted here that the location of the center of gravity can thus be determined without prior knowledge of the weight of the board, the $W_b$ term having been eliminated along the way.

Thus, when the following data were obtained for the subject depicted in Fig. 123

$$W = 590 \text{ N}$$

$$R_1 = 78 \text{ N}$$

$$R_2 = 388 \text{ N}$$

$$d = 1.8 \text{ m}$$

**Fig. 123.** *The reaction board method used to determine the location of the center of gravity in a simulated, erect standing position.*

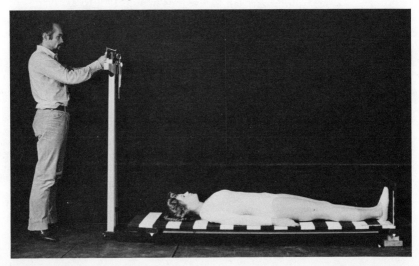

her center of gravity was located

$$X = \frac{(R_2 - R_1)d}{W}$$

$$X = \frac{(388 - 78)\ 1.8}{590}$$

$$= .95 \text{ m}$$

from the soles of her feet. Further, since she is 1.72 m tall, this distance of .95 m is equivalent to 55 percent of her standing height. (It is of some interest to note that because of differences in physique, the height of the center of gravity in women is usually a slightly smaller percentage of the standing height than it is in men (Fig. 124)).

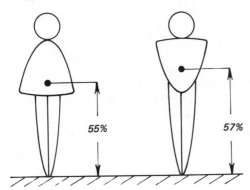

**Fig. 124.** *Differences in the physique of men and women—most notably, in the breadth of hips and shoulders—result in women having slightly lower centers of gravity than men when both assume an erect standing position.*

While simple reaction boards like that depicted in Fig. 122 are most frequently used to determine the location of the center of gravity of a subject in a simulated erect standing position (Fig. 123), they can also be used to determine its location, relative to the long axis of the board, in many other positions. For example, a reaction board can be used, as in Fig. 125, to analyze

1. how a baseball batter distributes his weight as he awaits the pitch,[6]
2. where the line of gravity falls in a well-executed handstand or headstand, and
3. where the line of gravity falls in various two- or three-person balances.

---

[6] Once the line of gravity has been established using the reaction board method, the conditions that characterize a body in equilibrium—the sum of the forces and the sum of the moments are each equal to zero—can be used to obtain estimates of the weight supported by each foot. For example, by taking moments of the forces about the center of the right foot, an estimate of the weight supported by the left foot can be obtained. An estimate of the weight supported by the right foot can be obtained in similar fashion or by simply subtracting the value for the left foot from the total weight of the batter.

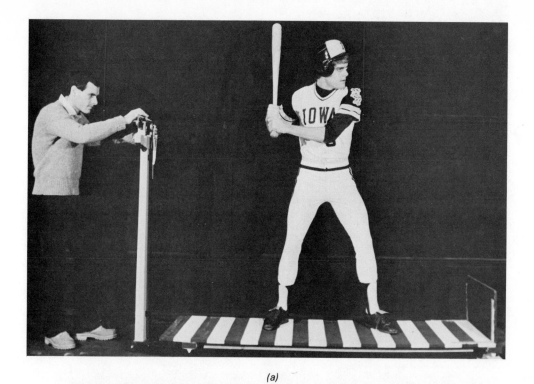

(a)

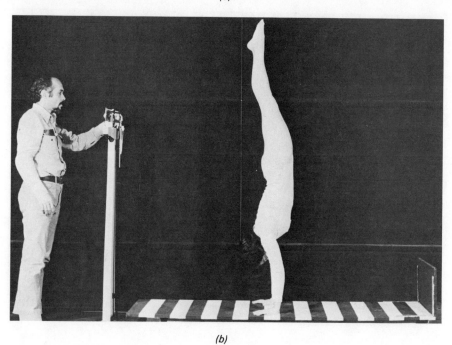

(b)

**Fig. 125.** *The reaction board method used to determine the location of the line of gravity in (a) a baseball batting stance, (b) a handstand,*

213

**Fig. 125.** *(cont.) (c) a headstand, and (d) a two-person balance.*

A rectangular block of wood has three readily identifiable dimensions—its length, its width, and its depth or thickness. When it is balanced on a sharp edge with its longest dimension—that is, its length—at right angles to the edge, a vertical plane containing its center of gravity is identified. This plane marks the location of the center of gravity in one dimension—that is, relative to the length of the body. Its locations in the other two dimensions—that is, relative to the left and right sides, and relative to the front and back of the block—remain unknown. The reaction board method similarly yields the location of the center of gravity of the human body in one dimension—relative to its length in Fig. 123, relative to its width in Fig. 125(a), and relative to its depth in Figs. 125(b) and (c). (Whether the determined location of the center of gravity is relative to the length, width, or depth of the body depends simply on its orientation on the reaction board.)

Several methods have been developed to determine the location of the center of gravity in more than one dimension. One of these involves the use of a large reaction board with three points of support, two resting on the platforms of scales and the third on a block of wood (Fig. 126). This method, an extension of the method already described, yields the location of the center of gravity in two

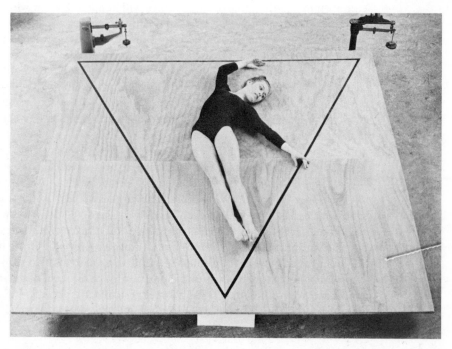

**Fig. 126.** *Large reaction board used to determine the location of the center of gravity in two dimensions.*

dimensions. Another method, and probably the most versatile yet devised, enables the location of the center of gravity of a human body to be determined from a photograph or, as is more frequently the case, from a single frame of a motion picture film.

This method, known as the *segmentation method*, involves three steps: (a) considering the body to be composed of a series of rigid segments—foot, shank, thigh, trunk, etc.; (b) measuring the positions of the ends of these segments; and (c) using known values for the weight of each segment and the location of its center of gravity—values often obtained from experiments with cadavers—computing the location of the whole body center of gravity. This procedure also yields the location of the center of gravity in two dimensions. However, if two or more cameras are used to record the position of the body, it can be extended to determine the location of the center of gravity in three dimensions. (Detailed discussion of the large reaction board and segmentation methods is beyond the scope of this book. Interested readers are referred to Hay[7] for further information on the subject.)

## STABILITY

If a body in equilibrium is given a small linear or angular displacement from its equilibrium position, one of three things will happen to it—it will keep going in the same direction, it will stop in the position to which is has been moved, or it will return to its original position. For example, consider a ball that is lying at rest on some fixed surface. If the surface is convex (Fig. 127(a)) and the ball is given a small displacement, the ball will keep moving in the direction of this displacement. It will simply roll farther and farther away from its original equilibrium position. If the surface is flat (Fig. 127(b)) and the ball is given a small displacement, it will remain in the position to which it has been displaced. Finally, if the surface is concave (Fig. 127(c)) and the ball is displaced, the ball will return to its initial equilibrium position. On the basis of these three different outcomes, the ball is said to have been initially in a state of *unstable equilibrium*, *neutral equilibrium*, or *stable equilibrium*, respectively.

The reason for these differences in behavior lies in the moments of the forces to which the ball is subjected. In each case, the ball is acted upon by two forces—its weight acting through its center of gravity and a reaction force exerted by the supporting surface through its point of contact with the ball. In the first instance, the moment of the weight about the point of contact acts to cause the ball to continue rolling away from its initial position. (The moment of the reaction force about the point of contact is, of course, zero and thus has no influence on the outcome.) In the second instance, the weight acts directly through the point of

[7]Hay, *The Biomechanics of Sports Techniques*, pp. 129–38.

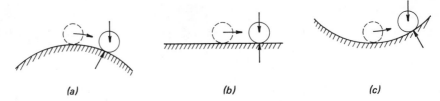

**Fig. 127.** *(a) Unstable, (b) neutral, and (c) stable equilibrium.*

contact and thus has no moment tending to cause the ball to move in one direction or another. The ball therefore remains in the position to which it has been moved. Finally, in the third instance, the moment of the weight about the point of contact is such as to cause the ball to return to its initial position.

Although the human body is infinitely more complex than a ball, it too can be in a state of unstable, stable, or neutral equilibrium. For example, a gymnast performing a handbalance is in a state of unstable equilibrium, and a gymnast hanging from a horizontal bar is in stable equilibrium.

The stability of a body in equilibrium is governed by the moment of the weight of the body about the axis of interest—usually the axis about which rotation is most likely to occur. This, in turn, is governed by the magnitude of the weight and of its moment arm. Since it is well known that it is more difficult to lift, push, or tip a heavy body than a lighter one, it should come as no surprise that the weight of a body is a factor in determining its stability. This, after all, is one of the principal reasons for having weight classes in combative sports like wrestling, judo, and boxing. The length of the moment arm of the weight depends on where the line of gravity falls with respect to the limits of the base of support. For example, in a parallel wrestling stance, the line of gravity passes much closer to the forward and backward limits of the base (Fig. 128(a)) than it does to the sideward limits (Fig. 128(b)). The moment arm and the stability of the body are thus much less in the former direction than in the latter. (It is important to emphasize here that, in referring to the stability of a body, it is usually necessary to make reference to the direction under consideration. A body may be very stable in one direction and quite unstable in another.)

Athletes in many sports make use of the relationship between stability and the distance from the line of gravity to the limit of the base to enhance their performances. Sprinters, swimmers, and football players move their centers of gravity close to the forward limits of their base so that only a small force will be necessary to upset their equilibrium and get them moving in the required forward direction; and defensive players in basketball and football often use stances in which the center of gravity is close to the backward limit of the base so that they can respond quickly to a forward movement by the opponent.

The length of the moment arm, and, thus, the stability of the body, may also be influenced by the height of the center of gravity. Consider the case of a wrestler who adopts two contrasting stances—one very erect and one a more orthodox

**Fig. 128.** *The distance from the line of gravity to the limit of the base is less in (a) the forward-backward direction than in (b) the sideward direction.*

crouch (Fig. 129(b)). If, while in the crouched stance, he is tipped clockwise through an angle $\theta$, his center of gravity remains to the left of the axis of rotation (Fig. 129(a)). In contrast, the same angular displacement of his body while he is in the erect stance causes his center of gravity to be moved to the right of the axis of rotation (Fig. 129(c)). The wrestler's position is thus much less stable when his center of gravity is high than when it is low. (The initial position of the line of gravity relative to the base and, of course, the weight of the wrestler are the same in both cases.)

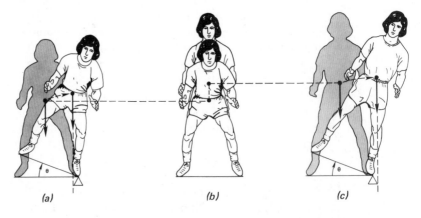

(a)                                    (b)                                    (c)

**Fig. 129.** *With all else equal, the higher the center of gravity, the less stable the equilibrium.*

According to Newton's first law (p. 154), bodies have a characteristic reluctance to change whatever they are doing. If they are at rest, they tend to stay at rest; if they are moving in a straight line, they tend to keep moving along that line. This characteristic of a body is called its *inertia*—a Latin word meaning idleness or laziness (p. 152). The mass of a body is a measure of its resistance to change in its state of rest or in its linear motion (p. 153).

In addition to resisting changes in its linear motion, a body also resists changes in its angular motion. Thus, to set a body rotating or to change its angular motion, one must also overcome its inertia, that is, its resistance to change. A body's resistance to change in its angular motion is measured not by its mass alone but, in addition, by how that mass is distributed relative to the axis about which the body is rotating or tending to rotate. If the mass is close to this axis, the resistance is small; if it is well away from the axis, the resistance is large.

The quantity that characterizes a body's resistance to changes in its angular motion is called its *moment of inertia*. The moment of inertia of any particle of a body—that is, of any very small part of that body—is formally defined as the product of its mass and the square of its distance from the axis of rotation. Thus, a particle of mass $m_1$ located at a distance of $r_1$ from the axis of rotation of the golf club in Fig. 130 has a moment of inertia

$$I_1 = m_1 r_1^2$$

Similarly, a particle of mass $m_2$ located at a distance of $r_2$ from the axis has a moment of inertia

$$I_2 = m_2 r_2^2$$

and so on. The moment of inertia of the entire golf club is obtained by adding up the moments of inertia of all the particles of which it is composed

$$I = I_1 + I_2 + I_3 + \cdots$$
$$= m_1 r_1^2 + m_2 r_2^2 + m_3 r_3^2 + \cdots$$
$$= \Sigma \, mr^2 \tag{23}$$

where $\Sigma$ is used to mean "the sum of all possible values of."

**Fig. 130.** *Defining the moment of inertia of a golf club.*

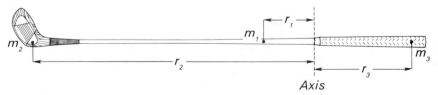

**219**

| Position | Axis | Moment of Inertia (kgm²) |
|---|---|---|
| 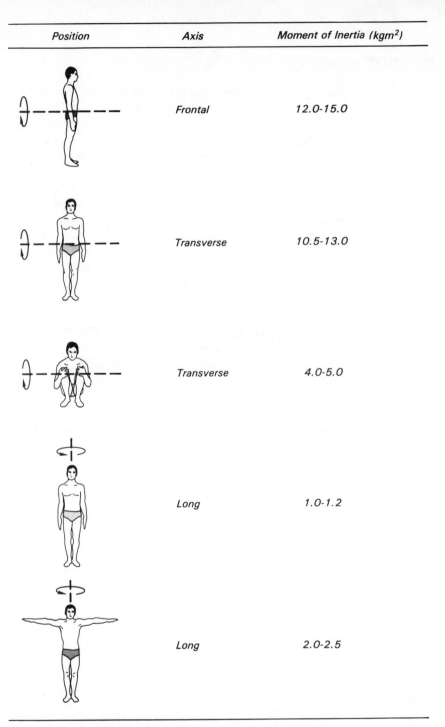 | Frontal | 12.0-15.0 |
| | Transverse | 10.5-13.0 |
| | Transverse | 4.0-5.0 |
| | Long | 1.0-1.2 |
| | Long | 2.0-2.5 |

**Fig. 131.** *The moment of inertia of the human body in various positions about various axes [Adapted from Gerhard Hochmuth, Biomechanik sportlicher Bewegungen, Frankfurt (Main): Wilhelm Limpert-Verlag, 1967, p. 60].*

While the formal definition of the moment of inertia presented here may be acceptable in theory, it is no doubt difficult to see how it can be used in practice—it is difficult, for example, to visualize how the mass of each and every particle of a body might be determined. Fortunately, however, mathematical methods can be used to determine the moments of inertia of bodies that are made up of regular geometric solids, and experimental methods can be used to obtain the same information for bodies, like the human body, that are irregular in shape. Experimental methods were used, for example, to determine the moments of inertia of the human body in the positions shown and for the axes specified in Fig. 131.

## PRINCIPAL AXES

An infinite number of axes might be considered to pass through the center of gravity of a body (Fig. 132(a)). Assuming that the body is not a solid of revolution—that is, a body like a cylinder or a sphere, which has the form a line would trace out when rotated or revolved about an axis—the moment of inertia of the body about one of these axes [AA' in Fig. 132(b)] will be larger than that about any other axis. Similarly, the moment of inertia of the body about one of the other axes [BB' in Fig. 132(b)] will be smaller than that about any other axis.

**Fig. 132.** *The principal axes of a block of wood and a woman in an erect standing position.*

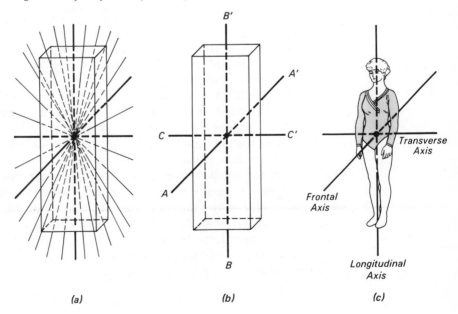

(a)                    (b)                    (c)

These two axes are at right angles to each other and together with a third axis [CC′ in Fig. 132(b)], which is at right angles to the other two, are the *principal axes* of the body.

When a woman assumes an erect standing position (Fig. 132(c)), the axis through her center of gravity about which her moment of inertia is a maximum runs from front to back. This is her *frontal* (or *anteroposterior*) *axis.* The corresponding axis about which her moment of inertia is a minimum runs down the length of her body. This is her *longitudinal axis.* Finally, the principal axis at right angles to these two runs across the length of her body and is known as her *transverse axis.* When the woman assumes other positions, the location of these three principal axes is often much more difficult to estimate. Their location can be determined, however, using appropriate mathematical methods.

## ANGULAR MOMENTUM

The quantity of motion of a body experiencing linear motion is called its momentum and is found by multiplying the mass of the body by its velocity (p. 153). The quantity of motion of a body undergoing angular motions is called its *angular momentum* and is found by multiplying the angular equivalents of mass and velocity—that is, the moment of inertia and the angular velocity of the body. Thus,

$$H = I\omega \qquad (24)$$

where $H$ = the angular momentum, $I$ = the moment of inertia, and $\omega$ = the angular velocity of the body, all relative to some carefully defined axis.

The angular momentum that the human body has in various physical education and athletic activities varies from zero, when it is simply not rotating about the axis in question, to values as large as $138\,kg \cdot m^2$ (Table 12). (While a knowledge of the angular momentum possessed by a human body and by its parts can be very useful in analyzing a movement, the computation of these values is often a very difficult task. For this reason, there is very little information available about the angular momentum possessed by the body in different activities, and most of what is available is with reference to motion about the transverse axis of the body—something that can be obtained with the film from one camera placed at right angles to the subject's "plane of motion." The data of Dapena on subjects performing the Fosbury flop and straddle styles of high jumping are among the extremely limited amount currently available on the angular momentum of the body about all three of its principal axes.)

**TABLE 12  Angular Momentum of the Human Body During Airborne Phases of Selected Athletic Activities**

| Activity | Principal Axis | Angular momentum $(kg \cdot m^2 \cdot sec^{-1})$ | Source |
|---|---|---|---|
| Long jump | | | |
|   Sail style | Transverse | 5 | Ramey[a] |
|   Hang style | Transverse | 14 | Ramey[a] |
|   Hitchkick style | Transverse | 20 | Ramey[a] |
| Forward dive | | | |
|   Pike | Transverse | 26 | Miller[b] |
|   Layout | Transverse | 29 | Miller[b] |
| Longhorse vault (Yamashita) | Transverse | | |
|   Flight | | 30 | Hay et al[c] |
|   Preflight | | 60 | Hay et al[c] |
| Front somersault, tuck | Transverse | 65 | Hay et al[c] |
| Forward $1\frac{1}{2}$ dive, pike | Transverse | 94 | Miller[b] |
| Forward $2\frac{1}{2}$ dive, pike | Transverse | 106 | Miller[b] |
| Double back somersault, layout | Transverse | 138 | Dapena[d] |
| High jump | | | |
|   Fosbury flop style | Transverse | 5-21 | Dapena[e] |
| | Frontal | 16-27 | Dapena[e] |
| | Longitudinal | 10-15 | Dapena[e] |
|   Straddle style | Transverse | 3-5 | Dapena[f] |
| | Frontal | 17-19 | Dapena[f] |
| | Longitudinal | 1-5 | Dapena[f] |

[a] M. R. Ramey, "The Use of Angular Momentum in the Study of Long Jump Take-offs," *Biomechanics IV*, ed. R. C. Nelson and C. A. Morehouse (Baltimore, Md.: University Park Press, 1973), pp. 144, 149.

[b] Doris I. Miller, "A Computer Simulation Model of the Airborne Phase of Diving," Ph.D. dissertation, (Pennsylvania State University, 1970).

[c] James G. Hay, Barry D. Wilson, and Jesus Dapena, "A Computational Technique to Determine the Angular Momentum of a Human Body," *Journal of Biomechanics*, 10 (1977), 269-277.

[d] Jesus Dapena, Unpublished report, (Biomechanics Laboratory, University of Iowa, 1976).

[e] Jesus Dapena, "A Simulation Method for Predicting the Effect of Modifications in Human Airborne Movements," Ph.D. dissertation, (University of Iowa, 1979).

[f] Jesus Dapena, Personal communication, March 31, 1980.

## NEWTON'S LAWS OF ANGULAR MOTION

Just as many quantities used to describe and explain linear motions have their direct equivalents in quantities used to describe and explain angular motions, Newton's three laws of motion have their directly equivalent angular forms.

## Newton's First Law

The angular form of Newton's first law may be stated as follows:

*A rotating body will continue to turn about its axis of rotation with constant angular momentum unless an external couple or eccentric force is exerted upon it.*

This law is also known as the *principle of conservation of angular momentum*—a name it was given because it states the conditions under which the angular momentum of a body is conserved or remains unaltered.

The principle of conservation of angular momentum is of particular significance in those athletic activities in which the performer is airborne. Assuming that the air resistance encountered is negligible, the only external force acting on an airborne performer is the performer's weight—a force that acts through the center of gravity and is thus incapable of altering the angular momentum about any axis through that point. The angular momentum of an airborne performer is thus conserved. (Whether somersaulting, side somersaulting, twisting, or performing some combination of these angular motions, an airborne performer rotates about an axis that passes through the performer's center of gravity.)

The fact that the angular momentum of an airborne performer is conserved has very important implications in practice. Consider the case of a gymnast who performs a round-off (cartwheel with a one-quarter turn) followed by a double back somersault in a floor exercise routine. As the gymnast takes off from the floor he has a certain amount of angular momentum about a transverse axis through his center of gravity. However, because his body is in a fairly extended position at this moment, his angular velocity is much too slow to allow him to complete the required double somersault in the time available. To overcome this problem, the gymnast tucks tightly soon after he leaves the floor. This reduces his moment of inertia about his transverse axis to approximately one fourth of its previous value (Fig. 131) and, because the product of his moment of inertia and his angular velocity must remain constant, increases his angular velocity to approximately four times its previous value. The gymnast's chances of successfully completing the double somersault are thus greatly improved.

For many gymnasts, success in performing a double back somersault and other similar movements depends on reducing the moment of inertia about the transverse axis to the absolute minimum. To this end, they adopt a tucked position with the knees apart and the head and trunk pulled downward and forward between the legs (Fig. 133). This thrusting of the upper body between the legs is known as "cowboying" the somersault, presumably because the straddled position of the legs is somewhat reminiscent of that used in riding a horse.

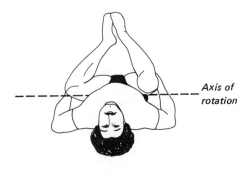

Axis of
rotation

**Fig.133.** *The "cowboy" position used by gymnasts to minimize the moment of inertia about a transverse axis when performing multiple somersaults.*

For purposes of demonstration, the conditions under which the angular momentum of a body is conserved are often simulated with the aid of a turntable. In one such demonstration, the subject stands on the turntable with his arms held sideward at shoulder level; the subject plus turntable is set in motion; and the subject moves his arms slowly in toward the midline of his body and out again (Fig. 134). Although the angular momentum is not conserved in such a case, the loss due to friction is generally so small that the interplay between the changes in the moment of inertia of the body about its longitudinal axis and its angular velocity about the same axis can be observed readily.

The conditions under which this demonstration is conducted—and, in particular, the gradual reduction of the body's angular momentum due to the friction in the turntable—closely parallel those under which dancers and skaters perform pirouettes on their toes or the tips of their skates. The techniques used to control the speed of rotation in these cases are also essentially the same as those

**225**

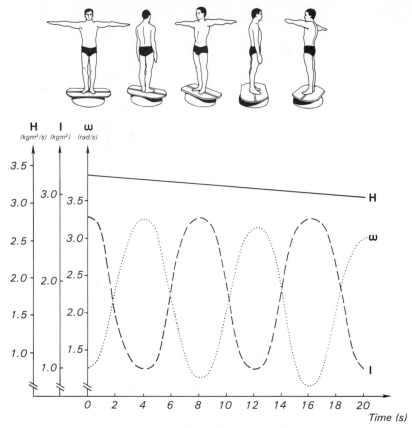

**Fig. 134.** *Turntable demonstration frequently used to show relationships among angular momentum (H), moment of inertia (I), and angular velocity (ω).*

used in the turntable demonstration—the dancers and skaters simply move the nonsupporting limbs toward and away from the axis of rotation. Trampolinists, gymnasts, and divers use similar techniques to control speed of rotation about their longitudinal axes in many airborne—and occasional nonairborne—twisting movements.

## Newton's Second Law

The angular form of Newton's second law may be stated as follows:

> *The angular acceleration of a body is proportional to the torque (or moment) causing it and takes place in the direction in which the torque acts.*

With an appropriate choice of units and the use of algebraic notation, this statement can be reduced to the form

$$T = I\alpha \qquad\qquad (25)$$

where $T$ = the torque exerted, $I$ = the moment of inertia, and $\alpha$ = the angular acceleration, all relative to some specified axis. (The terms in Eq. 25 represent instantaneous values of the respective quantities. The equation can be modified to allow the computation of average values by simply adding an overbar to each of the terms. The equation thus becomes $\overline{T} = \overline{I}\overline{\alpha}$.)

When a golfer strikes the ball off center in the course of making a putt, the force exerted by the ball on the putter causes it to rotate about a vertical axis through the center of gravity of the clubhead. This rotation alters the orientation of the blade of the putter and, thus, the direction in which the ball is propelled.

According to Eq. 25, there are just two factors that influence the angular acceleration of the putter during an off-center impact—the torque applied by the ball and the moment of inertia of the putter. The golfer thus has two ways in which to minimize the undesirable effects of an off-center hit. The first is to minimize the torque applied by striking the ball close to the center of the clubface—something the golfer presumably tries to do on every putt. The second is to use a putter whose mass is concentrated toward the heel and toe of the clubhead and whose moment of inertia about a vertical axis through its center of gravity is thus relatively large. Several manufacturers of golf clubs have developed such putters in recent years.

## Newton's Third Law

The angular form of Newton's third law may be stated as follows:

> *For every torque that is exerted by one body on another, there is an equal and opposite torque exerted by the second body on the first.*

Innumerable examples of the application of this law can be found in physical education and athletic activities. Of these, probably the most dramatic are those in which a contracting muscle or group of muscles exerts a torque on one part of the body, angularly accelerates it in the direction in which it acts, and simultaneously evokes a contrary angular acceleration of some other part of the body in reaction. A pole vaulter releases the pole and throws his head and arms in a counterclockwise direction to lift them clear of the bar. In reaction, his legs are accelerated in a clockwise direction (Fig. 135(a)). A ballet dancer performing a Saut de poisson gracefully moves her head and arms in a clockwise direction to the side, and her legs, in reaction, move in a counterclockwise direction (Fig. 135(b)). Finally, a softball player swings her upper body in a clockwise direction as she makes a left-handed throw, and, in reaction, her legs rotate in the opposite direction (Fig. 135(c)).

(c)

(b)

**Fig. 135.** *Angular action and reaction about (a) transverse, (b) frontal, and (c) longitudinal axes. [Photos (b) and (c) courtesy Ashley Collins and Drake Hokanson, respectively].*

Two important points should be noted with respect to the examples just given. First, the form of Newton's third law applies to angular motions about any axis. In the examples of Fig. 135, the angular motions took place about axes that roughly paralleled the principal axes of the body. This, however, is not a condition that must be met before the law is applicable. It applies regardless of the location and orientation of the axis. Second, care was taken to describe the direction of the angular motions involved in appropriate terms—namely, clockwise and counterclockwise. Errors in analysis often occur if terms used to indicate directions in linear motion—terms like forward and backward, upward and downward—are used to describe the direction of an angular motion. For example, describing the angular motion of the pole vaulter's arms and head in Fig. 135(a) as upward and backward might well lead to the conclusion that his legs should move downward and forward in reaction. This, however, would be precisely the same angular direction—clockwise—as that in which the arms and head are moving. The conclusion, therefore, would be patently incorrect.

When a torque is applied to a body, the angular acceleration that it acquires depends on its moment of inertia relative to the axis of rotation (Eq. 25). When equal and opposite torques are applied to different bodies, the angular accelerations that they experience are similarly dependent on their moments of inertia. Thus, although the torques exerted on two bodies that interact with each other are equal in magnitude, the effects that they produce are often quite different. In no case is this more evident than when the earth is one of the two bodies involved.

Consider, for example, a discus thrower who exerts large muscular torques to bring her body, and especially her upper body, around to the point at which she releases the discus. In reaction to these torques, her lower body tends to rotate in the opposite direction. This tendency is thwarted, however, because the thrower's feet are firmly fixed against the ground. The reaction is thus transmitted to the ground so that, instead of the upper body of the performer interacting with the lower body—as it did in the cases depicted in Fig. 135—her entire body is now interacting with the earth. The equal (and opposite) torques now exerted on the thrower and the earth have markedly different effects. Because of her small moment of inertia about the axis of rotation, the thrower experiences a clearly observable angular acceleration. The earth, on the other hand, appears to be singularly unaffected by the applied torque—its moment of inertia is so large that the angular acceleration that it experiences is quite imperceptible. For this reason, the earth is often said to "absorb the reaction" in such cases, and throwers, baseball batters, tennis players, and many others are instructed to "plant" their feet firmly against the ground so that the reactions to their movements can be absorbed in this manner.

## TRANSFER OF ANGULAR MOMENTUM

When a body is airborne and the angular momentum of one part of the body is changed, the angular momentum of some other part of it must also change if the total is to remain constant. This simple fact means that an airborne performer can stop a part that is rotating and cause the angular momentum it possessed to be taken up somewhere else in his or her body; conversely, the performer can set one part rotating and, by thus taking up angular momentum previously possessed by another part, cause this latter part to either stop rotating or to rotate in the opposite direction.

Many athletes make use of this ability to transfer angular momentum from one part of the body to another to improve their performances. Long jumpers who use the hitch-kick style circle their legs and arms to take up the angular momentum acquired at takeoff and thus prevent the trunk from being rotated into a poor landing position (Fig. 136). A gymnast performing a flyaway dismount does exactly the same thing when he circles his arms in one direction to rotate his body in the opposite direction and into the desired upright landing position (Fig. 137).

## EXERCISES

Give answers of magnitude less than 1.0—that is, $> -1.0$ and $< 1.0$—correct to two decimal places and all other answers correct to one decimal place.

1. When a marksman fires a pistol, a force is exerted on the bullet to drive it from the barrel of the weapon. The reaction to this force acts to drive the pistol in the opposite direction. This reaction is an eccentric force and thus tends not only to drive the pistol backward but to rotate it about a transverse axis through its center of gravity. To combat this latter tendency, the marksman applies a couple to the handle of the weapon. Draw a sketch of a pistol and superimpose on it three arrows—one to represent the reaction and the other two to represent the couple. Draw these arrows roughly to scale and so that the resultant moment about the transverse axis through the center of gravity of the pistol—a point that should also be included in the sketch—is in the same direction as the moment due to the reaction. (In pistol shooting, the tip of the barrel invariably rotates upward and backward a little despite the marksman's efforts to eliminate this motion.)

2. Is a trampolinist who is momentarily stationary at the peak of her flight during the execution of an upward jump in a state of equilibrium? If so, is this a state of static equilibrium or dynamic equilibrium? If she is not in a state of equilibrium, why is she not?

**Fig. 136.** *The angular actions of the legs and arms are used to produce a reaction that brings the trunk into a favorable position for landing.*

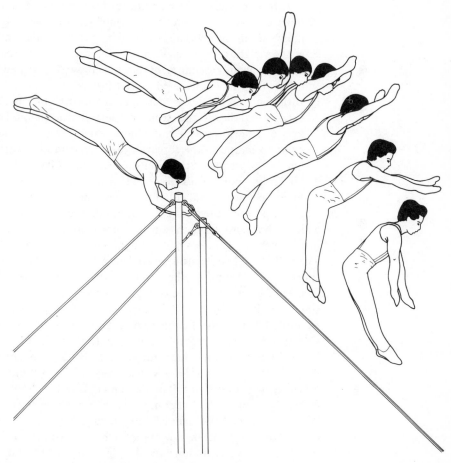

**Fig. 137.** *The angular actions of the arms produce the desired contrary reaction of the legs and trunk.*

3. How much force must be applied in a horizontal direction to the top of a high hurdle (1.07 m) to cause it to start rotating forward? The hurdle has two 32 N counterweights placed on the supporting legs of the hurdle at a distance of 0.6 m behind the axis about which it tends to tip. (Ignore the additional counterweight effect produced by the legs themselves.)

4. A weight lifter performing a curl with a 300 N barbell stops the lifting motion with his forearms in a horizontal position. (A curl is an arm exercise performed in the standing position and in which the barbell is raised from a position with the elbows extended and the barbell resting against the performer's thighs to one with the arms flexed and the barbell under his chin.)

   a. If the line of action of the weight of the barbell is 30 cm from the axis of rotation, what is the magnitude of the moment that the elbow flexors must exert to maintain the position? (Ignore the weight of the weight lifter's forearms and hands in making your calculation.)

   b. If the weight of the weight lifter's forearms and hands were taken into account, would the moment that the elbow flexors had to exert be more or less than that calculated in part (a)?

   c. If the only elbow flexor muscle active was the biceps—a rather unlikely event—and the distance from its insertion (the bicipital tuberosity) to the axis of rotation was 3 cm, what vertical force will the muscle have to exert to maintain the forearm in its horizontal position? (Once again, ignore the weight of the lifter's forearms and hands.)

   d. Under the conditions described, would the total force exerted by the biceps be more than, equal to, or less than that calculated in part (c)?

5. Compile a list of physical education, athletic, or recreational activities in which a body moves with constant linear and angular velocities—that is, in which the body is in a state of dynamic equilibrium.

6. The following statement appeared in a book on the use of mechanics in sports: "The weight of the mass on any one side of the center of mass [center of gravity] must be equal to the weight on the opposite side." This statement is wrong. Can you explain why it is wrong? Can you cite an example from physical education or athletics in which the weight on one side of the center of gravity is not equal to the weight on the opposite side?

7. A world record for a high jump by a lower-limb amputee was set by Arnie Bolt (Canada) in 1979 with a jump of 2.03 m. Following this jump, it was suggested that the lack of a complete lead leg (Bolt's lead leg has been amputated above the knee) may actually have been an advantage in one respect—it may have enabled Bolt to have his center of gravity higher at the instant of takeoff than would have been possible if he had had both legs intact. Under what circumstances is this suggestion valid?

8. The *Des Moines Register* for December 2, 1979 carried an account of a lengthy computer analysis conducted by scientists at Grinnell College for the purpose of determining the geographic center of the state of Iowa. A letter to the editor several days later correctly pointed out that the center of the state can be determined much more easily by pasting a map of the state on heavy card, cutting along the lines representing the state's border so that a cardboard template of the state is obtained, and then finding the point on which the template can be balanced in a horizontal position. This, of course, is the same as finding the center of gravity of the template using the balance method (p. 202).

   Obtain a large-scale map of your state—the kind issued free by oil companies is ideal—and carry out the same procedure to determine the geographic center of the state. Make a second estimate of the location of this point using the suspension method (p. 204) and compare the results you obtain with the officially designated (or recognized) center of the state.

9. A subject performing a vertical-jump-and-reach test flexes her knees as she reaches the peak of her flight. What effect does this have on the score she obtains?

10. Construct a mannikin according to the instructions in Appendix D (pp. 429–431) and use it in the following exercises:

   a. Put the mannikin in an erect standing posture (or as near as you can get it to such a posture) and note the position of the center of gravity relative to a line drawn to represent the ground. Then elevate the arms until they are directly overhead and note again the position of the center of gravity relative to the line used to represent the ground. Using the scale provided, determine how much the center of gravity of a 1.90 m man would be elevated if he were to move his arms in the same manner. How does this compare with the values reported by other investigators.[8]

   b. Put the mannikin in a simulated long jump landing position—head and trunk erect, hip angle approximately 120°, knee angle approximately 160°—with the arms extended horizontally forward. Note the *horizontal distance* from the heels to the line of gravity. Then move the arms so that they are extended horizontally backward and once again note the horizontal distance from the heels to the line of gravity. Which of these two positions of the arms yields the greater horizontal distance? If the two positions were adopted by a 1.90 m man, how large a difference would there be between the horizontal distances? How does this figure compare with the figure given by Ecker?[9]

[8]*Ibid.*, p. 131.
[9]Tom Ecker, *Track and Field Dynamics* (Los Altos, Calif.: Tafnews Press, 1974), p. 35.

    c. A defensive lineman in football jumps up to try and block a forward pass. Would he be able to reach higher if he flexed his knees and brought his heels up behind him at the peak of the jump? (Put the mannikin into a position with head and trunk erect, arms overhead, and legs extended and note the distance from the fingertips to the center of gravity. Then change the position of the lower legs so that the knees are flexed to approximately 90° and note again the distance from the fingers to the center of gravity.)

11. A woman who weighs 650 N lies on a reaction board. The reading on the scale supporting one end of the board is increased by 300 N compared with the reading before she took up her position. If the distance between the supporting "knife edges" is 1.8 m, what is the horizontal distance from the soles of the woman's feet to her center of gravity? If the woman is 1.54 m tall, is your answer to the previous question consistent with what the available research literature would lead you to expect?

12. What straight-body position might a gymnast adopt to be in a state of (a) stable equilibrium? (b) unstable equilibrium? (c) neutral equilibrium?

13. The masses and moments of inertia of a set of golf clubs are presented in Table 13. A close examination of this table reveals that, while the masses of the clubs increased from the No. 1 wood to the No. 10 iron, with only two exceptions, the corresponding moments of inertia decreased. Can you account for this seemingly strange result?

14. A tennis ball is hit with such severe topspin that, when it lands, the friction between the ball and the court acts to increase (rather than decrease) its horizontal velocity. How does the angular momentum of the ball after impact with the court compare with its angular momentum before impact?

**TABLE 13   Masses and Moments of Inertia of Golf Clubs\* About Transverse Axes 10 cm from the Grip End**

| Club | Mass (kg) | Moment of Inertia $(kg \cdot m^2)$ |
|---|---|---|
| 1 wood | 0.371 | 0.224 |
| 2 wood | 0.381 | 0.228 |
| 5 wood | 0.385 | 0.224 |
| 3 iron | 0.419 | 0.212 |
| 4 iron | 0.426 | 0.208 |
| 5 iron | 0.436 | 0.210 |
| 6 iron | 0.440 | 0.206 |
| 7 iron | 0.445 | 0.205 |
| 8 iron | 0.454 | 0.205 |
| 9 iron | 0.458 | 0.204 |
| 10 iron | 0.464 | 0.203 |

\*Wilson Sporting Goods Company, Sam Snead (1979 model).

15. A student who is performing a standing long jump in a physical education class takes off with zero angular momentum about her transverse axis. Then, shortly after takeoff, she flexes at the hip, knee, and ankle joints and assumes a tightly tucked position for the middle part of the airborne phase of the jump. (a) What is her angular velocity about her transverse axis at the instant she leaves the ground? (b) What effect will her actions after takeoff have on her moment of inertia about her transverse axis? (c) What effect will they have on her angular velocity about her transverse axis? [ *Hint*: Be very careful here.] (d) What effect will they have on her angular momentum about her transverse axis?

16. Figure 138 is a stick figure sequence taken from a film of a Russian circus

**Fig. 138.** *A Russian circus act.*

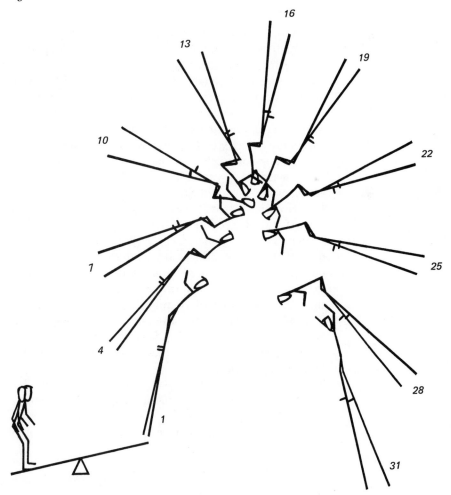

performer executing a back somersault on stilts. The trick is initiated by the performer who stands on one end of a teeterboard and is thrown into the air when two others in the troupe jump on the other end of the board. Assuming that air resistance is negligible, does the center of gravity of the man follow a parabolic path? Is the center of gravity of the man-plus-stilts system nearer his head than is his own center of gravity, or vice versa? About what axis does the man-plus-stilts system rotate?

Between frames 25 and 28, the man flexes sharply at the hips. What effect does this have on (a) the angular momentum of the man-plus-stilts system? (b) the moment of inertia of the system? (c) the angular velocity of the stilts? (d) the angular velocity of his trunk? (e) the angular velocity of the system? Why do you think he flexed his hips in this manner?

17. Some years ago, a well-known diver wrote "The body will . . . always be rotating slower on entry from a pike dive than from the same dive performed in a straight position." Although it may well seem to be wrong, this statement is actually correct. Can you explain why it is correct?

18. The gymnast in Fig. 139 is performing a straddle seat circle on the high bar as part of an uneven parallel bars exercise. (a) If her weight is 490 N, the distance between her line of gravity and the center of the bar is 0.47 m, and her moment of inertia relative to an axis through the center of the bar is 15.5 kg · m$^2$, what angular acceleration is she experiencing at the instant depicted? (Ignore the likely existence of a frictional torque exerted by the bar on the hands.) (b) As a practical matter, since she clearly would like as large an angular acceleration as she can get to complete the circle with ease, can you see any way in which she might increase her angular acceleration at the instant shown in Fig. 139?

19. In performing a front dive in the piked position, a diver leaves the board facing forward, rises high in the air, touches his hands to his toes, and then straightens out again for entry. While he is going into the piked (toe-touching) position, his legs appear to stop rotating, and, while he is coming out of it, his arms, head, and trunk appear to stop rotating. Can you explain, in mechanical terms, why first one half of his body and then the other appear to stop rotating?

20. The following statements have appeared in newspapers, coaching articles, and textbooks during the last few years. Examine each statement carefully and state whether it is true or false. If you decide that it is false, explain the basis for your decision.
    a. "Force must be applied to change the velocity of a body."
    b. "The force of attraction that the earth exerts on a body is called the body's weight."
    c. "If one point of an object is fixed, it will rotate regardless of where the force is applied."

**Fig. 139.** *An application of the angular version of Newton's second law.*

    d. "With angular motion the resistance offered to acceleration depends not only upon mass but also upon its distribution about the axis."

    e. "The center of gravity of the body is the axis of motion for the total body during airborne somersaults."

    f. "The somersault [long jump]er, because he is beginning to lean at take-off, is able to have his center of gravity increased."

    g. "Even though there is a definite flow of weight to the right side [during the backswing in a golf drive], the player's center of gravity remains over the left side."

## RECOMMENDED READINGS

Bishop, Robert D., and James G. Hay, "Basketball: the Mechanics of Hanging in the Air," *Medicine and Science in Sports*, 2 (Fall 1979), 274–277.

Dyson, Geoffrey H. G., *The Mechanics of Athletics*. New York: Holmes & Meier, 1977, pp. 34–73 (Forces (2)), 74–132 (Angular Motion).

Ecker, Tom, *Track and Field Dynamics* (2nd ed.). Los Altos, Calif.: Tafnews Press, 1974.

Faria, Irvin E., and Peter R. Cavanagh, *The Physiology and Biomechanics of Cycling*. New York: John Wiley, 1978, 89–108 (How Do I Pedal?).

Rackham, George, *Diving Complete*. London: Faber and Faber, 1975, pp. 141–150 (Body Movement and Free Fall).

Tricker, R. A. R. and B. J. K., *The Science of Movement*. London: Mills & Boon, 1966, pp. 28–47 (Simple Problems of Balance in Physics).

Wells, Katharine F., and Kathryn Luttgens, *Kinesiology: Scientific Basis of Human Motion*. Philadelphia: W. B. Saunders, 1976, pp. 328–352 (Motion and Force).

# Chapter 13

# Fluid Mechanics

In physical education and athletic activities, the motions of the performer and of the equipment being used take place in a fluid environment. This environment—the air in activities like badminton, cycling, and golf, the water in scuba diving, and a combination of air and water in canoeing, rowing, and swimming—influences the performance.

In some cases, like basketball, gymnastics, and wrestling, the effects produced by the fluid environment in which the activity takes place are so small that they can be disregarded in all but the most detailed analyses. In others, like badminton, golf, and ski jumping, the effects produced by the fluid environment have an enormous effect on the performance. The air resistance encountered by a soccer ball, for example, may reduce the range of a kick to almost half what it would be in the absence of air resistance,[1] whereas that encountered by a golf ball may actually serve to nearly double the range it would achieve in an air-free environment.[2]

## FLOTATION

When a swimmer assumes a horizontal position on the surface of the water, she is subjected to two vertical forces—her weight and the *buoyant force*, the resultant of all the vertical forces exerted on her by the water (Fig. 140(a)). Whether she floats or sinks depends on the relative magnitudes of these oppositely directed forces. If the buoyant force is equal to the weight, the swimmer floats; if it is less than the weight, she sinks.

[1] C. B. Daish, *The Physics of Ball Games* (London: English Universities Press, 1972), p. 41.

[2] Alastair Cochran and John Stobbs, *The Search for the Perfect Swing* (London: Heinemann Educational Books, 1968), p. 162.

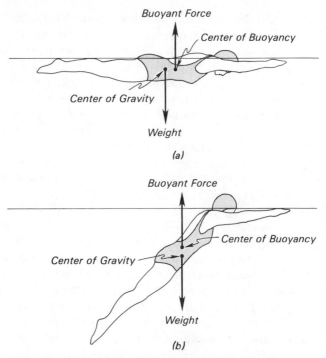

**Fig. 140.** *Whether a body floats horizontally or rotates to some inclined position is governed by the relative positions of the lines of action of the weight and the buoyant force.*

The magnitude of the buoyant force is, according to a principle discovered over 2,200 years ago by Archimedes, equal to the weight of the water that has been displaced by the submerged (or partially submerged) body. The maximum magnitude of the buoyant force is attained when the body is submerged entirely and is thus displacing the maximum possible volume of water—a volume equal to that of the body itself.

Whether the swimmer floats at the surface of the water or sinks to the bottom—there are no other alternatives despite a widespread belief that it is possible for her to "float" or be suspended at some intermediate depth—thus depends on two factors: her weight (or mass) and the volume of her body. The two quantities mass and volume together define the *density*, a quantity frequently used in describing the characteristics of a body:

$$\text{Density} = \frac{\text{Mass}}{\text{Volume}}$$

A body with a small mass and a large volume (or a low density) is likely to float, whereas one with a large mass and a small volume (or a high density) is unlikely to do so.

The density of a human body, and thus the likelihood that it will float, is governed by the amounts of bone, muscle, fat, and other tissues of which it is composed. Because muscle and bone each have a high density, anyone whose physique is "dominated" by these tissues—for example, a big-boned, heavily muscled wrestler or football player—is likely to have a high overall density and little chance of being able to float. Conversely, because fat has a low density, anyone with a physique in which fat predominates or is very well represented—as is often the case, for example, with channel swimmers—is likely to have a density sufficiently low to permit floating.

The proportions of the various tissues in the body vary with the age and sex of the individual. In early childhood, the muscles and bones are largely undeveloped, and the overall density of the body is thus relatively low. As a result, very young children generally have little difficulty in floating. With the demineralization of bone and the atrophy of muscle that occurs late in life, the density of the body once again becomes relatively low, and people who have been unable to float for many years often find that they can do so at this time in their lives. Because women generally have a higher proportion of fat in their physiques than do men—26 percent for women versus 11 percent for men, according to the findings of Wilmore and Behnke[3] and Sloan,[4] respectively—women are more likely to be able to float than are their male counterparts.

From the preceding paragraphs, it would appear that swimmers have little control over the factors that influence their ability to float. There is one factor, however, which has a considerable influence on this ability and over which one can exercise a certain measure of control—the volume of air in the lungs. If the swimmer takes a deep breath, the mass of the air inhaled is so small as to have no perceptible effect on the total mass of the body. The inflation of the lungs and the accompanying elevation of the thorax, however, produce a noticeable increase in the total volume of the body. By taking a deep breath, the swimmer thus decreases the density and improves the chances of being able to float. The importance of this factor in determining the ability to float has been clearly summarized by Whiting[5] who found that, while most men and most women will float in water if they have taken a full inhalation, the majority of the men will sink unless they have more than just residual air in their lungs.

While most people probably equate floating with the maintenance of a horizontal position at the surface of the water, only a very small percentage of floaters—that is, those who do not sink to the bottom—can maintain this

[3]J. H. Wilmore and A. R. Behnke, "An Anthropometric Estimation of Body Density and Lean Body Weight in Young Women," *The American Journal of Clinical Nutrition*, 23 (March 1970), 267–74.

[4]A. W. Sloan, "Estimation of Body Fat in Young Men," *Journal of Applied Physiology*, 23 (September 1967), 267–74.

[5]H. T. A. Whiting, *Teaching the Persistent Non-Swimmer* (London: G. Bell, 1970), p. 6.

position.[6,7] The position in which a person floats is dictated by the relative locations of the center of gravity of the body and of the point through which the buoyant force acts. This point, the center of volume of the submerged part of the body, is known as the *center of buoyancy* (Fig. 140(a)). If the centers of gravity and buoyancy lie on the same vertical line, the body remains in its present position. If the center of gravity lies nearer to the feet than does the center of buoyancy—as is almost invariably the case when a horizontal position is assumed—the weight and the buoyant force constitute a couple that serves to rotate the body, feet first, away from the horizontal position. In this case, the body eventually comes to rest—sometimes in an inclined position (Fig. 140(b)) and sometimes in a vertical one—with the center of buoyancy on the line of gravity.

## RELATIVE MOTION

Passengers sitting in a stationary train sometimes get the impression that they are moving when a train goes by on the next track. The reason for this is that when one train is stopped and a second train is moving, the visual effect is the same as if the first were moving and the second were stationary. In short, the motion of one relative to the other looks the same no matter which of the two is moving.

A similar situation exists with respect to motion in a fluid environment. In this case, the forces exerted on a body moving at a given velocity through a stationary fluid are the same as would be exerted on it if the body were stationary and the fluid were moving past it at the same velocity.

It is for this reason that the forces exerted on a body by the fluid through which it moves are often studied with the aid of wind and water tunnels in which the body is stationary and the fluid is made to move past it at a speed equivalent to that which pertains in the real-life situation (Fig. 141). The effects that air resistance has on the flight of discuses,[8,9,10,11] javelins,[12,13] cricket balls,[14] and golf

[6]H. T. A. Whiting, "Variations in Floating Ability with Age in the Male," *Research Quarterly*, 34 (March 1963), 84–90.

[7]H. T. A. Whiting, "Variations in Floating Ability with Age in the Female," *Research Quarterly*, 36 (May 1965), pp. 216–18.

[8]Richard V. Ganslen, "Aerodynamic Factors Which Influence Discus Flight," Research report (Fayetteville: University of Arkansas, 1958).

[9]Richard V. Ganslen, "Aerodynamic and Mechanical Forces in Discus Flight," *Scholastic Coach*, 28 (April 1959), 46, 77.

[10]Richard V. Ganslen, "Aerodynamic and Mechanical Forces in Discus Flight," *Athletic Journal*, 64 (April 1964), 50, 52, 68, 88–89.

[11]C. P. Kentzer and L. A. Hromas, Research report (Lafayette, Ind.: School of Aeronautical Engineering, Purdue University, July 2, 1958).

[12]Richard V. Ganslen and Kenneth G. Hall, *Aerodynamics of Javelin Flight* (Fayetteville: University of Arkansas, 1960).

[13]Juris Terauds, "Wind Tunnel Tests of Competition Javelins," *Track and Field Quarterly Review*, 74 (June 1974), 88.

[14]R. A. Lyttleton, "The Swing of a Cricket Ball," *Discovery*, 18 (May 1957), 188.

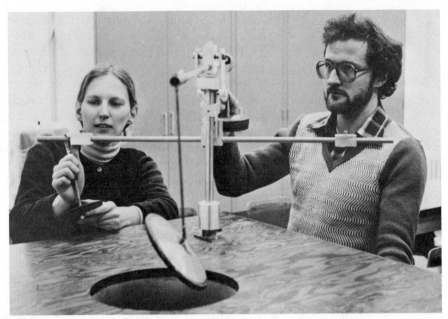

**Fig. 141.** *Wind-tunnel experiment designed to determine the forces exerted on a discus immersed in an airflow. The apparatus on which the discus is mounted is used to determine the forces exerted upon the discus when air is driven vertically upward past it.*

balls,[15] and on the motion of downhill skiers[16, 17] and ski jumpers[18] have been studied in this manner.

## FLUID RESISTANCE

If a javelin is mounted in a wind tunnel and air is made to flow past it, the resultant force that the air exerts on the javelin—the *air resistance*—usually acts at some angle to the direction from which the air approaches the javelin (Fig. 142). The component of the air resistance that acts in the direction of the approaching flow of air is called the *drag* and, in an actual throw, serves to oppose the forward motion of the javelin. The component that acts in a direction at right angles to that of the approaching air is called the *lift* and, in practice,

[15] John M. Davies, "The Aerodynamics of Golf Balls," *Journal of Applied Physics*, 20 (September 1949), 821–28.

[16] A. E. Raine, "Aerodynamics of Skiing," *Science Journal*, 6 (March 1970), 26–30.

[17] Kazuhiko Watanabe and Tatsuyuki Ohtsuki, "Postural Changes and Aerodynamic Forces in Alpine Skiing," *Ergonomics*, 20 (Nov. 2, 1977), 121–31.

[18] Itiro Tani and Matsusaburo Iuchi, "Flight-Mechanical Investigation of Ski Jumping," in *Scientific Study of Skiing in Japan* (Tokyo: Hitachi, 1971), pp. 33–53.

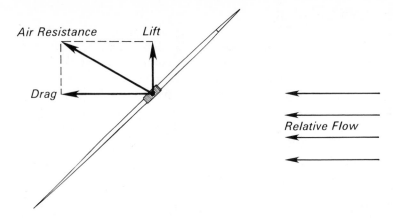

**Fig. 142.** *Lift and drag forces acting on a javelin.*

serves to lift the javelin above the path it would otherwise follow. In short, air resistance tends to simultaneously oppose the forward motion of the javelin and to lift it higher in the air.

## Surface Drag

When air is driven past the javelin in Fig. 143, those layers of air that come into direct contact with the javelin are slowed down. These layers tend to slow those next to them. These, in turn, tend to slow those next to them, and so on. This process of slowing the air in close proximity to the javelin is brought about because the javelin exerts forces on the air. The reaction to these forces—known as the *surface drag* because it arises as a result of the passage of the air across the surface of the body—contributes to the drag which the body experiences. (Because the motion of a fluid over the surface of a body is somewhat akin to the sliding of one body over another, the resulting surface drag is sometimes referred to as the *skin friction*.)

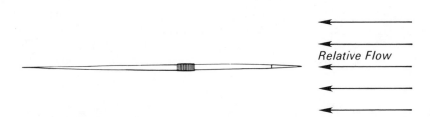

**Fig. 143.** *Surface drag slows the forward motion of the javelin.*

The magnitude of the surface drag depends, among other things, on the velocity of the flow relative to that of the body, the surface area of the body, and the smoothness of the surface—the higher the relative velocity, the greater the surface area; the rougher the surface, the greater the surface drag. This form of drag is especially important in activities like sailing, rowing, canoeing, tobogoning, and downhill skiing in which large surface areas and/or high velocities are involved.

## Form Drag

When air is driven past a smooth, spherical ball mounted in a wind tunnel (Fig. 144(a)), the part of it that strikes the front surface of the ball is diverted radially outward. It continues in this direction until it eventually reaches a point at which the neighboring layers of air are no longer capable of exerting the forces necessary to keep it in contact with the surface. At this point it breaks away (or separates) from the ball and continues on its outward path. In due course, the neighboring layers of air force the diverging parts of the flow back together. In this process of separating from the surface of the ball and reuniting downstream, a tapering "pocket" or zone is formed behind the ball. Because the pressure within this pocket is very low, the resultant force that acts on the rear half of the ball is relatively small. In contrast, and because of the "head-on collision" between the air and the ball, the resultant force exerted on the front half of the ball is relatively large (Fig. 144(b)). The difference between these two resultant forces is the *form drag*.

The magnitude of the form drag depends, among other things, on the cross-sectional area of the body relative to the flow, the shape of the body, and the smoothness of its surface—the greater the cross-sectional (or frontal) area, the less streamlined the shape and the smoother its surface, the greater the form drag.

Form drag plays an important role in many athletic activities, and teachers, coaches, and athletes correctly devote considerable attention to the use of techniques that maximize its benefits or minimize its detrimental effects. In cycling, for example, it has been found that variations in the frontal area from approximately $0.50 \text{ m}^2$ (upright position), to $0.42 \text{ m}^2$ (touring position), to $0.34$ $\text{m}^2$ (racing position) yield corresponding improvements in speed of 0.97 km/h and 1.13 km/h when cycling at approximately 24 km/h—and this, be it noted, with no change in the effort put forward by the cyclist.[19] Similar effects due to changes in frontal area and to streamlining of the body—it is virtually impossible

---

[19]Irvin E. Faria and Peter R. Cavanagh, *The Physiology and Biomechanics of Cycling* (New York: John Wiley, 1978), pp. 76–77.

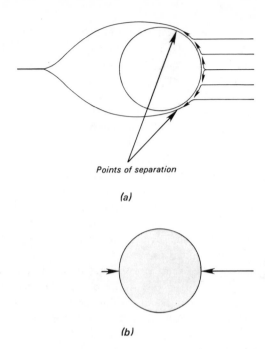

Points of separation

(a)

(b)

**Fig. 144.** *Form drag is due to an imbalance in the resultant forces acting on the front and rear halves of a body.*

to modify one without in some way altering the other—have been reported in a study involving downhill skiers.[20]

The role of surface smoothness in determining the magnitude of the form drag is deserving of special mention. The points at which the flow separates from the ball in Fig. 144 govern to a large extent the size of the low-pressure pocket behind the ball, the resultant force on the back of the ball, and, thus, the form drag. If the points of separation are near the front of the ball, the form drag is relatively large. If they are farther toward the back of the ball, the form drag is correspondingly less, assuming that all else is equal. The location of the points at which the flow separates from the ball is influenced by the nature of the air flow close to the ball. If the air is moving in parallel layers—*laminar flow*—the points of separation are nearer to the front than if the air is moving in an erratic, intermixing fashion—*turbulent flow*. Whether the flow is laminar or turbulent depends in large part on whether the surface of the ball is smooth or rough. If it is rough, the flow is more likely to be turbulent and the form drag less than if it is smooth. This fact is well known in golf where dimpled balls are used to minimize form drag and in games like baseball, cricket, and softball where the seams and variations in the

---

[20] A. E. Raine, "Aerodynamics of Skiing," *Science Journal*, 6 (March 1970), 26–30.

roughness of the surface of the ball are used to produce variations in the points of separation on different parts of the ball. With skilled performers this, in turn, produces dramatic effects on the flight path followed by the ball.

The effect that surface smoothness (or roughness) has on the drag experienced by a body is often a source of confusion. As the observant reader may have noticed, the smoother the surface of a body, the less the surface drag it is likely to experience (p. 244); the rougher the surface, the less the form drag it is likely to encounter (p. 245). Whether efforts should be made to smooth or roughen the surface of a body involved in a given case depends on which of these two effects dominates. In rowing, where the surface area of the shell is very large and where a very streamlined shaped is involved, the surface effects are much more important than are the form effects. Under such circumstances, a smooth surface is preferable to a roughened one—a fact that is well-understood in practice if the highly polished surfaces of rowing shells and of sailing boats, canoes, and kayaks are any indication. In games like baseball, cricket, golf, and softball—which involve the use of a spherical (or nearly spherical) ball, the small surface area of the ball and its essentially blunt, nonstreamlined shape ensure that the form effects have a greater influence on the drag encountered than do the surface effects. For minimum drag, and thus maximum velocity, the surface of the ball should thus be roughened—as it is in golf and, less routinely, in some other sports.

## Lift

Contrary to what might be expected, and to what is certainly implied by the term, lift does not necessarily act vertically upward and, thus, elevate the body involved. Depending on the circumstances, it can act in any direction including vertically upward, vertically downward (!), and horizontally. When a javelin is released point upward in a horizontal direction—that is, with an angle of release of 0° (Fig. 142)—the lift component of the air resistance acts vertically upward. When it is released at an angle of 35°, an angle close to that used by most top-class throwers, the lift component is directed upward and backward at an angle of 35° to the vertical (Fig. 145). (The angle at which a body is inclined to the horizontal, the so-called *attitude angle*, does not necessarily coincide with the angle of release. A javelin may thus be released at some angle other than that indicated by the direction of its long axis, as shown in Fig. 142.) When the positions of stabilizers mounted on a Grand Prix car are adjusted as the car goes into a corner, the lift components of the air resistance they encounter are directed vertically downward (Fig. 146). The stabilizers thus serve to improve the car's traction by increasing the forces holding it against the track. Finally, when a

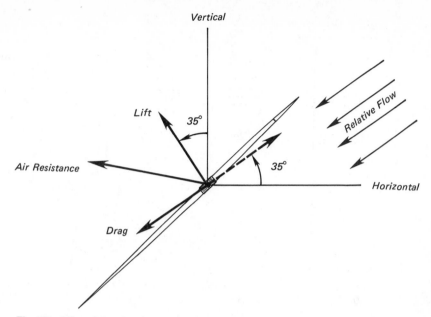

**Fig. 145.** *Lift and drag forces acting on a javelin released with an angle of attack of 35°.*

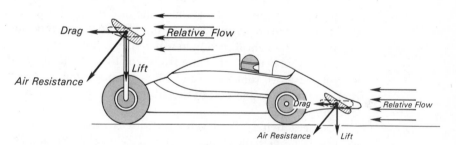

**Fig. 146.** *The stabilizers on Grand Prix cars are used to increase the force holding the tires against the road and thus to improve traction.*

breaststroke swimmer moves her hands laterally during the pull phase of the stroke, the lift component of the water resistance experienced by each hand acts in the forward horizontal direction—that is, in the direction in which she is swimming (Fig. 147).

In any given case involving a nonspherical body, the magnitude and direction of the fluid resistance encountered—and the magnitudes and directions of the lift and drag components—vary according to how the body is oriented relative to the flow. The angle between the body and the approaching flow is called the *angle of attack* and is positive if the flow strikes the undersurface of the body and negative

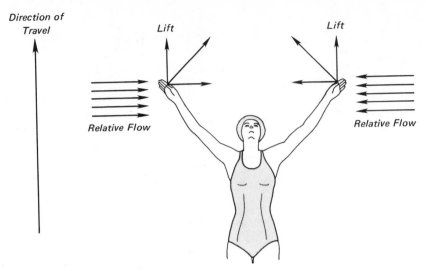

**Fig. 147.** *The lift force propels the swimmer forward in the breaststroke.*

if it strikes the upper surface. If the angle of attack is zero (Fig. 143), the resistance encountered is relatively small and acts in approximately the same direction as the approaching flow. There is thus little, if any, lift. If the angle of attack is 90°—that is, if the body is broadside on to the flow—the resistance encountered is relatively large. The direction in which it acts, however, is still approximately the same as that of the approaching flow, and there is once again little, if any, lift. Between these two values—0° and 90°—the magnitude of the resistance varies from small to large and that of the lift from zero to some maximum value and back to zero again.

In some activities like javelin and discus throwing and ski jumping, the performers' best interests would be served if they could simultaneously minimize the drag and maximize the lift acting on their bodies. In this way, the forward motion would be retarded as little as possible and the time of flight prolonged as much as possible. As might already be apparent from the preceding discussion, these two objectives are basically incompatible because the value of the angle of attack that yields the least drag invariably yields less than the maximum possible lift. The performer must therefore try to find the compromise solution that affords the best overall result.

Some attempts have been made to resolve this problem by using a wind tunnel to determine the lift and drag forces acting on a body at different angles of attack. The values obtained in this manner are then expressed as a ratio—the so-called *lift–drag ratio*—and the angle of attack that yields the highest ratio, the most lift for the least drag, is taken as the best angle of attack. This approach, however, ignores one very important fact—the fact that the angle of attack, the orientation of the body to the flow, is continually changing during the course of the flight. Thus, the angle of attack that yields the highest lift–drag ratio may not be the one with which the body should be projected into the air to obtain the best performance overall. In a few cases, advanced mathematical techniques have been used to try and determine the way in which the body should be oriented to the flow at the instant of release or takeoff so that the best overall result is obtained. Cooper, Dalzell, and Silverman[21] have found, for example, that discus throwers who throw in the 45–60 m range should use an angle of release between 35° and 40° and an attitude angle of between 25° and 30°. This means that the angle of attack with which the discus should be released is between 0° and −15°.

## Magnus Effect

When topspin is applied to a tennis ball, the air that comes into direct contact with the ball is carried around with it. The layer of air that becomes effectively attached to the ball in this process is called the *boundary layer*.

As the ball travels forward, the air on the top of the ball at any given instant has a forward velocity whereas that on the bottom of the ball has a backward velocity *relative to the center of gravity of the ball* (Fig. 148(a)). The difference in the velocity with which the air in different parts of the boundary layer meets the air through which the ball passes creates a corresponding difference in the pressure exerted on the two halves of the ball (Fig. 148(b)). The "head-on" collision of the air on the top of the ball with the air through which the ball is passing causes an increase in pressure in the area immediately above the ball. The motion of the air on the bottom of the ball in the same direction as the relative flow causes a decrease in pressure in the area immediately below the ball. As a result of these two changes in pressure, the ball experiences a downward force that diverts it from the path it would otherwise follow (Fig. 148(c)). The ball thus returns to the court much more quickly than it would if no top-spin were applied to it. Since this naturally decreases the time that an opponent has to react to the shot, the topspin drive is widely used in tennis—and in table tennis—as an offensive stroke.

The influence that rotation has on the path followed by a body traveling through a fluid is called the *Magnus effect*, after a German scientist who is generally credited with its discovery—and this, it might be noted, despite Newton having reported the same phenomenon nearly two hundred years earlier!

In addition to tennis and table tennis, the Magnus effect also plays an important role in baseball, where the skilled pitcher applies spin to the ball to cause it to move disconcertingly in the air (Fig. 149); in golf, where unwanted sidespin imparted to the ball causes it to hook or slice (Fig. 150); and in soccer, where sidespin can be applied judiciously to allow the ball to be kicked around the "wall" of opponents set up to defend a free kick (Fig. 151).

[21]Leonard Cooper, Donald Dalzell, and Edwin Silverman, "Flight of the Discus," Research report (Lafayette, Ind.: Division of Engineering Science, Purdue University, May 18, 1959).

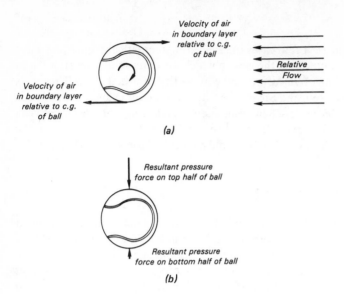

Velocity of air
in boundary layer
relative to c.g.
of ball

Relative
Flow

Velocity of air
in boundary layer
relative to c.g.
of ball

*(a)*

Resultant pressure
force on top half of ball

Resultant pressure
force on bottom half of ball

*(b)*

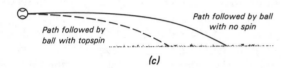

Path followed by ball
with no spin

Path followed by
ball with topspin

*(c)*

**Fig. 148.** *Differences in pressure cause a spinning ball to deviate from the path it would follow if it were not spinning.*

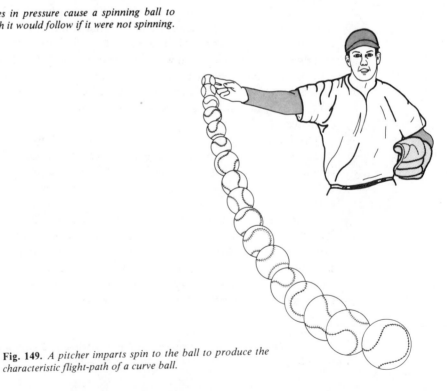

**Fig. 149.** *A pitcher imparts spin to the ball to produce the characteristic flight-path of a curve ball.*

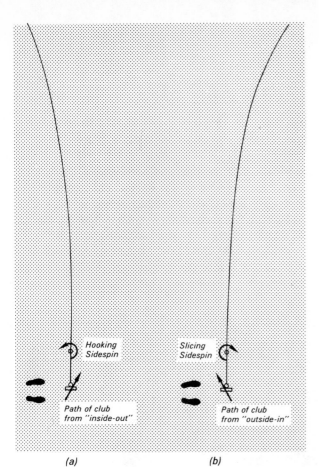

(a)                                    (b)

**Fig. 150.** *Motion of the clubhead across the intended line of flight produces (a) a hook and (b) a slice.*

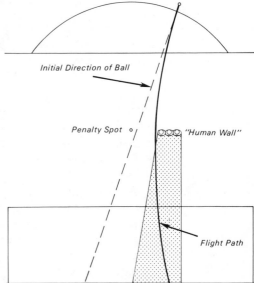

Initial Direction of Ball

Penalty Spot

"Human Wall"

Flight Path

Goal

**Fig. 151.** *Spin imparted to the ball during a free-kick in soccer may cause the ball to pass around a wall of players set up to block the kick. (Benno Nigg, Einige mathematische und physikalische Grundlagen der Biomechanik. Mimeographed lecture notes, Federal Institute of Technology, Zurich, Switzerland.)*

## EXERCISES

Give answers of magnitude less than 1.0—that is, $>(-1.0$ and $<1.0)$ correct to two decimal places and all other answers correct to one decimal place.

1. Figure 152 shows a boat with a hull of steel and concrete—a so-called ferrocement hull—under construction. At first, the idea of building a boat with concrete might well seem absurd, but, in certain parts of the world, such boats are quite common. (a) How do you account for the fact that a boat made of such a heavy material can float? (b) If the completed boat displaced 32.5 m³ of fresh water, how heavy would it be? [*Hint*: You'll need to find out how much a cubic meter of water weighs before you can answer this question.] (c) Would the volume of sea water the boat displaced be the same, more than, or less than the volume of fresh water it displaced?

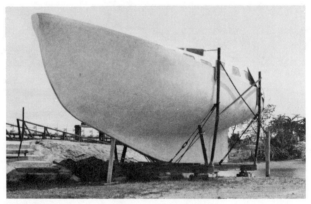

**Fig. 152.** *A ferro-cement boat under construction.*

2. The forces acting on a body placed in a fluid are its weight acting vertically downward and the buoyant force acting vertically upward. Under these circumstances, is it possible for a body that is unable to float at the surface of water to "float"—that is, to maintain a stationary position—at some level between the surface and the bottom?

3. A wind-tunnel experiment was conducted to determine the effects that changes in the angle of attack had on the lift, the drag, and the lift–drag ratio recorded for a discus immersed in the airflow. Unfortunately, at the end of the experiment, the four sets of data became mixed up so that it was no longer clear which set of data was associated with which quantity. The four sets of data were as follows:

| (1) | (2) | (3) | (4) |
|------|------|------|------|
| 1.2 | 0 | 0 | 0 |
| 7.0 | 2.0 | 27 | 13.8 |
| 2.0 | 0.7 | 49 | 8.6 |
| 15.5 | 0.4 | 65 | 5.6 |
| 17.5 | 0.1 | 88 | 0.9 |

Which quantity was associated with each set of data?

4. Obtain a large, narrow-necked jar and glue a crutch tip (or something of similar size and shape) to the center of the inside bottom of the jar. Allow the glue to set and then fill the jar with water. Take a penny and, keeping its two flat surfaces vertical or nearly vertical, try to drop it into the crutch tip. Use your knowledge of fluid mechanics to explain why it is difficult to get the penny to drop into the crutch tip.

5. Take a piece of elastic or rubber cord and tie one end to the uppermost part of a table leg and the other to the corresponding part of another leg on the same side of the table. Fasten one end of a narrow (5–8 cm), long (30–35 cm) piece of cloth to the center of the elastic and wind the other end several times around a cardboard cylinder. Holding one end of the cylinder in each hand, draw it back across the surface of the table until the elastic is taut and then release it.[22] Use your knowledge of fluid mechanics to explain what happens to the cylinder.

## RECOMMENDED READINGS

Daish, C. B., *The Physics of Ball Games*. London: English Universities Press, 1972, pp. 41–56 (The Flight of Balls Through the Air), 57–72 (The Swerve of Balls in Flight).

Faria, Irvin E., and Peter R. Cavanagh, *The Physiology and Biomechanics of Cycling*. New York: John Wiley, 1978, pp. 71–88 (What is Still Slowing Me Down?, Drag and Friction).

Hay, James G., *The Biomechanics of Sports Techniques* (2nd ed.). Englewood Cliffs, N.J.: Prentice-Hall, 1978, pp. 165–183 (Fluid Mechanics).

Rogers, Eric M., *Physics for the Inquiring Mind*. Princeton, N.J.: Princeton University Press, 1977, pp. 154–169 (Fluid Flow).

Tricker, R. A. R. and B. J. K., *The Science of Movement*. London: Mills & Boon, 1966, pp. 131–140 (Hydrodynamics).

[22]Eric M. Rogers, *Physics for the Inquiring Mind* (Princeton, N.J.: Princeton University Press, 1977) p. 165.

Whiting, H. T. A., "Variations in Floating Ability with Age in the Male," *The Research Quarterly*, 34 (March 1963), 84–90.

———, "Variations in Floating Ability with Age in the Female," *The Research Quarterly*, 36 (May 1965), 216–218.

# Chapter 14
# Free Body Diagrams

Analyses of human motions are usually conducted with a view to gaining a greater understanding of how a particular movement is performed and/or predicting what would happen if certain changes were made in the pattern of movement. These analyses are often flawed, however, because the person making the analysis has failed to take all the factors involved—and, in particular, all the external forces and torques to which the body is subjected—into account.

One way to overcome this problem is to draw what is known as a *free body diagram* (FBD). To do this, one first draws a simple sketch of the body as if it had been totally removed from its surroundings—that is, without including anything other than the body itself. The next step is to superimpose on the sketch arrows that represent the external forces acting upon the body. These include the weight of the body acting downward through its center of gravity, the reaction forces acting through the body's points of contact with other bodies, and, in certain cases, buoyancy and fluid resistance. (In some instances it is useful to represent one or more of these forces in terms of their components rather than as a single force. Thus, the reaction force exerted by the ground against the foot of a runner might be represented by a normal component and a frictional component. Similarly, the air resistance force encountered by a discus in flight might be represented by its lift and drag components.)

Once this task has been completed, the next step is to superimpose curved arrows on the diagram to represent the external torques exerted on the body at those points where it makes contact with other bodies. (It is often difficult to decide whether an external torque is being applied to a body at some point where it makes contact with another. In such cases, it is better to assume that such a torque exists and to include it on the FBD than to risk leaving out a torque that may be crucial to the analysis. Where numerical computations are involved, a torque included on the FBD when no torque actually exists works out to have zero magnitude and thus no effect on the results obtained.) The final step is to place labels on the arrows superimposed on the sketch. These are usually letters, when the magnitudes of the forces and torques are unknown, and the appropriate quantities—for example, 600 N, 450 N·m, etc.—when they are known.

A series of FBDs is shown in Figs. 153, 154, and 155 to illustrate the procedures just described. In each case the spatial relationships of the bodies involved are shown in the upper figure—sometimes referred to as a *space diagram*—and the completed FBD is shown in the lower figure.

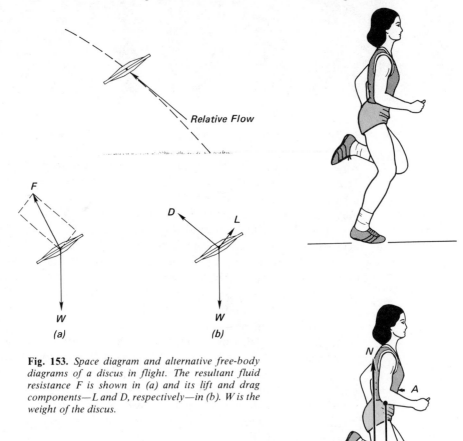

**Fig. 153.** *Space diagram and alternative free-body diagrams of a discus in flight. The resultant fluid resistance F is shown in (a) and its lift and drag components—L and D, respectively—in (b). W is the weight of the discus.*

**Fig. 154.** *Space and free-body diagrams of a woman during the support phase in running. (A = air resistance, F = friction, N = normal reaction, and W = weight of the woman.)*

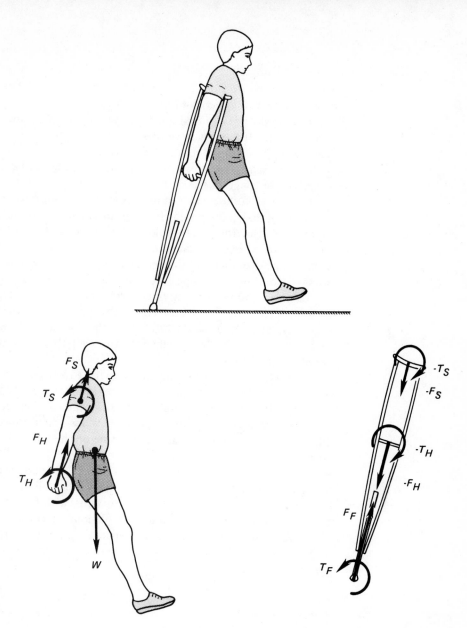

**Fig. 155.** *Space and free-body diagrams of crutch walking. ($F_F$ and $T_F$, $F_H$ and $T_H$, and $F_S$ and $T_S$ are the forces and torques at the floor, hand, and shoulder, respectively, and $W$ is the weight of the man.)*

**EXERCISES**

1. Draw free body diagrams of the following:
   a. a shot at the peak of its flight through the air. (Consider the shot to be moving from left to right as viewed in the diagram.)
   b. a billiard ball lying at rest on a billiard table.
   c. a bowling ball rolling down a lane. (Consider the ball to be moving from left to right in the diagram.)
   d. a basketball at the instant of impact with the floor during a bounce pass. (Consider the ball to be moving from left to right and assume that no spin was imparted to it by the player making the pass.) Be sure to label each of the vectors included in each diagram.

2. A skier in a downhill event is acted on by four forces—the skier's weight, the reaction from the ground perpendicular to the slope (the normal reaction), friction (also known as snow resistance), and air resistance. Draw a space diagram and a free body diagram of the skier and answer the following questions: (a) Under what conditions will the skier travel down the slope at a constant velocity? (b) Under what conditions will the skier's motion down the slope be accelerated?

3. A highly skilled water-skier holds on to the tow rope with one foot while skimming over the water supported on one ski by the other foot. Draw a space diagram and a free body diagram of the skier and consider how the moments of the several forces acting on the skier about his or her transverse axis might differ from those that were acting when he or she held the tow rope in the normal manner.

4. A gymnast places his left hand on a stall bar at a little above head height and his right hand, vertically below the left, on another stall bar at about midthigh level. He then levers his feet off the floor until, with both arms straight, his body is in a horizontal position. For obvious reasons, this is sometimes referred to as the flag position. Draw a space diagram and a free body diagram of the gymnast. [*Hint*: Include any external torques that may be acting on the gymnast.]

# Part III

# Practical Application

# Chapter 15
# Qualitative Analysis

A knowledge of basic anatomical and mechanical concepts is of relatively little use to the physical education teacher or coach unless this knowledge can be applied to solving problems encountered in the practical teaching or coaching situation. The purpose of this final part of the text is to demonstrate how the material presented in the preceding parts may be used by teachers and coaches to improve the results obtained by those in their charge.

## DEFINITION OF TERMS

Three terms—*motor skill* (or skill), *performance*, and *result*—will be used repeatedly in this chapter. These terms are defined as follows:

A *motor skill* is a series of voluntary movements of the human body designed to achieve a specific goal. A golf drive, a tennis serve, a vault in gymnastics, and a punt in football are examples of motor skills.

A *performance* is the manner in which all the movements comprising a motor skill are executed. In a long jump, for example, it is the manner in which the performer moves the various segments of his body as he sprints down the runway, leaps into the air, and lands in the pit.

A *result* is the measure of the outcome of a performance. It is the time, height or distance recorded in a track or field event, the number of points scored in a dive, and either success (2 points) or failure (0 points) in a jump shot in basketball.

The examples given here are obvious ones. The result for a motor skill used in a team game, in a dance, or in a number of other activities may be much less obvious. The process of identifying the result is discussed in detail later in this chapter.

## METHODS OF ANALYSIS

There are basically two methods by which motor skills can be analyzed—the *qualitative method* and the *quantitative method*. With the qualitative method, the performance is evaluated subjectively on the basis of direct, visual observation. With the quantitative method, the performance is first recorded using photography, cinematography, electromyography, or some other recording technique and then is evaluated objectively on the basis of measurements taken from the record(s).

The qualitative method is used by physical education teachers, coaches, athletes, and spectators, among others. In its simplest form it consists of an evaluation of the performance based on a simple, visual observation of the result. Thus, when a spectator says, "Nice catch!" or "Good shot!" or "Dumb move!" in reaction to something that has occurred in a game, that person is using a simple, and somewhat incomplete, form of qualitative analysis to arrive at a conclusion. In its more complete form, the method consists of a systematic evaluation of not only the result but also of all the various factors that have contributed to the result. Thus, when a diving coach tells a diver, "Drive your chest forward and upward as you takeoff," the advice is probably based on the coach's knowledge of the results obtained on the diver's most recent attempts to perform the dive in question and a consideration of the factors that caused them to be less than perfect. As in the case of the spectator, the coach's evaluation is based on simple visual observation of the performance. (In some cases, observation of the performance is facilitated by the use of a Polaroid sequence camera, a videotape recorder, or a motion picture camera. The method of analysis—that is, the qualitative method—is not altered by the use of such aids.)

The quantitative method is used extensively by researchers, occasionally by coaches and only very rarely by physical education teachers. Researchers use the quantitative method for a wide variety of purposes. It is used, for example, to examine questions concerning the effectiveness of alternative techniques for performing the same basic task. Thus, in a study of the relative merits of various high jump styles, a researcher might take high-speed motion picture films and analyze these frame by frame to determine such quantities as the horizontal and vertical velocities of the center of gravity, and the angular momentum of the body about selected axes, at critical instants through the sequence of motion—at the instants of touchdown and takeoff for each ground support phase during the run-up and takeoff, for example. On the basis of these measures and, usually, some form of statistical analysis, the researcher would then reach conclusions concerning the relative merits of the styles under consideration. A researcher working directly with a coach might also use the quantitative method to gather data for the purpose of improving the performances of the members of a particular team. Since expensive equipment, highly trained people, and many hours are required for such work, analyses of this type are generally conducted only at the highest levels of competition—in preparation of athletes for Olympic competition, for example.

In summary, the qualitative method is based on visual observation of the performance, whereas the quantitative method is based on measurements taken from a record of the performance. The qualitative method is subjective throughout. The quantitative method, on the other hand, is basically objective.

## QUALITATIVE ANALYSIS

Physical education teachers and coaches are repeatedly called upon to evaluate the performance of a motor skill and to suggest ways in which the performance may be improved. Since they almost invariably lack the background, equipment, and time to conduct quantitative analyses, and since their effectiveness as teachers and coaches depends in part on the accuracy of the analyses they make, it is imperative that they develp some proficiency in qualitative analsis. The remainder of this chapter is devoted to a discussion of a procedure for conducting qualitative analyses.

## PREREQUISITE KNOWLEDGE

To conduct a qualitative analysis requires some prior knowledge of the sport or activity concerned and, in particular, of the motor skill to be analyzed. At the very least, one *must* know what the performer is trying to achieve and what restrictions are imposed by the rules governing the event. Consider, for example, how difficult it would be to evaluate the performance of an athlete who is competing in a sheep-shearing event (Fig. 156) if the observer did not know whether the aim of the contest were to remove the wool from the sheep in the shortest time possible, or to remove it so that the fleece were intact and the sheep uninjured, or to remove it using some ideal form specified in the rules. Consider, too, how difficult it would be to suggest changes that might improve the performance if one did not know what was permissible within the rules governing the event.

Our present knowledge of the motor skills used in physical education and athletics has derived from teachers, coaches, and athletes who have experimented in the practical setting; from researchers in biomechanics, motor learning, and related areas who have conducted scientific experiments in their laboratories and in competitive situations; and from scholars who have performed theoretical analyses based on their knowledge of anatomy, mathematics, and mechanics. The knowledge gleaned from these various sources is an essential prerequisite to the use of the qualitative analysis system presented in this chapter. It is, therefore, important that those seeking to develop proficiency in the use of this system acquaint themselves as completely as possible with the available knowledge concerning the motor skills in which they are interested and that they endeavor to stay abreast of new developments. To this end, they should seek out those books, journals, and other materials that contain information on the motor skills of interest.

**Fig. 156.** *A competitor in a sheep-shearing contest.*

It is important to note that difficulties frequently arise when outstanding coaches present their views on how a motor skill should be performed. In many instances, the advice given is consistent with what teachers, coaches, researchers, and others have learned about that skill over the years. In other instances, it is in direct conflict with what is presently known about the skill. There are two obvious reasons for such conflict. In a few cases—as, for example, when Brown and Counsilman[1] first presented their revolutionary theory of propulsion in swimming—the conflict arises because the person concerned has discovered something that was not previously known. In other cases—and these, regrettably, constitute a large proportion of all that is written about motor skills—it arises because the person concerned is simply ignorant of the present state of knowledge in the field. Consider, for example, this statement made by one of the nation's most successful and widely recognized track and field coaches:

> *A good sprinter stays high up on his toes. As a result, his heel will never be seen hitting the ground.*

[1] Ron M. Brown and James E. Counsilman, "The Role of Lift in Propelling the Swimmer," *Selected Topics on Biomechanics: Proceedings of the C.I.C. Symposium on Biomechanics*, ed. John M. Cooper (Chicago: The Athletic Institute, 1971), pp. 179–188.

This statement, and many of the others made in the same article, is incorrect. Except during the start, the heel of a sprinter almost invariably makes contact with the ground during each supporting phase of the running cycle. This simple but important characteristic of the technique used by skilled runners has been repeatedly noted in the literature on sprinting[2,3,4] and can readily be observed in slow-motion films of sprinters in action. It can even be seen in the photo sequences accompanying the article that contained the statement quoted!

Given such circumstances, those who seek information on the performance of motor skills should read the available material, and especially that written by coaches who may have been successful for reasons other than their knowledge of the subject, with a great deal of caution. Certainly it is folly to assume that coaches must be authorities on the techniques used in their sports because they have coached many outstanding individuals or teams. Coaches may, instead, be excellent recruiters or motivators and know next to nothing about the techniques involved.

The methods used by skilled performers must also be kept in correct perspective. Many highly successful performers owe their success to factors other than the actual technique they employ in performing the skill (or skills) involved in their sport. They are strong, or fast, or flexible, or fiercely competitive, for example, and succeed in their chosen sport not because of the technique they use but in spite of it. It is a mistake, therefore, to use such a technique as a basis for comparison when training a young athlete.

Some performers use techniques that, although very sound, can only be used effectively by people with specific physical characteristics. Edwin Moses, the 1976 Olympic champion in the 400 m hurdles, provides a simple case in point. Because of his speed and leg length, Moses is able to take 13 strides between the hurdles whereas athletes less favorably endowed must take 15 or more. It is important, therefore, that the technique used by a champion not be used as a model for a young athlete who lacks the critical physical characteristics that might otherwise make that technique suitable. To put it in the words of a well-known track coach, it is important to recognize that "You can't build a cathedral on the foundations of a house!"

## BASIC STEPS

People involved in the teaching of motor skills are repeatedly faced with the task of observing a performance and offering suggestions as to how that performance

---

[2]Toni Nett, "Foot Plant in Running," *Track Technique*, 15 (March 1964), 462–463.

[3]Fred Wilt, "Fundamental Mechanics of Running and Hurdling," in *Olympic Track and Field Techniques*, ed. Tom Ecker, Fred Wilt, and Jim Hay (West Nyack, N.Y.: Parker, 1974), p. 24.

[4]Geoffrey H. G. Dyson, *The Mechanics of Athletics* (New York: Holmes & Meier, 1977), p. 136.

might be improved. Although they are often not fully aware of it, the procedure they follow in carrying out this task consists of three basic steps. First, they observe the performance and identify the faults in that performance; second, they establish an order of priority among these faults; and, finally, they give instructions to the performer.

The qualitative analysis system presented here follows the same general pattern with one major exception. It includes the development of a theoretical model as a basis for identifying faults and judging their relative importance. This model serves to supplement whatever experience the teacher or coach might have and to channel or direct the analysis in a logical, systematic fashion.

The system thus consists of four basic steps:

1. Development of a model (or block diagram) showing the relationships between the result and the factors that produce that result (Fig. 157).

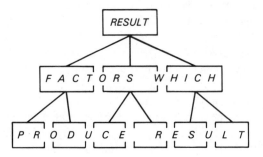

**Fig. 157.** *Basic structure of a model.*

2. Observation of the performance and identification of faults.
3. Evaluation of the relative importance of these faults.
4. Instruction of the performer in accord with the conclusions reached in the course of the analysis.

These four basic steps are discussed, with reference to a simple example, in the remaining sections of this chapter. Other examples in the use of the qualitative analysis system are presented in Chapter 16.

## DEVELOPMENT OF MODEL

### *Result*

The first step in the development of a model consists of identifying the result. For this purpose, it is instructive to classify skills into two broad categories according to the manner in which the result is measured:

***Objective measures.***    For many skills, the result is an objective measure of the performance. In the shot put, it is the *distance* thrown; in running, swimming, and cycling, it is the *time* taken to cover the specified distance; in ten-pin bowling, it is the *number of pins* knocked down; and in shooting with a bow, pistol, or rifle, it is the number of *points scored*.

In shooting for goal in basketball, hockey, and soccer and in kicking field goals and points after touchdown in football, the result is also the number of *points scored*. However, since there are only two possible outcomes of the performer's attempt to score—one either succeeds or fails—the number of points that can be scored is similarly limited. It is either the number associated with a successful attempt—1, 2, 3 points, as the case may be—or the zero points associated with an unsuccessful one.

A similar situation exists in the high jump and pole vault where the athlete attempts to clear a bar set at a given height. If one succeeds, he or she receives credit for that height. If one fails, he or she receives no credit. The result is thus the *height* of the bar for that attempt or zero, depending on whether the attempt was a success or failure.

***Subjective measures.***    Some skills are evaluated by means of a rating scale specified in the rules governing the competition. For such skills—those used in diving and in the vault in gymnastics, for example—the result is a subjective measure of the performance. It is the number of *points awarded* by the judges.

Many skills are not evaluated separately in the normal competitive situation. In team and racket games, for example, the manner in which the various skills are performed is reflected only in the final score—an overall measure of many performances of many different skills. The same is true in gymnastics and figure skating, where different skills are linked together to form complete exercises. Such skills are evaluated subjectively by teachers, coaches, and others in terms of the contribution they make to the final result. Consider, for example, a player executing a long one-handed pass to a fast-breaking teammate in basketball. If the pass is thrown so that the receiver is able to catch the ball without interruption in his forward motion, his chances of scoring are enhanced. Conversely, if it is thrown so that he must slow down, stop, or come back to catch the ball, his chances of scoring are reduced. Under such circumstances, a pass in basketball must be evaluated on whether it was executed in such a way as to permit the maximum advantage to be extracted from the situation in which it occurred. The same is true for passing in games like soccer (where passes are made by throwing, kicking, and heading) and hockey (where they are made by hitting, pushing, and flicking); for catching in baseball and football; and for hitting in baseball, tennis, and squash. The skills used in gymnastics and figure skating exercises are evaluated in similar fashion in terms of whether they are executed in such a manner as to make the maximum possible contribution to the aesthetics of the overall result. In all these cases, and in many others like them, the performance is thus evaluated subjectively to determine whether the skill was executed to best advantage. For the sake of brevity and convenience, the result in these cases is simply stated as *advantage*.

The two categories of result and a selection of the skills that fall within each category are shown in Table 14.

Once the result has been identified for the skill of interest, the word (or words) used to describe it are printed in a rectangular box drawn near the top of a sheet of paper (Fig. 157). The rest of the model can then be arranged in appropriate order in the space below.

## Division of Result

The second step in the development of the model consists of determining whether the result may be divided readily into a series of distinct, consecutive parts. This can usually be done only in the case of those skills for which the result is a measure of length or time. In the long jump, for example, the total distance for which the performer receives credit—the distance from the front edge of the board to the nearest mark made in the sand—can be divided conveniently into three lesser distances (Fig. 158). These are the horizontal distance from the performer's center of gravity to the front edge of the board at takeoff (the takeoff distance), the horizontal distance traveled by the performer's center of gravity while airborne (the flight distance), and the horizontal distance between the performer's center of gravity at touchdown and the mark made in the sand (the landing distance)—three distances that together add up to the total distance, or result. (If the athlete's center of gravity is behind the front edge of the board at takeoff, as is often the case if the approach run is not well practiced, the takeoff

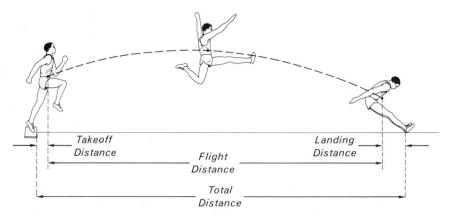

**Fig. 158.** *The distance recorded in the long jump may be divided into takeoff, flight, and landing distances.*

**TABLE 14  Classification of Results for Selected Skills**

| Category | Result | Skill |
|---|---|---|
| Objective measures | Distance | Discus throw |
| | | Hammer throw |
| | | Javelin throw |
| | | Long jump |
| | | Shot put |
| | | Triple jump |
| | Height | High jump |
| | | Pole vault |
| | Number of pins | Bowling |
| | Points scored | Archery |
| | | Curling |
| | | Kicking field goals (Rugby) |
| | | Kicking field goals or PATs (Football) |
| | | Shooting (Basketball) |
| | | Shooting (Hockey) |
| | | Shooting (Pistol and Rifle) |
| | | Shooting (Soccer) |
| | | Shooting (Water polo) |
| | Time | Alpine skiing |
| | | Canoeing |
| | | Cross-country skiing |
| | | Cycling |
| | | Rowing |
| | | Running |
| | | Speed skating |
| | | Swimming |
| | Weight | Weight lifting |
| Subjective measures | Points awarded | Diving |
| | | Vaulting (Gymnastics) |
| | Advantage | Blocking and tackling (Football) |
| | | Catching (Baseball) |
| | | Catching or receiving (Football) |
| | | Exercise parts (Figure skating) |
| | | Exercise parts (Gymnastics) |
| | | Hitting (Baseball) |
| | | Hitting and serving (Tennis) |
| | | Lifting and leaping (Dance) |
| | | Passing (Basketball) |
| | | Passing (Soccer) |

distance is negative. The landing distance is similarly negative if the performer falls backward and makes a mark in the sand behind the position of his or her center of gravity at touchdown.)

The height recorded in the pole vault and the distance recorded in the shot put can be similarly divided into distinct, consecutive parts (Figs. 159 and 160).

The time taken to cover the specified distance in swimming can be divided into the time spent starting, the time spent stroking, and the time spent turning—three times that together add up to the total time. (The total time consists of the time spent starting plus consecutive, alternating times spent stroking and turning

$$t_{Total} = t_{Starting} + t_{Stroking} + t_{Stroking} + t_{Turning} + \cdots$$

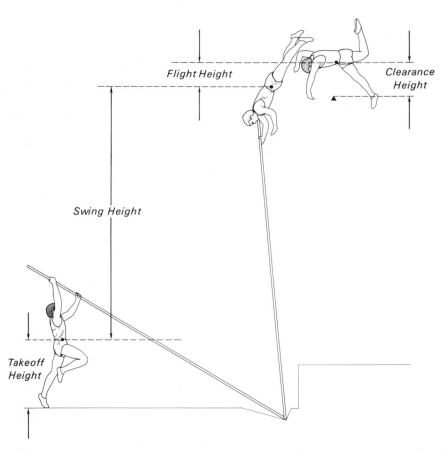

**Fig. 159.** *The height recorded in the pole vault may be divided into takeoff, swing, flight, and clearance heights.*

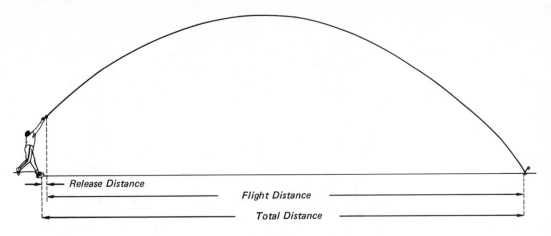

Release Distance

Flight Distance

Total Distance

**Fig. 160.** *The distance recorded in the shot put may be divided into release and flight distances.*

However, because of the very close similarity that normally exists between the stroking and turning techniques employed at different stages in a race, it is sufficient for purposes of analysis to sum all the times spent stroking and, similarly, all the times spent turning.)

Although the second step in the procedure can only be taken in a limited number of cases, it is nonetheless an important step because, where it can be used, it greatly simplifies the subsequent development of the model. This is especially true when the skill involves some form of projectile motion. In these cases, the division of the result is made so that the part of the result associated with the airborne motion of the body is separated from the part (or parts) associated with the nonairborne motion (Figs. 158–160). The further development of the model then proceeds with relative ease because factors that determine the result of the airborne motion—usually the most important part of the total motion—are well known. They are, of course, the speed, angle, relative height of release and the air resistance encountered in flight.

Once the result has been divided as described, additional rectangular boxes are drawn to accommodate the contributing, or component, parts of the result. These boxes are drawn below the box containing the result and are linked to it as shown in the examples of Fig. 161.

### Factors Influencing Result

The third and final step in the process of developing the model is to determine those factors that produce the result or, if the result has been divided, the respective parts of that result. Two important rules should be followed in identifying these factors.

*Where possible, the factors included in the model should be mechanical quantities.* (Appropriate mechanical quantities—for example, velocities, masses, coefficients of restitution, and moments of inertia—were defined and discussed

271

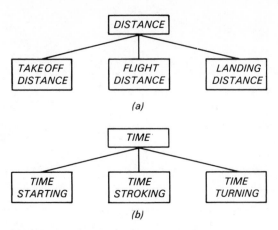

**Fig. 161.** *Models for (a) the long jump, and (b) swimming, showing division of result into distinct, consecutive parts.*

in Chapters 9–13.) This rule should be disregarded only when the use of another term conveys the same meaning in a more concise way. Thus, for example, it would be more appropriate to use the term body position at touchdown than the more precise but much more long-winded terms coordinates of the heels at touchdown, lengths of body segments, and relative angles of the body segments at touchdown when listing the factors that determine the landing distance in the long jump. All the factors that are summarized in the more concise statement must, however, be considered in the final analysis.

*Each of the factors included in the model should be completely determined by those factors that appear immediately below it.* Two methods are used to ensure that this condition is met. The first involves the use of simple addition. Consider the case of the incomplete model shown in Fig. 161(a). In this case, the total distance or result is simply the sum of the takeoff, flight, and landing distances. It is, therefore, completely determined by the three factors that appear immediately below it in the model. The same simple addition is also evident in the other incomplete model shown in Fig. 161(b).

Simple addition is also used in another situation. In developing a model for the high jump, the vertical velocity of the athlete at the instant of takeoff—a very important factor in determining the result of the jump—can be considered equal to the sum of his vertical velocity at the instant of touchdown and the change in vertical velocity that takes place during the takeoff phase:

$$\text{Vertical velocity at takeoff} = \text{Vertical velocity at touchdown} + \text{change in vertical velocity}$$

When the corresponding boxes of a high jump model are drawn, it can be seen that the required condition is met. The factor at one level in the model is completely determined by the factors in the level immediately below it. The simple relationship involved in this case (Final value = Initial value + change in value) is often used in the development of models.

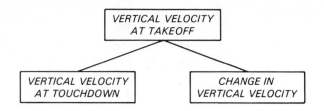

The second method involves the use of a known mechanical relationship. Consider locomotor skills like cycling in which the performance is evaluated in terms of the time taken to cover some set distance. The average speed at which the performer moves in such cases is given by Eq. 1 (p. 120):

$$\text{Average speed} = \frac{\text{Distance}}{\text{Time}}$$

This equation can be rearranged to show (even more clearly than does Eq. 1 itself) that the time is completely determined by just two factors—the distance involved and the average speed of the performer over that distance:

$$\text{Time} = \frac{\text{Distance}}{\text{Average speed}}$$

The first two levels of a cycling model can thus be written

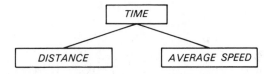

The same three quantities can also be included in the models for running, swimming, canoeing, rowing, and various other locomotor skills.

Several other mechanical relationships are used with some frequency in the development of models. For example, the horizontal distance that a projectile travels while it is airborne is governed by the speed, angle, and relative height at which it is released and by the air resistance it encounters in flight.

## Summary

The procedure used in developing a model is perhaps best summarized with the aid of an example. Suppose one wished to develop a model to be used in the qualitative analysis of a high jumper's performance.

The first step in the procedure consists of identifying the result and this presents no real difficulty. In the high jump, the result is simply the height cleared. The word *height* is thus entered in a box at the top of the page on which the model is to be drawn.

The second step consists of determining whether the result can be divided into distinct, consecutive parts. In the case of the high jump, the height of the jump can be divided into three parts. These are (Fig. 162) the height of the athlete's center of gravity at takeoff (the *takeoff height*), the height that the athlete's center of gravity rises from takeoff to the peak of its flight (the *flight height*), and the difference between the peak height of the athlete's center of gravity and the height of the bar (the *clearance height*).

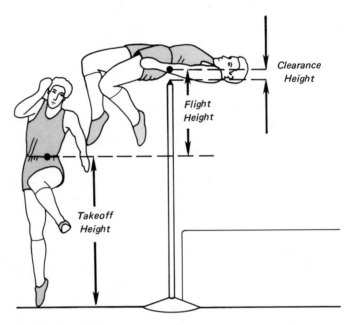

**Fig. 162.** *The height recorded in the high jump may be divided into takeoff flight and clearance heights.*

If the athlete's center of gravity passes beneath the bar while his or her body passes above it (a possible, but unlikely, occurrence), the height cleared is given by

Height = Takeoff height + flight height + clearance height

Alternatively, if the athlete's center of gravity passes above the bar, the height is given by

Height = Takeoff height + flight height − clearance height

In either case, the result is entirely determined by the three parts into which it has been divided. This condition must be met any time a result is divided into smaller parts. Having divided the result into parts, the words or symbols used to designate these parts are entered in boxes below the box containing the result.

The third step consists of determining those factors that produce the result or its respective parts. In the present case, this step involves identifying all the factors that determine the magnitudes of the takeoff, flight, and clearance heights.

The takeoff height is determined in part by (a) the lengths, (b) the masses, and (c) the location of the centers of gravity of the athlete's body segments. These factors are represented in the model by the term *physique*. The takeoff height is also determined in part by how the segments of the athlete's body are positioned at that instant. The factors that describe the position of the segments—the coordinates of the toe of the athlete's takeoff foot, the lengths of the segments, and the relative angles of the segments—are represented in the model by the term *position*. When the factors that determine the takeoff height are included, the model for the high jump is as shown in Fig. 163.

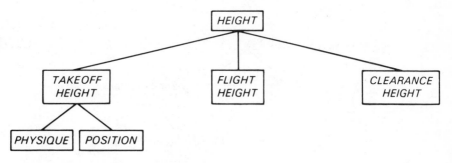

**Fig. 163.** *Incomplete model for the high jump.*

The amount that any projectile rises above its height of release is governed by its vertical velocity at release—the greater the vertical velocity, the greater the height to which the body rises—and by the air resistance it encounters in flight. The flight height in the high jump is thus completely determined by the athlete's *vertical velocity at takeoff* and by the *air resistance* encountered as the athlete rises to his or her peak height. The vertical velocity at takeoff is equal to the sum of the *vertical velocity at touchdown* and the *change in the vertical velocity* during the takeoff (p. 273). The factors that determine how much the vertical velocity changes during the takeoff can be identified with the aid of the impulse-momentum relationship (p. 165). This relationship states that the change in momentum is equal to the impulse that produced the change. That is, in algebraic terms,

$$mv_f - mv_i = J$$

When this equation is rearranged,

$$v_f - v_i = \frac{J}{m}$$

It becomes obvious that the change in velocity experienced by a body is determined by its mass and by the impulse that is exerted upon it. In the specific case of the high jump, therefore, the change in the athlete's vertical velocity during the takeoff depends on his or her *mass* and on the *vertical impulse* exerted during the takeoff. The vertical impulse is equal to the sum of all the *vertical forces* transmitted to the ground via the athlete's jumping leg and the *time* for which they act. These *vertical forces* are, in turn, the result of the muscular actions associated with the movements of the athlete's arms ($A$), trunk ($T$), lead leg ($LL$), and jumping leg ($JL$). When the various factors that determine the flight height are added, the model takes the form shown in Fig. 164.

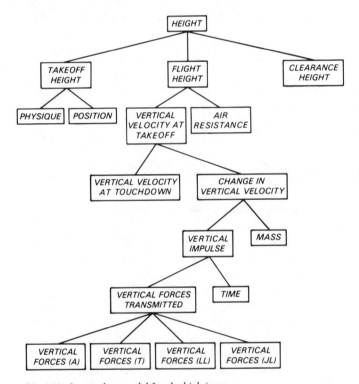

**Fig. 164.** *Incomplete model for the high jump.*

The clearance height, like the takeoff height, is governed by the *physique* of the athlete and by his or her *position* at the moment the center of gravity reaches its peak height. It is also governed by the path followed by the center of gravity relative to the bar and the motions of the athlete's body segments, relative to the center of gravity, during their passage across the bar. If the athlete's center of gravity does not reach its peak height directly above the bar, the clearance height is a greater negative quantity, and thus less favorable, than if it did (Fig. 165). The clearance height also suffers if the segments of the athlete's body are not moved in such a way as to extract the maximum credit from the height to which the athlete has elevated the center of gravity. In the interest of brevity, these two factors are represented by the term *actions during clearance*. With these remaining factors included, the complete model for the high jump is as shown in Fig. 166.

Several additional points can now be made about the models developed for use in qualitative analyses:

1. Different models can be developed for the same skill. It would be possible, for example, to develop a model for the high jump in which the result is divided into four parts instead of three. These might be the height of the athlete's center of gravity at touchdown, the height that the center of gravity is raised during the takeoff, and the flight and clearance heights, as previously defined.

2. While some models might be better suited for their intended purposes than others, there is no such thing as "the correct model" for a given skill. Any model that satisfies the basic conditions described earlier (p. 266) is a correct model.

3. Models should be developed only to the point where their further development seems likely to be unrewarding. Unless some restraint is exercised, an otherwise simple model can become much too large and complex to be really helpful in the subsequent analysis. Consider the factors that influence the vertical velocity at touchdown in the high jump model. These are the speed, angle, and relative height of takeoff for the last stride of the approach and the air resistance encountered during the flight phase of that stride. The speed of the takeoff for the last stride can be considered as the sum of the speed of the preceding touchdown and the change in speed during that grounded phase. The speed of the previous touchdown is determined by . . . and so on. While one could follow such a process of development all the way back to the first stride of the approach, there is little or nothing to be gained by doing so. For this reason, the development of the model in this direction is halted once the vertical velocity at touchdown has been included.

4. For the sake of clarity, the relationships between factors shown at the same levels of a model are not indicated. Thus, although the magnitude of the forces transmitted via the high jumper's takeoff leg is governed in part by the time available for this purpose, no effort is made to indicate the relationship between these two factors in the model. It is important to remember, however, that such relationships not only exist but are often of considerable practical significance. (While the relationship cited here as an example could be indicated by simply linking the appropriate two boxes, serious difficulties arise when four or five factors shown at the same level are interrelated.)

5. A model for a given skill can be used in analyzing the performances of any number of individuals performing that skill.

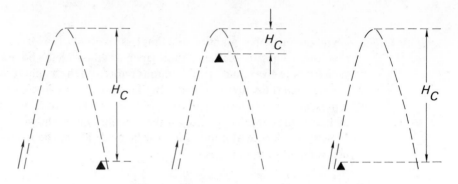

**Fig. 165.** *Clearance height in the high jump ($H_C$) is influenced by the relationship between the trajectory of the jumper's center of gravity (---) and the bar (▲). The height and shape of the trajectory—dictated by the height, speed, and angle of takeoff—is the same in all three cases.*

**Fig. 166.** *Model for the high jump.*

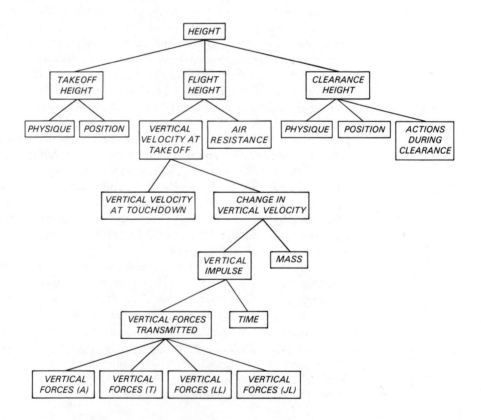

## OBSERVATION OF PERFORMANCE

Success in the use of any system of qualitative analysis depends largely on the accuracy with which the movements involved are perceived. If the teacher or coach conducting the analysis is able to perceive exactly what the performer is doing, appropriate conclusions are more likely to be reached than if this ability were lacking. Developing the ability to perceive movements accurately is, thus, a critical step in acquiring proficiency in qualitative analysis.

The process of observing the performance of a motor skill is normally considered to be visual. The observer simply watches the performance and gathers information about it via his or her eyes. There are, however, other sources of useful information. There are instances, for example, when useful information may be obtained via the sounds made by the performer and/or the equipment being used, via direct contact between the observer and the performer, and via the performer's perception of his or her own movements. These additional forms of observation are referred to here as *aural*, *tactile*, and *kinesthetic* observation, respectively.

Since motor skills are often highly complex, qualitative analyses of motor skills are often similarly complex. It makes no sense, therefore, to reduce the chances of reaching appropriate conclusions by using only part of the information available. Good qualitative analyses thus take into account all the information from all the available sources—visual, aural, tactile, and kinesthetic.

### Visual Observation

Two basic forms of visual observation may be used in analyzing motor skills. These are direct visual observation of the movements involved as they are performed (or, if they have been recorded, as they are displayed at some later time) and indirect visual observation of the movements using the various marks or "clues" that remain after the performance has been completed.

**Direct Visual Observation.** The accuracy with which one can perceive the performance of a motor skill using direct visual observation depends on a number of factors. These include the setting in which the skill is performed, the position of the observer relative to the performer, the point on which the observer focuses his or her attention, and the availability of aids to assist the observer.

**Setting.** The setting in which a motor skill is performed plays a large part in determining the accuracy with which the movements involved can be perceived. If the skill is performed in a class setting, the activities of the others in the performer's class can serve to distract the attention of the observer. This is especially likely to happen where the observer is the class teacher rather than an outside observer with no particular responsibility for the class. The observer can also be distracted if the skill is performed in a situation in which his or her view of the performance is obstructed from time to time. This occurs frequently when observing skills being performed in games, races, and other competitive situations.

Most teachers and coaches have only limited control over the settings in which they must make their observations. They are, therefore, obliged to exercise this limited amount of control to their best advantage and then do the best they can under the circumstances imposed by the setting. It should be obvious, therefore, that physical education classes and athletic competitions are not usually the best settings in which to acquire proficiency in the observation of human movements. While one may eventually work in settings that are less than ideal from an observational standpoint, the process of developing proficiency in observation is best obtained in a setting that is as near to the ideal as can be arranged.

The ideal setting is one that closely simulates the setting in which the skill is normally performed, that is devoid of external activities that might distract the attention of the observer and the performer, and that places no important restrictions on the positions that may be used by the observer.

***Protocol.***   Assuming that a setting that meets these requirements has been selected and prepared, the observer must instruct the performer as to how to proceed. Several important points should be considered here. First, the performer should be instructed to warm up as he or she would prior to performing the skill under normal conditions. Thus, if a golfer would normally do some loosening exercises and hit a bucket of balls on the driving range before stepping up to the first tee, the golfer should go through this same precompetition procedure before an analysis of his or her drive is conducted. Second, once ready to begin, the performer should be asked to perform a given number of repetitions of the skill in his or her own time. This number is a function of the skill involved—some skills can only be performed a few times in succession while others can be repeated almost indefinitely. It is also a function of the ability and experience of the observer. The more able and experienced the observer, the fewer the number of repetitions that should be necessary. However, as a general rule, the greater the number of repetitions—up to a maximum of fifteen to twenty—the better the chances are that a thorough, accurate, and efficient analysis will be conducted. Third, the performer should be told that the observer will not make any comments about the performance until all repetitions have been completed. This eliminates any expectation of immediate feedback on the performer's part and, thus, any pressure that the observer might feel to make snap judgments about what he or she has seen. It also means that the performer is more likely to be consistent from one repetition to another than if he or she were able to introduce variations based on the observer's comments. Finally, the observer should move into position to observe the first trial(s) and instruct the performer to begin whenever ready.

*Position.* The accuracy with which movements can be perceived depends also on the extent to which the observer is free to choose the positions from which to observe the performance. If the observer is free to take up the best possible positions, the chances of obtaining an accurate perception of the performance are enhanced. If, however, the observer is unable to use the best vantage points—because of limitations in the physical arrangement of the setting in which the performance takes place or because of constraints imposed by the rules or the organizers of the competition—the chances of obtaining an accurate perception of the performance are reduced.

The optimum position(s) from which to view the performance of a motor skill varies not only from skill to skill but also according to the specific feature(s) of the skill that are being observed. There are, however, general rules that should be followed when selecting the position(s) from which to observe a performance.

In most cases, the best vantage point is one that is at right angles to the general direction of the body's motion and opposite the point of interest. Thus, if one wished to observe the performance of a gymnast during takeoff for a vault over the horse, the best vantage point would be on a line at right angles to the gymnast's approach run and directly opposite the takeoff board (position 1 in Fig. 167). Similarly, if one wished to observe the gymnast during the hand support or landing phases, the best vantage points would be opposite the positions on the horse and landing mat at which these occurred (positions 2 and 3 in Fig. 167, respectively). Where the motion of interest extends over some distance, the best vantage point is one opposite the midpoint of this distance. Thus, in the example of the gymnastic vault already mentioned, the best point from which to observe the vault as a whole would be opposite the midpoint of the distance from takeoff to landing (position 4 in Fig. 167).

The speed at which the performer's body moves across the observer's field of view influences the latter's ability to perceive exactly what movements are taking place. This speed depends on the rate at which the performer is moving and the distance between the performer and the observer. As the speed at which the performer moves increases or the distance between the observer and the performer decreases, the speed at which the latter moves across the observer's field of view gets progressively faster. To overcome the problems of perception inherent in this situation, the observer should take up a position sufficiently far from the plane of action to allow as clear a view of the performance as possible. For skills involving slow movements or a very limited range—skills like shot putting, place kicking, and weight lifting, for example—a distance on the order of

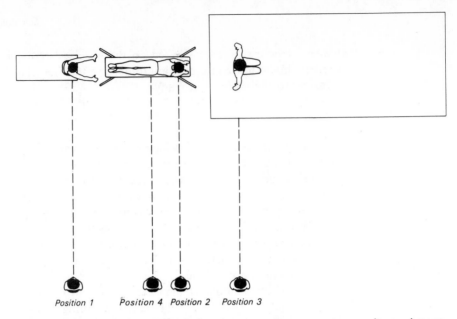

Position 1    Position 4  Position 2    Position 3

**Fig. 167.** *The optimum position from which to observe a performance varies according to what part of the performance is to be observed.*

10–15 m is usually appropriate. For skills involving fast movements and/or skills that take place over an extended range, this distance should be increased to 20–40 m.

Finally, it is very unlikely that any one vantage point will permit the observer to see all that should be seen. For this reason, it is essential that the observer move from one vantage point to another as attention is focused first on one aspect of the performance and then on another.

*Focus.* The point (or points) on which the observer focuses attention have a marked bearing on what is seen. If an attempt is made to get a general impression of the whole performance, the observer is unlikely to gain a very distinct impression of how any one part of the body is moving. Conversely, if attention is focused on one part, the movements of the other parts are not likely to be seen very clearly. It is important, therefore, that the observer decide exactly what he or she is going to look for before each trial. In this respect, it is usually wise to devote the first two or three trials to obtaining a general, overall impression of the performance and subsequent trials to a systematic examination of the separate movements that contribute to the whole.

*Aids.* There are a few observational techniques and a wide variety of recording devices that the observer may use to improve the accuracy with which a performance is perceived.

In some cases, the observer's attention is distracted by the movements of body parts other than those on which he or she is trying to focus. One technique that is

282

occasionally used to overcome this problem is to use a card (or a hand) to mask the observer's view of these distracting movements. Thus, walking judges, whose sole concern is whether the walker is maintaining unbroken contact with the ground, sometimes use a card to mask their view of the walker's upper body. Track coaches occasionally use the same device to determine whether a hurdler's head is at the same level as he crosses the hurdle as it is during the running strides before and after the hurdle.

Many skills contain movements that are so complex and that occur at such high speed that it is very difficult for all but the most highly trained observer to perceive them accurately. It is very difficult, for example, to see a golfer's body position at the instant of impact, a hammer thrower's position at the end of the third turn, and a sprinter's position at the instant of touchdown. There is, however, a very simple and effective technique that may be used to gain an accurate impression of the performer's position at a given time. This technique consists of closely following the sequence of movements involved in the performance and then quickly closing the eyes at the moment of interest. When this is done, the last image to reach the eyes is retained unaltered for some time and can readily be examined to obtain the required information.

A camera or videotape recorder may also be used to assist the observer in obtaining an accurate perception of performance. Equipment of this kind is usually quite expensive, is often inconvenient to use (especially where film processing may take as long as seven to ten days), and is occasionally difficult to operate. However, because its use greatly increases the likelihood that the observer will obtain an accurate perception of the performance, it should nonetheless be used whenever possible.

**Indirect Visual Observation.**   Almost all human activities produce changes in the environment in which they occur. These changes may often be used as a basis for deducing the nature of the activity and of the individuals who took part in it. Thus, a detective will examine the scene of a crime for bullet holes, marks, scratches, and other indicators of what took place during the commission of the crime. Anthropologists, working many centuries after the events, also make use of such methods when trying to deduce the nature of prehistoric peoples and the manner in which they lived.

Teachers, coaches, and others involved in the analysis of human movements have also made use of such methods. In track and field, for example, an examination of the marks left in a cinder track has often been used to provide valuable information about an athlete's performance. This point is well illustrated by Webster's account[5] of what might be learned from an examination of the marks left by a hurdler.

---

[5]F. A. M. Webster, *Why? The Science of Athletics* (London: John F. Shaw, 1937), p. 238.

*The hurdler, by observing the imprint of his take-off foot, can tell clearly whether he is driving straight at the hurdle, while his landing mark will tell him if he is chopping the leading leg down close enough to the fence to ensure a good knee pick-up action in the next stride, after landing. He can also work out accurately the proper proportioning of his three-stride action between landing and take-off. Another important point is that he will see if he is turning his foot out on landing from hurdle clearance, a fatal fault which will cause him to deviate badly from his true course.*

Similarly useful information can also be obtained by examining the marks left by sprinters, jumpers, and vaulters.

There are numerous other instances in which this form of indirect visual observation can be used to good effect. The divot taken out of the ground by a golf club as it swings through the ball can be used as a source of information about a golfer's performance. The direction of the divot provides an indication of whether the clubhead was moving along the intended line as it was brought into contact with the ball. The position of the divot gives an indication of whether the low point of the swing was correctly located with respect to the ball.

The point at which a ball or a throwing implement lands is usually an indication of how it was projected into the air. Thus, for example, there are usually important differences between the sequence of movements that yields a hit to left field in baseball and the sequence that yields a hit to right field. In some cases, the manner in which an implement lands is also a useful indication of how it was released. For example, the angle to the horizontal and to the direction of the throw at which a javelin lands provides a very clear indication of the manner in which it was released.

The marks left in the snow by a skier who is executing a turn and those left on the ice by a figure skater are an excellent source of information about a performance. Such marks are often used by ski instructors and coaches of skating to aid them in their analysis of a performance and their subsequent instruction of the performer.

Because important information can often be gained from an examination of the marks made during the course of a performance, observers should be continually alert to the possibility that such marks might be used as an aid in their analyses of a given skill.

### Aural Observation

The performance of a motor skill is usually accompanied by a certain amount of noise. In some cases, this is little more than an outward sign of the effort involved. The tennis player grunts when smashing the ball and the thrower shouts when releasing the implement. Such noices are generally of little value as a supplementary means of gaining information on how the skill was performed. In other cases, however, the sounds that accompany the performance can be used to good effect. This is particularly true for those motor skills in which rhythm plays an important role. Thus, an experienced coach of the jumping events in track and field can often obtain important information about a performance by simply listening to the rhythm of the athlete's feet during the final strides of the approach to the takeoff. The coach can also obtain useful information about the performance of a triple jump by listening to the rhythm beaten out as the athlete executes the hop, step, and jump phases. Useful information can also be obtained by listening to the rhythm of a basketball player's feet during the performance of a lay-up shot, to the sounds of an athlete's breathing during the execution of a clean and jerk in weight lifting, and to the sound made as the ball makes contact with the hands during the performance of a set of volleyball. The intelligent observer makes use of such auditory information whenever possible to supplement the information obtained from other sources.

## Tactile Observation

Teachers and coaches occasionally find themselves in direct contact with the performer. Thus, a gymnastics coach might place his or her hands on a performer to assist the performer in completing a given movement successfully. Alternatively, a coach might use a hand-held spotting belt or an overhead spotting belt to maintain direct contact with the performer. In such cases, the coach is very ill equipped to observe the performance visually—being much too close to the performer to see clearly what is happening and too concerned with assisting the performer to be able to devote much attention to observing precisely what the performer is doing. The coach can, however, gain a sound, general impression of the performance via the direct contact with the performer.

## Kinesthetic Observation

The analysis of a motor skill should be a cooperative venture in which both the performer and the observer participate. The observer is usually well placed to gather information by visual and auditory observation and, on some occasions, by tactile observation. The performer, on the other hand, is generally in a poor position to gather information in these ways. The performer is, however, in a unique position to gather information in another way. The person who is performing the skill is the only one who can say how the performance felt. The performer is, in short, the only person who is able to observe the performance from a kinesthetic standpoint.

The completeness and accuracy of the kinesthetic observations made by a performer are a function of his or her familiarity with the skill involved. Just as a person seeing an unfamiliar skill performed is unlikely to gain more than a fleeting impression of the movements involved, so an inexperienced performer of a skill is unlikely to be able to accurately perceive the movements kinesthetically. An experienced performer, however, can often gain a very accurate perception of his or her own movements. This is particularly true if the performer has developed a reasonably analytical approach toward the practice of the skill. In cases like this the observer is well advised to make a practice of asking the performer "How did it feel?" after each trial. Although the answer given will not always be an accurate indication of what happened, it is very likely to contain information of value to the observer. In some cases, the information given by the performer will merely serve to confirm the impressions gained from other sources; in others, it will resolve a question in the observer's mind; in still others, it will raise doubts about the tentative conclusions already reached by the observer. Considering the very real benefits to be gained, it is unfortunate that teachers and coaches make so little use of this valuable source of additional information.

## Conclusion

The ability to obtain an accurate perception of the movements involved in the performance of a motor skill is developed through extended and intelligent practice. People wishing to develop such ability should take every available opportunity to observe human movements. They should watch people moving in the street, in shops, in airports, in factories, and on farms. They should watch them, in short, wherever and whenever they possibly can. They should experiment with the various positions that might be taken to observe a motor skill and note what aspects of the performance can be readily observed from each of these positions. They should note how different people perform the same skills. They should try, for example, to identify how the walking gaits of the very young and the very old differ from each other and from the gaits of people who are neither very young nor very old. They should observe as wide a range of different skills as possible. They should be ever alert to the possibility of supplementing the information available to them via direct visual observation with that available via other forms of observation. It is only by such prolonged and well-directed efforts that they will become truly proficient in the observation of human movement.

## IDENTIFICATION OF FAULTS

In the qualitative analysis system described in this chapter, the observational techniques of the preceding section are used to discriminate between those parts of a performance that are executed correctly and those that are not. At least two general procedures may be used for this purpose. In the first—referred to here as the *sequential method*—the total performance is considered to be made up of a series of consecutive parts. For example, the performance of a batter in baseball might be considered to consist of four parts: the stance, stride, swing, and follow-through. The positions that the performer adopts and the movements executed within each part are then compared, in sequential order, with the observer's mental image of an ideal form for these positions and movements. Special attention is devoted to differences between the observed and ideal forms, and, where judged appropriate, instruction is given with a view to eliminating these differences.

Although this method is used widely by teachers and coaches and is often taught to students, it has one major limitation that makes it less than ideal for its intended purpose. The lack of any valid basis for determining the ideal form usually leads to the assumption that the form used by an outstanding athlete is the ideal form. This practice, while understandable, is a dangerous one because many outstanding athletes, even Olympic champions and world recordholders, have forms that are demonstrably less than ideal. Furthermore, there is almost certainly no one form that is ideal for everyone. Instead the ideal form for one person is likely to be different from the ideal form for another simply because of differences in their physical and mental attributes.

The second method—referred to here as the *mechanical method*—is based on the theoretical model described earlier in this chapter. In this method, the observer examines the performance systematically to determine whether there is any way in which the contributions made by the various factors included in the model for that skill might be improved. For example, in analyzing the performance of the straddle-style high jumper in Fig. 168, the observer would look for ways in which the takeoff, flight, and clearance heights might be improved. To do this, the observer would consider each of these heights—and the factors that govern them—in turn.

**Takeoff height.**    The only factors that influence the magnitude of the takeoff height are the physique of the performer and the position at the instant of takeoff (Fig. 166). Of these, the only one that might conceivably be improved under normal circumstances is the athlete's position at takeoff. A close examination of the position shown in Fig. 168(d)—the best available indication of the athlete's position at the instant of takeoff—suggests that he could increase his takeoff height by assuming a more erect position of his head and trunk and by increasing the height to which he elevates his arms and leading leg.

**Flight height.**    The magnitude of the flight height depends on the vertical velocity at takeoff, which, in turn, depends on the vertical velocity at touchdown and on the change in vertical velocity during the takeoff (Fig. 166). Although most high jumpers have a downward vertical velocity at the instant of touchdown, the need to arrest this downward motion before the body can again

*(a)*

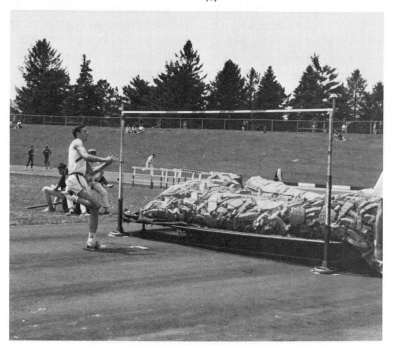

*(c)*

**Fig. 168.** *Sequence for analysis of a straddle-style high jump.*

(b)

(d)

(e)

(g)

*(f)*

*(h)*

be driven upward into the air means that better results are likely to be obtained if this motion is kept to a minimum. This is usually achieved by keeping the body low during the preceding support phase, driving the hips forward and upward into the last stride, and "grounding" the takeoff foot before the body can acquire much downward velocity during the flight phase of the last stride.[6] Figs. 168(a) and (b) suggest that the athlete shown does not perform this sequence of movements effectively and that his vertical velocity at touchdown is thus larger than is desirable.

The change in vertical velocity during the takeoff depends, ultimately, on the vertical forces exerted against the ground as a result of the actions of the athlete's arms, lead leg, and jumping leg and the times for which these forces act (Fig. 166). The actions of the arms and lead leg suggested by the positions shown in Figs. 168(b)-(d) appear to be capable of substantial improvement. The arms appear to have been lifted with little vigor and through a very limited range of motion. The lead leg appears to have been brought forward from behind the body with a similar lack of vigor and, due to the flexion of the knee, through a limited range. The contribution that the jumping leg makes to the flight height is decreased if the body is leaning forward at takeoff, as suggested by Fig. 168(d). When the body is vertical, the forces exerted on the ground by the jumping leg are primarily directed vertically downward. The contribution that these forces make to the flight height is thus maximized. When the body is inclined forward, these same forces are directed at an angle to the vertical and their contribution to the flight height is reduced.

The forces exerted by the muscles of the performer during the takeoff produce two effects. They cause each of the joints involved to move through a given range and they cause forces to be transmitted via the foot of the jumping leg to the ground. If the forces exerted by the muscles are large, the joints move rapidly and the forces exerted against the ground are large. Conversely, if the forces exerted by the muscles are small, the joints move slowly and the forces exerted against the ground are small. In short, the duration of the joint motions—and thus the times over which forces are exerted against the ground—are inversely related to the magnitudes of these forces. Since the ways in which these forces might be increased have already been considered, there is obviously little to be gained by considering the times over which they act as independent factors.

**Clearance height.**    The clearance height is governed by the athlete's physique, position at the peak of the flight, and actions during clearance (Fig. 166). Of these, only the last two can be altered under normal circumstances.

From the photo sequence of Fig. 168, it is impossible to be sure what part (or parts) of the athlete's body dislodged the bar. It would appear, however, that it

[6]Vladimir M. Dyatchkov, "The High Jump," *Track Technique*, 34 (December 1968), 1059–1074.

was struck by the inside of the athlete's leading leg—(Figs. 168(f) and (g)). This would certainly be consistent with the forward-diving takeoff in which the position of the trunk would make it difficult for the athlete to lift his leading leg. Aside from this problem with the leading leg, the athlete's position at the peak of his flight appears to be capable of relatively little improvement. His position in Fig. 168(g) suggests that he draped his body around the bar very well and that he was reasonably effective in extracting credit from the peak height to which he had elevated his center of gravity.

The athlete's movements during his passage across the bar appear to be deficient in at least one respect. On the basis of the positions shown in Figs. 168(g) and (h), it appears that the athlete made little effort to externally rotate the thigh, shank, and foot of his trailing leg to lift it clear of the bar. This lack of external rotation is especially evident in the position of the left foot. Thus, if the leading leg had not dislodged the bar (assuming again that it did), the trailing leg may well have done so.

The systematic consideration of the factors that contribute to the result has yielded a number of possible avenues for improvement in the performance of the athlete in Fig. 168. These may be summarized as follows:

Takeoff height: Assume more erect position of head and trunk
                   Increase height to which arms and leading leg elevated

Flight height: Decrease downward vertical velocity at touchdown
                 Increase range and vigor of arm and lead leg actions
                 Increase backward lean of body at touchdown
                 Decrease forward lean of body at takeoff

Clearance height: Externally rotate trailing leg to avoid contact with bar

Several points should be made with respect to the mechanical method for identifying faults and with respect to the example used here to illustrate its application.

1. The method proceeds in a systematic manner dictated by the structure of the model and thus demands that the model be referred to repeatedly during the course of the analysis.
2. Since only those factors that have a role in determining the result are included in the model, and the method calls for each of these factors to be considered in turn, it is very unlikely that an important factor will be overlooked or that an unimportant one will be given undue attention.
3. The method does not rely on the notion of an ideal form. It does, however, make use of the form employed by skilled performers as a guide to what movements others have attempted and found effective and, thus, to what changes might be contemplated in a given case.

4. Although the performance depicted in Fig. 168 provided a useful basis for discussion of the mechanical method, the analysis was necessarily limited by the amount of information available. It was limited in particular by the absence of information about the athlete's body positions at critical instants that fell between consecutive photographs, by the absence of information that is best obtained from other vantage points, and by the lack of any clear indication of the timing or rhythm of the movements. These limitations should be borne clearly in mind when conducting analyses based on photographic sequences like that of Fig. 168.

5. The ability of observers to identify ways in which the factors governing the result may be improved depends on their prior knowledge of the skill concerned, their observational abilities, and their experience in observing that skill. Thus, if two people reach different conclusions following their analyses of the performance of a skill, it is more than likely to be because they differ in one or more of these respects.

## EVALUATION OF FAULTS

Because it is virtually impossible for a performer to work on correcting more than one or two faults at a time, it is necessary to establish an order of priority among the various faults identified.

The first step in this process is to exclude from further consideration at this time all those faults that appear to be the effects produced by other faults rather than faults in themselves. There are numerous instances in which such faults (or effects) are observed. For example, when a pole vaulter reaches bar level with his body nearly parallel to the bar, rather than perpendicular to it, the cause is almost certainly an error in the manner in which he left the ground. He may have failed to get the pole directly overhead, to get his takeoff foot directly behind the pole, or to drive straight forward and upward into the takeoff. Any one of these errors can cause the vaulter's body to swing "around" the pole (rather than forward and upward in the desired vertical plane) and thus produce the fault observed at bar level. In cases like this—and in scores of other examples that could have been cited—it makes little sense to try and correct the observed effect (or fault) directly. Rather, one should attempt to identify the cause of the problem and work to correct that fault at its source.

The next step is to establish an order of priority among those faults (or avenues for improvement) that remain. Before this can be done, however, it is necessary to establish some basis for deciding that one fault is more important (or should be corrected sooner) than another. As a general rule, the aim of the performer (and of the teacher or coach) is to obtain the largest improvement in the result that is possible in the time available. The order of priority among a number of faults should, thus, be based on *the improvement in the result that could be expected if each fault were corrected to the extent that the available time permitted.*

The effective use of this criterion for establishing an order of priority requires that the teacher or coach have some knowledge of (a) the physical and psychological characteristics of the performer and (b) the nature of the faults involved.

Some performers are very much aware of the movements of their various body segments when they perform a motor skill; others are not. Some have the ability to concentrate their attention on a particular instant or phase in a sequence of movements and to modify it almost at will; others can only make such modifications with the greatest difficulty. Some are so well coordinated that they can not only make such modifications but can also, without apparent effort, make the adjustments in the rest of the movement sequence necessary to accommodate them. For others, the modification of one part of a performance often leads to a total disruption of the remainder of that performance. The extent to which a performer has these various abilities obviously has a considerable influence on the length of time required to correct faults in his performance.

Some faults can be corrected by drawing them to the performer's attention and simply asking him or her to make the appropriate change. Others can only be corrected by breaking the movements involved into their component parts, practicing these parts with the aid of special drills, and then carefully reassembling the movements. The length of time required to correct a fault may thus vary from one to two practice sessions in the former case to as much as several months in the latter.

To show how an order of priority is established in a specific case, consider once again the case of the athlete in Fig. 168 and assume that the photographs were taken during a competition some two weeks before the conference championships for which he has been aiming all season. Assume, too, that the athlete has indicated that his jumping on the day the photographs were taken did not differ markedly from his jumping in the preceding few weeks. Two important conclusions can be reached on the basis of this information.

First, the faults identified in the analysis of Fig. 168 are likely to be well ingrained and thus fairly resistant to change. Second, there is insufficient time to make substantial changes in the athlete's performance unless his ability to make such changes is well above average. These conclusions must be kept clearly in mind during the process of establishing an order of priority among the faults (or avenues for improvement) identified.

Several of the faults observed in the performance of the high jumper in Fig. 168 can be eliminated from further consideration because they are likely to be the effects of other faults rather than faults themselves. The athlete's failure to externally rotate his trailing leg to ensure that it does not make contact with the bar is no doubt due in large part to his body position at the peak of the jump.

This, in turn, is largely dictated by his forward lean at takeoff and by his failure to drive upward rather than forward and upward. Finally, the faulty position of the athlete's head and trunk at the instant of takeoff is largely due to his lack of backward lean at the instant of touchdown. The action of the athlete's trailing leg during the clearance and the faulty position of his head and trunk at takeoff should therefore be excluded from further consideration at this stage. (In the unlikely event that they persist once the other faults in his performance have been corrected, these faults must be regarded as faults in themselves rather than as effects produced by other faults and must be treated accordingly. For the present, however, there is little to indicate that they warrant such status.)

With the deletion of those faults judged to be effects rather than faults in themselves, the list of possible avenues for improvement is substantially reduced:

Takeoff height: Increase height to which arms and leading leg elevated
~~Assume more erect position of head and trunk~~

Flight height: Decrease downward vertical velocity at touchdown
Increase range and vigor of arm and lead leg actions
Increase backward lean of body at touchdown
~~Decrease forward lean of body at takeoff~~

Clearance height
~~Externally rotate trailing leg to avoid contact with bar.~~

Two of the athlete's remaining four faults—his excessive downward vertical velocity and his lack of backward lean at touchdown—are almost certainly due to a failure to correctly execute the movements involved in making the transition from the run-up to the takeoff. These movements involve a gradual lowering of the athlete's center of gravity over the last three to four strides of the approach; a driving forward and upward of his hips during the takeoff for the last stride—which action produces the desired backward lean at touchdown; and an early grounding of his takeoff foot—which serves to minimize his downward vertical velocity at that instant. These movements are invariably difficult to learn and even the most highly skilled athlete requires several months of concentrated practice to master them. In view of this, it would be unwise to suggest that the athlete attempt to correct these faults in the two weeks remaining before the conference championships. Such an attempt would be much more likely to reduce, rather than improve, the athlete's chances of success.

The remaining faults—the limited range and vigor of the arm and lead leg actions and the failure to have these limbs high at the instant of takeoff—are closely associated with each other and to some extent with the previous two faults. The height of these free limbs at the instant of takeoff depends on their position at the instant of touchdown and on how vigorous and wide ranging their action during the takeoff. Thus, the lack of range and vigor evident in the arm and leg actions is undoubtedly responsible in large part for their lack of elevation at the instant of takeoff. The range through which the athlete can swing his free limbs is, in turn, governed to some extent by the position of his body at the instant of touchdown. If he is not leaning backward at this instant, the range of motion of his free limbs is restricted accordingly.

The actions of the free limbs obviously depend also on the extent to which the athlete seeks to use these actions to advantage. If he swings his free limbs hard through a large range, and in an appropriately coordinated fashion, he derives as much benefit from these actions as the other aspects of his performance will permit. If he does not, the result he obtains suffers accordingly. In the present case, the athlete is apparently making very little effort to swing his arms and lead leg vigorously forward and upward during the takeoff and to have them high at the instant of takeoff. Since these are things that could most certainly be improved upon in the time available, it would be appropriate to suggest to the athlete that he direct his attention to the simultaneous correction of these closely related faults in the remaining two weeks. (It should be noted, incidentally, that any improvement in the range and vigor of the free limb actions is likely to lead to improvements elsewhere. It is likely, for example, to rotate the athlete upward and backward and, thus, place him in a more nearly erect position at takeoff. This, in turn, could be expected to increase both his takeoff and flight heights.)

The process of establishing an order of priority among the faults identified in the performance of the high jumper in Fig. 168 has proven to be a relatively simple one. Three of the seven faults identified previously were judged to be effects produced by other faults rather than faults in themselves. Two of the remaining faults were found to require more time than was available to bring about changes that would produce an improved result. Finally, the remaining two faults were judged to be capable of significant improvement in the time available and, since they could be corrected simultaneously, were together given top priority.

To further illustrate the process of establishing an order of priority, consider once again the high jumper of Fig. 168 and assume that this time he is about to begin training for his conference championships one year away. The three faults previously judged to be the effects of other faults are, again, excluded from further consideration in the first step of the ordering process. This leaves four faults, as before. These must now be placed in an order of priority. (Since there is ample time for the correction of all four faults, none can be excluded, as in the previous example, on the grounds of insufficient time.)

It is often difficult to estimate the improvement that might result from the correction of each fault in a performance and, thus, to establish an order of priority. This is not the case, however, in the present instance, because two of the faults—the unfavorable vertical velocity at touchdown and the lack of backward lean at the same instant—are so closely related that the correction of either one would necessarily result in the correction of the other, so these two can be considered as one. Further, because the correction of these two faults would almost certainly have a much more pronounced effect on the athlete's entire performance than any improvement that might be effected in the actions of his arms and lead leg, there is little doubt that they should be assigned top priority. (If one is unable to distinguish among the amounts of improvement that might result if a number of faults were corrected, these faults should be assigned an order of priority based upon the order in which they occur in the performance of the skill in question. Thus, a fault that occurs early in the sequence should be given preference over one that occurs later. In this manner, the likelihood of mistakenly attempting to correct a fault that is actually the effect of an earlier fault, rather than a fault in itself, is reduced. Incidentally, if a reader were unwilling to accept the conclusions reached here concerning the ordering of the last four faults and simply proceeded as described, the ordering of priorities would be identical to that already obtained.)

## Conclusion

The process of establishing an order of importance among the observed faults in a performance should be based on (a) a knowledge of the mechanical principles underlying that performance, (b) a knowledge of pertinent research findings, where applicable, and (c) practical experience in the observation, teaching, and coaching of the motor skill involved. While the process is necessarily subjective, and thus subject to error, the magnitude and frequency of error are likely to be small if the teacher or coach has the required theoretical and practical knowledge and makes appropriate use of this knowledge in reaching his conclusions.

## INSTRUCTION OF PERFORMER

Once the observed faults in a performance have been placed in an order of importance, the next task of the teacher or coach is to give instructions that will lead to the systematic elimination of these faults. As might be expected, the manner in which these instructions are given is critical to the overall success of the qualitative analysis procedure.

The extent to which the instructions given by a teacher or coach serve their intended purpose depends on many factors. Among these are a number that research and practical experience suggest are of particular importance. These factors, discussed in the following paragraphs, should be given careful consideration in the final step of the analysis procedure.[7]

The immediate objective of this final step is to convey information to the performer. To this end, the teacher or coach may simply talk to the performer, may demonstrate what is required, or may manually guide the performer's body through the desired sequence of movements. In short, the teacher or coach may use any one of several methods—or, as is most common, some combination of these methods—to communicate with the performer.

No single method of communication is intrinsically superior to all others. The nature of the situation in which it is to be used dictates the best method or combination of methods for that situation. In any given case the teacher or coach must, therefore, select that method that appears to be most appropriate.

Having selected and used an appropriate method, the teacher or coach should use whatever means are available to check whether the information transmitted is being received completely and without distortion by the performer. This can often be done with one or two carefully phrased questions. If the answers to such questions, or the feedback from other sources, are unsatisfactory, the method of communication must be reevaluated and modified before proceeding. The value of the entire analysis rests on the quality of the communication at this point. If it is flawed, the preceding analysis can be wasted and, worse yet, the future results of the performer can be impaired rather than improved. It is imperative, therefore, that every effort be made to obtain complete understanding between the teacher or coach on the one hand and the performer on the other.

The effectiveness of the communication depends in large part on the teacher's (or coach's) knowledge of the physical and psychological characteristics of the performer. In this respect, it is especially important to have some knowledge of the performer's motivation. If the performer is poorly motivated, any attempts to modify the way in which a skill is executed are almost certainly doomed to failure. For best results, the performer must first have, or be given, a genuine desire to improve the performance.

The intelligence of the performer has a bearing on his or her ability to understand instructions and, thus, on the methods of communication that can be effectively employed. For this reason, teachers (and coaches) should take care to ensure that the instructions they give are compatible with the intellectual capacity of the performer with whom they are working.

---

[7]Readers interested in a more comprehensive discussion of these factors than is either possible or desirable here, might like to consult Ronald G. Marteniuk, *Information Processing in Motor Skills* (New York: Holt, Rinehart and Winston, 1976), and Robert N. Singer, *Motor Learning and Human Performance* (New York: Macmillan, 1975).

The setting in which instructions are given often has a decided influence on the effectiveness of the communication between the people involved. If the sun is shining in the performer's eyes, the teacher or coach is much less likely to receive the performer's full attention than would otherwise be the case. The performer's attention is also likely to be distracted if there are interesting activities taking place nearby. Teachers and coaches should thus be alert to the presence of such environmental factors and take appropriate steps to minimize their negative effects.

The information conveyed to the performer can be in one or two fundamental forms. It can be a specific instruction ("Keep your body erect as you take off") or a question directed toward the same end ("How can you increase your height at takeoff?"). The relative merits of these contrasting approaches—known, respectively, as the *drill approach* and the *problem-solving approach*—have been widely debated. Although the research evidence is far from conclusive on the issue, there appears to be general agreement that the drill approach is appropriate in those cases where the sole (or principal) objective is to teach the performer how to perform a specific skill and that the problem-solving approach is appropriate when much broader educational objectives are also involved. The drill approach is used extensively in athletics coaching where, despite occasional protestations to the contrary, the overwhelming concern of most coaches is to improve the skill levels of those with whom they work and, thus, to increase their chances of winning. The problem-solving approach is more frequently used in physical education classes than in athletics coaching, and this, of course, is consistent with the broad objectives normally sought in such classes.

Teachers or coaches using the drill approach to correct a specific fault have a wide range of instructions that they may give the performer. They can simply tell the performer *directly* to correct the fault ("Keep your body erect as you take off."), they can ask the performer to do something that will lead *indirectly* to the correction of the fault ("Drive your head upward at takeoff."), they can use some form of imagery to convey the same idea ("Imagine you are a giraffe reaching for a high branch."), and so on. Some of these instructions are more likely to be effective than others. Although research provides little guidance as to the likely effectiveness of different types of instruction, practical experience suggests that the direct form of instruction should be tried first. Only when this proves unsuccessful should indirect forms be used. In this way, the performer is likely to get a more accurate appreciation of the faults in the performance than would be the case if they were not confronted directly from the outset.

Care must be taken to ensure that the volume and nature of the information conveyed to the performer does not exceed his or her ability to comprehend, retain, and use such information. Because they are naturally concerned that the performer understands what should be done to correct the observed faults, inexperienced teachers and coaches often give so many instructions that the performer becomes more confused than enlightened. This natural tendency to say more than is necessary, to ask for more than the performer can possibly produce, must be strongly resisted.

Unless the performer has had extensive experience with the skill in question or is exceptionally gifted physically, it is unlikely that he or she will be able to direct his or her attention to more than one aspect of the performance at a time. The high jumper in Fig. 168 is unlikely, for example, to be able to attempt a correction in his takeoff position and in his actions over the bar during the same jump. The teacher or coach is well advised, therefore, to confine instructions to one point at a time. To do otherwise is to invite failure or, at least, to limit success.

Where it is possible for the performer to comply, he or she should be instructed to correct faults while executing the complete skill at the speed that would be used in a competition. If the movement involved is too complex to permit this, it should be broken down into the least number of parts that will permit the necessary corrections. If the movements occur too quickly to allow the performer to make the required corrections, they, or the relevant parts of them, should be performed at less than normal speed while corrections are being made. Finally, where the movements involved *must* be broken down or slowed down so that faults may be corrected, the performer should be instructed to return to practicing the complete, full-speed movement as soon as possible.

## EXERCISES

1. What is the result—the measure of the outcome of the performance—for (a) a standing long jump? (b) a handspring vault in gymnastics? (c) a block in football? (d) a giant slalom in skiing?
2. The result in a triple jump is the distance from the front edge of the takeoff board to the nearest mark that the performer makes in the landing pit. How might this distance be divided for the purposes of qualitative analysis?
3. Using the initial steps described on p. 268 and the general direction suggested on p. 271, develop a model to be used in the qualitative analysis of a running long jump. What changes would be required to make the model suitable for use in the analysis of a standing long jump?
4. Develop a model to be used in the qualitative analysis of a discus throw.
5. Develop a model that might be used in the qualitative analysis of any trampoline skill. [*Hint*: There is a very close similarity between diving and trampolining.]

## RECOMMENDED READINGS

Barret, Kate R., "Observation for Teaching and Coaching," *Journal of Physical Education and Recreation*, 50 (January 1979), 23–25.

Chapman, A. E., "Mechanical Analysis as an Aid to Coaching," *Track Technique*, 73 (Fall 1980), 2317–2318.

Drowatzky, John N., *Motor Learning: Principles and Practices*. Minneapolis: Burgess, 1975.

Magill, Richard A., *Motor Learning: Concepts and Applications*. Dubuque, Iowa: Wm. C. Brown, 1980.

Nelson, George, *How to See*. Boston: Little, Brown, 1977.

# Chapter 16

# Examples in the Use of Qualitative Analysis

The performance of a high jumper was analyzed to illustrate how the method presented in the last chapter might be applied in a specific case. To further illustrate the application of the method, and to increase the reader's familiarity with the procedural steps involved, three additional analyses are presented in this chapter.

## RUNNING

The photo sequence of Fig. 169 shows a runner competing in a 100 m event. The sequence shows four strides taken in the second half of the race when the runner was moving at, or close to, top speed. (The film was exposed at a rate of approximately 24 frames/s.)

### Development of Model (Fig. 170)

In running, the result is the *time*[1] taken by the runner to cover the specified distance. This time is completely determined by two factors—the *distance* involved and the *average speed* of the performer over that distance (p. 273).

---

[1] The factors to be included in the model are shown in italics. These factors should be included in the order in which they are identified and linked to those other factors with which they have causal relationships.

(a)

(b)

(e)

(f)

**Fig. 169.** *Sequence for analysis of sprint-running.*

(c)

(d)

(g)

(h)

(i)

(j)

(m)

(n)

(k)

(l)

(o)

(p)

(q)

(r)

(u)

(v)

(s)

(t)

(w)

(x)

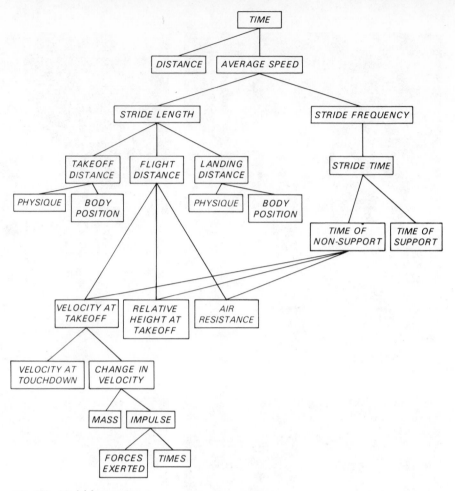

**Fig. 170.** *Model for running.*

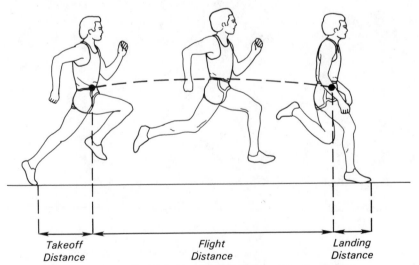

Takeoff
Distance

Flight
Distance

Landing
Distance

**Fig. 171.** *The stride length may be divided into takeoff, flight, and landing distances.*

The distance that the runner travels is generally subject to only minor variation. It is the specified distance of the race plus any additional distance that the runner may cover by deviating from the measured course. Under such circumstances, there is little to be gained by developing the model further in this direction.

The average speed of a runner is equal to the product of the runner's average *stride length* and average *stride frequency.*[2] That is,

Average speed = Average stride length × average stride frequency

Thus, for example, if the average length of the stride is 2.0 m and the runner takes an average of 4.0 strides per second,

$$\text{Average speed} = 2.0 \text{ m} \times 4.0/\text{s}$$

$$= 8.0 \text{ m/s}$$

The stride length, like the distance of a long jump (p. 268), may be considered as the sum of three separate distances (Fig. 171):

1. the horizontal distance between the runner's center of gravity and the toe of the "takeoff" foot at the instant of takeoff (the *takeoff distance*),
2. the horizontal distance that the center of gravity travels while the runner is in the air (the *flight distance*), and
3. the horizontal distance between the center of gravity and the toe of the leading foot at the instant the runner lands (the *landing distance*).

The takeoff distance for a given runner is determined completely by his or her *physique* and by the *body position* at the instant of takeoff. The landing distance is similarly governed by *physique* and by *body position* at the instant of landing.

The flight distance of a runner, like the range of any projectile, is determined by the speed and angle at which the runner is projected into the air, by the height at which the runner is projected relative to the height at which he or she lands, and by the air resistance encountered in flight. For the sake of brevity, these factors are expressed as the *velocity at takeoff*, the *relative height at takeoff*, and *air resistance*.

The velocity of the runner at takeoff is equal to the sum of the *velocity at touchdown* (or landing) and the *change in velocity* experienced during the support phase:

Velocity at takeoff = Velocity at touchdown + change in velocity

---

[2] Although the word "average" should strictly be included in the model in each of the above cases and in the case of all the factors that contribute to these two, it is omitted here in the interests of brevity.

According to the impulse–momentum relationship (p. 165), the change in velocity depends on the *mass* of the runner and on the *impulse* the runner exerts on the ground—and that the ground exerts on the runner in reaction. The latter, in turn, depends on the *forces exerted* on the ground via the supporting foot and the *times* for which they act.

The average stride frequency of a runner is determined by the average length of time taken to complete one stride—that is, by the average *stride time*. If a half a second is taken to complete one stride, the average stride frequency is 2/s; if a third of a second is taken, it is 3/s; if a fourth of a second is taken, it is 4/s; and so on. The stride frequency is, in short, the inverse of the stride time.

The stride time may be considered as the sum of

1. the time during which the athlete is in contact with the ground (the *time of support*) and
2. the time during which the athlete is in the air (the *time of nonsupport*).

The support phase in running is generally of very short duration—approximately .07–.09 s for a top-class sprinter running at full speed and .15–.20 s for a top-class middle-distance runner. Differences in the time of support due to differences in running action are, naturally, even smaller—so small, in fact, as to be virtually imperceptible to the human eye. Since the time of support is thus of little value for purposes of qualitative analysis, there is little to be gained by developing the model further at this point.

The time of nonsupport, like the time of flight for any projectile, is determined by the velocity at takeoff, the relative height at takeoff, and the air resistance encountered in flight. Since these factors have already been included in the model, all that remains to be done is to indicate their causal relationship with the time of nonsupport. This is done by simply linking the respective boxes.

With all the identified factors included, the complete model for running is as shown in Fig. 170.

Several general observations should be made with respect to this model:

1. The upper levels of the models for many other cyclic skills—that is, skills in which the performer executes the same basic sequence of movements over and over again—have the same general form as shown in the running model. For example, the top five boxes in the running model could be used without alteration in models for race walking, speed skating, and cross-country skiing. With the word "stride" replaced by the word "stroke," the same five boxes could also be used in models for canoeing, rowing, and swimming.
2. Reference has already been made to the similarity between the manner in which the stride length was divided into three lesser distances and that in which the distance of a long jump might be divided (p. 268). This similarity goes much farther. Indeed, with only a few alterations and additions, that

part of the running model that shows the stride length and the factors that influence it could serve as a model for the long jump.

3. There are other instances in which similar groupings of factors appear in more than one model. For example, the part of the running model that shows the factors influencing the velocity of takeoff is very similar to a part of the high jump model shown in Fig. 166 (p. 278).

4. The term "velocity at takeoff" is used in preference to the terms "speed at takeoff" and "angle of takeoff," for the sake of brevity. It must be remembered, therefore, that both the magnitude (speed) and the direction (angle) of the velocity are to be considered whenever the term "velocity" appears in a model.

5. Three factors (the velocity at takeoff, the relative height at takeoff, and the air resistance) influence both the stride length and the stride frequency. This explains, in part, why it is difficult to increase the stride length without decreasing the stride frequency, and vice versa. Consider, for example, the case of a runner who increases the angle of takeoff and keeps the speed at takeoff and relative height at takeoff unaltered. Such a change would almost certainly result in an increase in stride length (due to an increase in flight distance) and a decrease in stride frequency (due to an increase in the time of nonsupport).

### Identification of Faults

The time taken to run a given distance depends on that distance and on the average speed of the runner (Fig. 170). Assuming that the runner of Fig. 169 ran the shortest possible distance from start to finish, the only way he can improve his time in future races is to bring about an improvement in his average speed. (In the practical situation, there is no need to make any assumptions about the distance covered by the runner. One can simply take up a position facing along the lane and observe the path directly.)

The average speed of a runner is equal to the product of the average stride length and average stride frequency (Fig. 170). For purposes of analysis, the runner's stride length is considered to be the sum of three lesser distances—the takeoff, flight, and landing distances, respectively (Fig. 170).

**Takeoff distance.** The first of these is determined completely by the runner's physique and by the position at the instant of takeoff (Fig. 170). Of these factors, only the latter can conceivably be improved under normal conditions. In the present case, the position of the runner at the instant of takeoff from his left foot is recorded in (a) and, again, in (m). His position at the instant of takeoff from his right foot must be deduced from his positions immediately before and after that instant—that is, from the positions shown in (f) and (g) and, again, in (r) and (s).

A close examination of these photographs reveals several ways in which the runner's position at takeoff might be altered to increase his takeoff distance. He might swing the thigh of his recovery leg more forward and upward (and, thus, move the center of gravity of the leg farther forward) than is shown in the photographs. Although one cannot be sure, in the absence of photographs taken at exactly the right moment, it would seem that the position of the left thigh in particular is capable of improvement in this respect—see (f) and (g) and (r) and (s). The runner might also modify the position of his head, neck, and trunk so that they are inclined slightly forward rather than being held erect or, as it seems in some photographs—(r) and (s), for example—inclined slightly backward. Finally, he might be able to more completely extend the knee of his driving leg. It is often difficult to tell whether a runner has completely extended the knee of his driving leg. This difficulty arises because of differences between runners in the range of motion possible at the knee—complete extension for one may be less than complete for another—and because the contours of the hamstrings and triceps surae often give the impression that the knee is incompletely extended even though it has reached the anatomical limit of its motion in that direction.

Two of the faults identified in the runner's position at takeoff may well be closely related to one another. The height to which the knee is lifted depends on a number of factors. It depends, for example, on the hip mobility of the runner. If the runner has a limited range of flexion at the hip (presumably because of "short" hamstrings), the height to which the knee can be lifted is similarly limited. The lack of knee lift evident in the action of the runner in Fig. 169 might thus be due to a lack of hip mobility. Although one would be most unwise to assess an athlete's hip mobility solely on the basis of the way in which the athlete runs, the erect (or back arching) posture of the runner's trunk (a logical compensation for an inability to lift the knees by other means) provides further encouragement for the view that the lack of knee lift evident in the present case is due to a lack of hip mobility.

***Flight distance.***    The flight distance, which accounts for approximately 50–65 percent of the stride length at top speed,[3] depends on the velocity of the runner at takeoff, on the relative height of takeoff, and on the air resistance encountered in flight (Fig. 170). The runner's velocity at takeoff—without question the most important of these three factors—depends on the velocity at touchdown and on the change in velocity during the support phase (Fig. 170). The velocity at touchdown is completely determined by what took place in the preceding stride and cannot, thus, be altered directly. The change in velocity depends, ultimately, on the forces that the runner exerts against the ground (or, more precisely, the

---

[3]Anne E. Atwater, "Kinematic Analysis of Striding During the Sprint Start and Mid-Race Sprint," Paper presented at 26th Annual Meeting of American College of Sports Medicine, Honolulu, Hawaii, May 1979.

forces that the ground, in reaction, exerts against the runner) and on the times for which they act (Fig. 170).

The horizontal forces exerted on the supporting foot of a runner act first to decrease and then to increase the runner's forward velocity (p. 164). There are, therefore, two basic ways in which a runner may improve the flight distance by changing the performance during the support phase. The runner can reduce the amount by which the horizontal velocity is decreased during the first part of the support phase, or increase the amount by which it is increased during the latter part.

The horizontal forces exerted against the foot as it comes into contact with the ground depend on its horizontal velocity *relative to the ground.* If it is moving forward, as is almost invariably the case, it will evoke horizontal forces that oppose not only the forward motion of the foot but that of the body as a whole. The magnitude of these braking (or retarding) forces depends primarily on how fast the foot is moving forward at the moment it makes contact with the ground. If it is moving forward relatively fast, the body's forward motion will be opposed by relatively large horizontal forces, and vice versa. Thus, if a runner wishes to improve his or her performance by reducing the amount by which the forward velocity is decreased during the first part of the support phase, the runner must concentrate on reducing the forward velocity of the leading foot at the instant of touchdown. This is usually done by simply allowing the foot to drop once it has reached a position directly below the knee, or in some cases, by actively swinging it downward and backward *relative to the athlete's center of gravity.* In either case, the foot lands on, or just slightly forward of, the athlete's line of gravity. (Deshon and Nelson[4] found a significant positive relationship between the angle the leg made with the ground at the instant of touchdown and the speed of running. On the basis of this finding, they concluded that "efficient running is characterized by . . . placement of the foot as closely as possible beneath the center of gravity of the runner.")

With these general comments as background (or prerequisite knowledge), it is appropriate to return to the case of the runner in Fig. 169. An examination of the photographs showing the runner immediately prior to and at the instant of touchdown reveals that the runner does not allow his foot to simply drop to the ground once it reaches a point directly below the knee but, instead, swings it forward in an apparent effort to increase the length of his stride—(b) and (c), (h) through (j), (n) and (o), and (s) through (u). This leads—as is clearly evident in (j) and strongly suggested by (c) and (d), (o) and (p), and (u) and (v)—to his foot landing well forward of his line of gravity. It also leads, almost certainly, to a braking, horizontal reaction from the ground that is larger than necessary. The runner could thus reduce the extent to which his horizontal velocity is decreased

[4]Deane E. Deshon and Richard C. Nelson, "A Cinematographical Analysis of Sprint Running," *Research Quarterly*, 35 (December 1964), 453–454.

during the first part of the support phase by altering the manner in which he brings his foot into contact with the ground.

The horizontal forces that act, during the latter part of the support phase, to increase the runner's horizontal velocity are the result of the vigorous extension of the driving leg and the accompanying actions of arms and recovery leg. There are several ways in which the runner of Fig. 169 might improve his performance in this respect. If he is not completely extending the knee of his driving leg (and this possibility has already been discussed), he might work to obtain the additional horizontal force from this source.

The position of the runner's foot, relative to the direction in which he or she is running, has some bearing on the forces that the runner exerts against the ground and that the ground, in reaction, exerts on the runner. The runner in Fig. 169 places his feet on the ground with the toes pointing forward and to the side rather than directly forward. He is, in other words, "toeing out." This is most clearly apparent in (d) through (f) and (p) through (r), where the right foot is in contact with the ground. (The position from which the photographs of Fig. 169 were taken is ill suited for the purpose of determining whether a runner is toeing out. The fact that the right foot can be seen to be toeing out and not the left does not necessarily mean that the right is more at fault than the left. It may simply mean that the position is less ill suited for observing the right foot than it is for observing the left!)

Finally, if the runner were to swing his arms more directly forward and backward and less across the line of his body than is shown, for example, in (b) through (d), (n) and (o), and (v) through (x), the resulting forces transmitted to the ground via the driving leg might well be more favorable.

There is little scope for improving the runner's stride length by changing the relative height of his takeoff or the air resistance that acts on him in flight. The first of these would be slightly and positively affected by the changes in the runner's position at takeoff described earlier with reference to the takeoff distance. The second is, for practical purposes, beyond the runner's control— unless, of course, the athlete is running on a very windy day.

*Landing distance.*    The seemingly obvious way to increase stride length is to increase the landing distance by reaching well forward with the foot immediately prior to touchdown. This leads, however, to a pronounced braking effect and a concomitant decrease in speed (p. 164). The optimum landing distance is thus very close to zero. The ways in which the runner of Fig. 169 might modify his position at touchdown to obtain a nearly zero landing distance were discussed earlier with reference to the flight distance.

*Stride frequency.*    The stride frequency of a runner is very difficult to evaluate qualitatively under any circumstances and well-nigh impossible when the only available record of his performance is a series of still photographs. Any attempt to evaluate the stride frequency of the runner in Fig. 169 by qualitative means—

that is, by simple visual inspection—is thus almost certain to be unproductive.

The preceding analysis has yielded a number of possible avenues for improvement in the performance of the runner in Fig. 169. These, in summary, are as follows:

Increase height to which knee of recovery leg is lifted.
Assume posture with head, neck, and trunk inclined slightly forward.
Fully extend knee of driving leg.
Eliminate forward and downward swing of lower leg and foot immediately prior to touchdown.
Eliminate "toeing out" of feet.
Swing arms more forward and backward than across body.

## Evaluation of Faults

The first step in the process of establishing an order of priority is to eliminate from further consideration those faults that are likely to be effects produced by other faults. With this in mind, it would be appropriate to determine whether the runner has an acceptable range of motion at the hip joints and, in particular, whether he can flex his hips to a greater extent than is evident in Fig. 169. Suppose, for the purposes of this analysis, that one of the well-known "toe-touching" tests[5] were used for this purpose and the runner found to be below average on the test. This would suggest that his lack of knee lift and his less than ideal posture of head, neck, and trunk are probably effects rather than faults in themselves.

The actions of a runner's arms serve to balance the actions of the legs and, if performed correctly, to increase the magnitude of the forces evoked from the ground during the driving phase. Because of their role in balancing the leg action, the pronounced cross-body action of the arms apparent in Fig. 169 is likely to be due to the observed faults in the runner's leg action—that is, to his toeing out and/or his reaching forward with his foot prior to touchdown.

With these three faults eliminated from further consideration at this point, the next task is to place the others in an order of priority. For this purpose, suppose that the runner is preparing for an important meet in five to six weeks, that he has been running in the manner shown in Fig. 169 for a long time, and that his ability to learn new movements or modify old ones is no more than average.

Of the remaining faults, the forward and downward swing of the lower leg and foot appeals as the one whose correction is likely to yield the greatest improvement in the time available. This correction could conceivably be brought about simply by telling the runner not to reach forward with his foot or, more

[5]Barry L. Johnson and Jack K. Nelson, *Practical Measurements for Evaluation in Physical Education* (Minneapolis: Burgess, 1979), pp. 78–79.

positively, by asking him to "sweep" his foot downward and backward as it approaches the ground. (It is important, of course, that this "sweeping" action not be exaggerated to the extent that in correcting one fault it creates others.)

As indicated earlier, the apparently incomplete extension of the knee joint of the driving leg may or may not be a fault depending on whether the runner is capable of obtaining a more complete extension. However, even if he is, the effect that correction of this fault would have on his takeoff and flight distances would almost certainly be small. This, coupled with the likelihood that eliminating the "toeing out" of the feet will produce larger improvements in the runner's performance in the time available, leads to the conclusion that fully extending the knee of the driving leg should be placed third in the order of priority.

With these decisions made, the order of priority established for the three faults, not judged to be primarily the effects of other faults, is thus

1. Eliminate forward and downward swing of lower leg and foot immediately prior to touchdown.
2. Eliminate "toeing out" of feet.
3. Fully extend knee of driving leg.

It might be noted that no mention was made of the runner's hip mobility during the process of establishing an order of priority among observed faults in his running action. This is because there is a fundamental difference between training directed toward changing the physical attributes of a performer and that directed toward changing the way in which he or she performs a motor skill. Training directed toward changing an athlete's physical attributes is usually undertaken with the aid of special exercises and drills performed in a separate session or in a separate part of the training session from that directed toward changing the way in which a specific motor skill is performed. Thus, a coach working with the runner in Fig. 169 might prescribe a series of exercises designed to improve his hip mobility and have him perform these at set times outside normal training sessions and/or at the beginning of each training session. The athlete could thus work to improve his hip mobility and to improve faults in his running action without the one interfering with the other. There is, therefore, no need to consider his lack of hip mobility—or, in other cases, corresponding limitations in the physical makeup of the performer—when establishing an order of priority.

## DIVING

The photo sequence of Fig. 172 shows a diver performing a $2\frac{1}{2}$ front somersault dive in tuck position from the 3 m board. (The film was exposed at a rate of approximately 24 frames/s. The number above each photograph refers to the corresponding frame of the film from which the photographs were taken.)

As in other skills that are evaluated by a panel of judges, the result in diving is the number of *points awarded* for the performance. This number is determined by multiplying the *judges' score* (specifically, the sum of the middle three scores awarded by the judges) and the *degree of difficulty* for the dive involved. Thus, for example, if the five judges awarded 5.8, 6.1, 6.1, 6.2, and 6.5 points for a dive with a degree of difficulty of 2.0, the diver would receive

$$(6.1 + 6.1 + 6.2) \times 2.0 = 36.8 \text{ pts}$$

Disregarding deliberate bias, a judge's score is based on his or her perception of three characteristics of the dive. These, expressed where possible in mechanical terms rather than in the more general terms normally used by judges, are

1. the linear motion of the diver, reflected in the path followed by the diver's center of gravity,
2. the angular motion of the diver, reflected in the angular distances through which the diver rotates about the transverse (or somersault) and longitudinal (or twist) axes, and
3. the extent to which the diver's performance has the aesthetic qualities expected.

These characteristics, or factors, are simplified and abbreviated for inclusion in the model as *trajectory of center of gravity, angular distance,* and *form.*

The path followed by a diver's center of gravity from takeoff to entry is governed by the diver's *velocity at takeoff,* by the diver's *relative height at takeoff*—that is, the height of the center of gravity at takeoff relative to its height at the instant of contact with the water—and by the *air resistance* encountered in flight.

As in the case of a runner, the velocity of takeoff of a diver may be considered as the sum of the *velocity at touchdown* following the preceding step and the *change in velocity* that occurs during the takeoff. This latter is determined by the *impulse* that the diver exerts on the board (and that the board, in reaction, exerts on the diver) during the takeoff and by the *mass* of the diver (impulse-momentum relationship). The impulse exerted on the diver is, by definition, determined by the *forces exerted* and the *times* for which these forces act.

The velocity with which the diver lands on the board at the conclusion of the hurdle step is governed by the *velocity at takeoff* into the hurdle step, *the relative height at takeoff*—that is, the height of the center of gravity at takeoff for the hurdle step relative to its height at touchdown—and the *air resistance* to which the diver is subjected during the flight. (To distinguish between the instant of takeoff for the dive and that for the hurdle step, the latter is indicated in the model with an "H" set in parentheses.)

The angular distance through which a diver's body rotates while airborne depends on the diver's *angular momentum at takeoff;* on how the body position

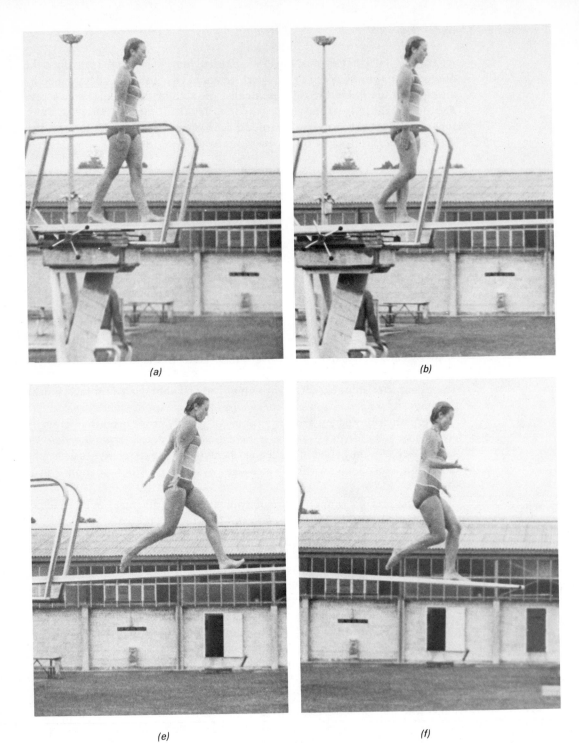

(a)  (b)

(e)  (f)

**Fig. 172.** *A good position at the instant of takeoff.*

(c)

(d)

(g)

(h)

*(i)*

*(j)*

*(m)*

*(n)*

(k)

(l)

(o)

(p)

(q)

(r)

(u)

(v)

(s)

(t)

(w)

(x)

(y)

(z)

(c')

(d')

(a')                                                    (b')

is adjusted to increase or decrease angular velocity once the diver has left the board (*moment of inertia during flight*); and on the length of time the diver has in the air (*time of flight*). If all else is equal, the greater the angular momentum at takeoff *or* the smaller the moment of inertia during the flight *or* the longer the time of flight, the greater will be the angular distance through which the diver rotates.

The angular momentum at takeoff can be considered as the sum of the *angular momentum at touchdown* following the hurdle step and the *change in angular momentum* that occurs during the takeoff.

The impulse–momentum relationship has been used in all three models developed to date as a basis for identifying the factors that produce a change in velocity. This relationship applies only to linear motion and so cannot be used in a similar fashion to identify the factors that produce a change in the angular motion of a body. However, a corresponding relationship—not discussed previously in this text—can be used for this purpose. This is the angular impulse–angular momentum relationship. It states that the sum of all the torques that act on a body multiplied by the times for which they act—that is, the angular impulse—is equal to the change in angular momentum that results. In other words, the change in angular momentum is determined completely by the *angular impulse*. This latter quantity is, in turn, determined completely by the *torques exerted* on the body and the *times* for which they act. These torques may be produced in two different ways—by a couple (*couples exerted*) or by a force whose line of action does not pass through the axis of rotation. (If the "forces exerted" factor entered previously in the model is considered to refer not only to the magnitudes and directions of the forces involved but also to their lines of action, there is no need to include an additional factor in the model at this point. All that needs to be done is to indicate the causal link between the appropriate factors.)

The time of flight is determined by the velocity at takeoff, the relative height at takeoff, and the air resistance encountered in flight. The causal relationships among these factors (already included in the model) and the time of flight are indicated by linking the respective boxes.

A diver's form—that is, the aesthetic characteristics of the diver's performance—depends on a number of factors. These include the positions of the various body parts during the dive and the manner in which these parts move from one position to another. Although these, and the other factors involved, are important in determining the overall success of a dive, the effort required to identify all of them and to define them precisely is much greater than any benefit that might be gained from doing so. For this reason, the model is not developed further in this direction.

With all of these identified factors included, the complete model for diving is as shown in Fig. 173.

## Identification of Faults

The score awarded by a judge for a given dive is based upon the trajectory of the diver's center of gravity, the angular distance through which the diver rotated, and the aesthetic characteristics of the dive (Fig. 173).

**Trajectory.**   To the extent that safety and the other requirements of a dive permit, the ideal trajectory is one in which the peak height attained by the diver's center of gravity is as high as possible and the horizontal distance that it travels from takeoff to entry is as short as possible.

The performance of the diver shown in Fig. 172 could be improved in both these respects. At the peak of her flight—which occurred at or very close to the instant recorded in (t)—her center of gravity was approximately 0.85 m higher than at the instant of takeoff. This is markedly less than the 1.14–1.50 m commonly attained by highly skilled divers.[6] (Without sophisticated methods to

---

[6]Dale P. Mood, "A Mechanical Analysis of Six Twisting Dives," M.A. thesis (University of Iowa, 1968), p. 163.

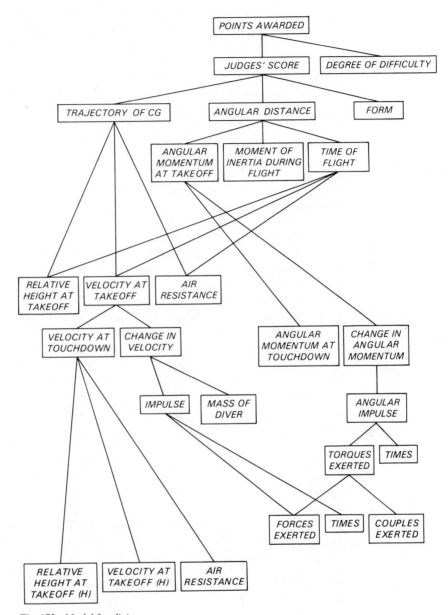

**Fig. 173.** *Model for diving.*

locate the center of gravity of the diver and an appropriate linear scale to convert measured to real-life distances, it is impossible to obtain anything more than a very rough estimate of the distances traveled by the center of gravity from one position to another.) The diver's position in (c′), some 1.80–1.85 m clear of the board, implies that the horizontal distance that she traveled from takeoff to entry was somewhat more than necessary to ensure her safe passage past the board. This horizontal distance would undoubtedly have been even greater if the board from which she was diving were perfectly level—as required by the rules—rather than sloping upward toward the takeoff end. A board like the one she was using tends to "throw" the diver upward and backward compared with one that is horizontal.

The path followed by the center of gravity of the diver from takeoff to entry is determined by her velocity at takeoff, by her relative height at takeoff, and by the air resistance she encounters in flight (Fig. 173).

Since the height that a projectile rises above its height at takeoff (or release) is governed primarily by its vertical velocity at takeoff, the lack of height evident in the dive of Fig. 172 implies a corresponding lack of vertical velocity at takeoff. Similarly, the excessive horizontal travel evident in Fig. 173 implies a larger horizontal velocity at takeoff than is desirable.

The position of the diver's body determines whether the height of her center of gravity at takeoff (and, thus, her relative height at takeoff) is capable of improvement. If her body is not erect, and her arms extended overhead at this instant—and they most certainly are not in Fig. 172(p)—her relative height at takeoff is, at least theoretically, capable of improvement.

The air resistance the diver encounters during the flight is effectively beyond her control and, thus, not subject to improvement.

The diver's velocity at takeoff is determined by her velocity at the instant of touchdown following the hurdle step and by the change in velocity that occurs during the takeoff (Fig. 173). The velocity at touchdown is determined by her velocity and relative height at takeoff into the hurdle step and by the air resistance she encounters during the subsequent flight (Fig. 173), the last of which can again be dismissed as a possible source of improvement.

A close examination of Fig. 172 reveals one way in which the diver's velocity at takeoff into the hurdle step might be improved. If she keeps her arms straight and swings them vigorously forward and upward during the takeoff rather than simply lifting them, flexed at the elbows, in an outward and upward motion (e–g), the downward force exerted against the board in reaction will be increased. This, in turn, will result in an increase in the upward force exerted on her by the board and an increase in her vertical velocity at takeoff.

The height of the diver's center of gravity at takeoff into the hurdle step (and, thus, her relative height of takeoff) could be improved marginally if she had her arms fully extended overhead rather than in the position shown in (h). These heights would also be increased slightly if she held her head erect and dorsiflexed her right ankle. Although a completely erect position of the head would be desirable for a number of reasons, the diver's need to keep her eyes focused on the end of the board—an acknowledged prerequisite to a successful takeoff[7]— probably dictates that she keep her head inclined forward slightly throughout the hurdle step. Thus, while some improvement is almost certainly possible, a totally erect position of the head is probably unattainable. Finally, dorsiflexion of the right ankle would be ill advised because the diver needs it plantar-flexed for aesthetic reasons.

The change in the diver's velocity during the takeoff is determined ultimately by the forces that act on her during the takeoff and by the times for which they act (Fig. 173). The forces that act on her are those applied by the board in reaction to the movements (strictly, the accelerations) of her arms, legs, and trunk. There are at least two ways in which the diver of Fig. 172 might favorably alter her actions and, thus, the change in her velocity during the takeoff. She could keep her arms straight and swing them vigorously forward and upward (as already mentioned with respect to the takeoff for the hurdle step), and she could reduce the forward motion of her center of gravity. This latter could be achieved by keeping her head and trunk more nearly erect and, in particular, by reducing the extent of her hip and shoulder flexion during the latter stages of the takeoff. In addition to reducing her forward motion, the assumption of a more erect position would likely result in the forces exerted by the legs being directed more vertically upward than, as in (k) through (p), forward and upward. These various changes in the diver's actions could be expected to produce an increase in her vertical velocity and a decrease in her horizontal velocity at the instant of takeoff.

***Angular distance.*** The position of the diver at entry (c′) is somewhat short of the vertical (or nearly vertical) position desired. This is a clear indication that the angular distance through which she rotated between takeoff and entry—roughly 800°—was less than required for best results.

The angular distance through which the diver rotates in the air is determined by her angular momentum at takeoff, by her moment of inertia relative to the axis of rotation during the flight, and by her time of flight (Fig. 173).

The diver in Fig. 172 could decrease her moment of inertia relative to the transverse axis through her center of gravity (her "somersaulting" axis) and thus increase the angular distance through which she rotates between takeoff and entry. Assuming that she has sufficient strength and flexibility for the purpose, she could achieve this by pulling hard on her lower legs and forcing a more pronounced flexion of the hips and knees than is evident in Fig. 172. She might

[7]Charles Batterman, *The Techniques of Springboard Diving* (Cambridge, Mass.: M.I.T. Press, 1968), pp. 11–12.

also flex at the neck and bring her head into a more tightly tucked position. (Some divers appear to have difficulty timing their "come out," or body extension, prior to entry when they tuck the head in this manner. In these cases, tucking the head is likely to be more of a liability than an asset and should, thus, be avoided.)

The diver's time of flight—and, thus, the angular distance through which she rotates—could be increased by increasing her vertical velocity and/or her relative height at takeoff (Fig. 173). The means by which these two factors might be increased were discussed previously, on p. 330. (As in several earlier instances, the diver has no real control over the last of the three factors that determine her time of flight, namely, the air resistance that opposes her motion.)

The diver came reasonably close to completing the required $2\frac{1}{2}$ somersaults despite a takeoff that limited her time of flight and a position in the air that might have been improved. This suggests that she had ample angular momentum at takeoff.

**Form.**   A knowledge of anatomy and mechanics is little, or no, basis for evaluating those aesthetic aspects of a dive that are separate and distinct from the mechanical characteristics already discussed. Evaluation of the form—the uniquely aesthetic characteristics of the performance—shown in Fig. 172 is thus properly left to those with the needed qualifications.

The preceding analysis has yielded several possible avenues for improvement in the performance of the diver in Fig. 172. These are listed as follows in rough, sequential order and are numbered for ease of reference:

Takeoff to hurdle:
1. Keep arms straight and swing them vigorously forward and upward during the takeoff.
2. Have arms fully extended overhead at instant of takeoff.
3. Hold the head more nearly erect at instant of takeoff.

Takeoff:
4. Increase the vertical velocity at takeoff.
5. Decrease the horizontal velocity at takeoff.
6. Keep arms straight and swing them vigorously forward and upward during the takeoff.
7. Assume takeoff position with body more nearly erect and arms extended overhead.

Flight:
8. Increase peak height attained during the flight.
9. Decrease horizontal distance traveled between takeoff and entry.
10. Increase the angular distance through which the body rotates between takeoff and entry.

11. Decrease the moment of inertia during the flight.
12. Increase the time of flight.

## Evaluation of Faults

Of the twelve faults identified in the preceding analysis, those referred to in 4, 5, 8, 9, 10, and 12 can be eliminated from further consideration because they are not faults in themselves but rather are the effects produced by other faults.

For the purposes of establishing an order of priority among the remaining faults identified, suppose that the diver has no intention of using the $2\frac{1}{2}$ front somersault dive in competition until she has attained a high level of proficiency in its execution. She has, therefore, a reasonably unlimited amount of time in which to bring about the required changes.

Of the six faults still under consideration, the very pronounced forward lean at the instant of takeoff and the lack of range and vigor in arm swing during the takeoff appeal, in that order, as the ones whose correction is most likely to yield the largest improvement in her performance. The corresponding faults in the takeoff to the hurdle step appeal, but in reverse order, as those most likely to yield the next largest improvements. (The magnitude of the change that might be effected in the diver's moment of inertia during the flight appears to be relatively small and thus unlikely to produce any substantial change in the result obtained.)

The most appropriate order of priority for the six possible sources of improvement listed earlier would, thus, appear to be 7, 6, 1, 2 and 3, 11. There is one other point that should be considered, however. Because the takeoff follows immediately after the hurdle step, the manner in which the latter is executed has a direct influence on what the diver does during the takeoff. Thus, the diver's ability to assume a more erect body position and to execute an effective arm swing *may* by impaired by the faulty hurdle step that preceded it. Under such circumstances, it is only logical that attention be focused first on correcting the faults in the hurdle step and then on the related, and apparently more important, faults associated with the takeoff. (It is perhaps worth noting here that, due to the very close similarity between the faults evident in the takeoff to the hurdle step and in the takeoff to the dive itself, efforts directed toward the correction of the one are likely to have a similar effect on the other.)

The revised order of priority would thus read

1. Keep arms straight and swing them vigorously forward and upward during the takeoff to the hurdle step.
2. Have arms fully extended overhead and head more nearly erect at instant of takeoff into the hurdle step (2 and 3, p. 332).
3. Assume takeoff position with body more nearly erect and arms extended overhead.

4. Keep arms straight and swing them vigorously forward and upward during the takeoff.
5. Decrease the moment of inertia during the flight.

## SERVING IN TENNIS

The photosequence of Fig. 174 shows a player executing a first service in a tennis match. (The film was exposed at a rate of approximately 24 frames/s. The number above each photograph refers to the corresponding frame of the film from which these photographs were taken.)

### *Development of Model (Fig. 175)*

As in most other sports where a serve is used to begin play—badminton, handball, racquetball, squash, and table tennis, for example—the performance of a player serving in tennis is judged in terms of the *advantage* gained as a result of the player's efforts.

Before any advantage can be gained, the server must first satisfy the requirements for a legal serve—that is, he must throw the ball into the air, hit it with the racket, and cause it to pass over the net and into the service court. The server must, in short, cause the ball to be displaced from his hand to the service court, in accord with the rules—*displacement of ball (hand to court)*.

Assuming that the player has qualified to gain some advantage by executing a legal service, the advantage he actually gains depends on the path followed by the ball after it has bounced in the service court (*trajectory of ball after bounce*). If the trajectory of the ball is such that the receiver has ample time to assume an appropriate position and to execute the return stroke, the server is likely to gain little, if any, advantage. Conversely, if the trajectory is such that the receiver is unable to assume an appropriate position or is forced to rush the return shot, the server may gain sufficient advantage to win the point.

The displacement of the ball from hand to service court may be considered as the sum of (1) the displacement it experiences from the moment it is released from the hand until it is struck by the racket and (2) the displacement it experiences from the moment it leaves the racket until it strikes the service court. That is,

$$\frac{\text{Displacement of ball}}{\text{(hand to court)}} = \frac{\text{Displacement of ball}}{\text{(hand to racket)}} + \frac{\text{Displacement of ball}}{\text{(racket to court)}}$$

(The displacement experienced by the ball during the approximately .004 s it is in contact with the racket is ignored.)

The displacement of the ball from the moment it is released from the hand until it is struck by the racket is determined by its *velocity at release*, by its *height at release*, and by the *air resistance* that opposes its motion. The relative height at release is equal to the difference between the height of the ball at release from the hand and at the instant of contact with the racket (*height of ball above ground at release* and *height of ball above ground at contact*).

The displacement that the ball undergoes between the time it leaves the racket and the time it lands in the service court is similarly governed by its *velocity at release*, by its *height of release*, and by the *air resistance* that acts upon it during its flight.

The velocity of the ball at release—that is, immediately following the elastic impact between ball and racket—is governed by the *velocity of the ball at contact*, the *velocity of the racket at contact*, the *mass of the ball*, the *mass of the racket*, and the *coefficient of restitution* for the ball–racket impact.

Because it is simply another characteristic of the same projectile motion, the velocity of the ball at contact is governed by the same factors that govern the horizontal and vertical displacements of the ball from hand to racket.

The velocity of the racket at the instant it makes contact with the ball is determined—according to the impulse–momentum relationship—by three factors. These are the forces that act on the racket as the player performs the various serving movements that precede contact (*forces on racket*), by the *times* for which these forces act, and by the *mass of the racket*.

The path followed by the ball after it bounces in the service court depends on where the ball landed (*location of ball at contact*), on its *velocity at release* following impact with the court, and on the *air resistance* exerted upon it during its flight. Since the impact between ball and court is an elastic one, the velocity at which the ball leaves the court is governed by the *velocity of (the) ball at contact*, the *velocity of (the) court at contact*, the *mass of (the) ball*, the *mass of (the) court*, and the *coefficient of restitution* of ball and court.

Finally, because they are all characteristics of the same projectile motion, the point at which the ball makes contact with the court and its velocity at that time are governed by the same factors that determine the displacement of the ball from racket to court.

The various factors that determine the result obtained when serving in tennis are summarized in the model of Fig. 175.

This model is without question the most complicated of all the models developed in this chapter. This complexity stems, in part, from the three-phase nature of the skill involved and, in part, from the very large number of interacting factors that contribute to the result obtained. The model is, however, neither as complicated as it might be nor as complicated as it might appear.

The alert reader will have noted that, except for its indirect influence on the air resistance encountered, no account has been taken of the effects produced by the spin imparted to the ball. The addition of those factors necessary to fully account for the effects of spin, while a simple enough task in itself, would greatly add to the complexity of the model.

*(a)*

*(b)*

*(e)*

*(f)*

**Fig. 174.** *Sequence for analysis of a tennis serve.*

(c)

(d)

(g)

(h)

(i)

(j)

(m)

(n)

(k)

(l)

(o)

(p)

(q)                                                                  (r)

There is a rough parallel between the sequence of movements that comprise the total skill and the development of the model that should be of assistance in understanding and using the model. The factors that govern the initial toss of the ball and swing of the racket appear to the left and bottom of the model. Those that govern the contact between ball and racket and the flight of the ball over the net and into the service court appear in the middle of the model. Finally, those that govern the ball's trajectory after it has bounced in the service court are shown in the upper right of the model. These rough subdivisions are shown in Fig. 175.

### Identification of Faults

In evaluating the performance of a motor skill, it is desirable to have some knowledge of the result obtained. In the present case, the ball left the racket with good speed, considering the age and experience of the player, but at an angle that caused it to land in the right-hand doubles lane well beyond the service court.

The advantage that a player obtains from a service is determined by the displacement that the ball experiences from the moment it leaves the hand until the moment it lands in the service court and by the trajectory that it follows after it has bounced (Fig. 175). The player of Fig. 174 gained no advantage from his service simply because he was unable to complete the displacement of the ball

(s)                                                    (t)

from his hand to the service court and thus satisfy the requirements for a legal serve. Evaluation of his performance must, therefore, be focused primarily on those factors that govern the magnitude and direction of this displacement.

The displacement of the ball from hand to service court can be considered as the vector sum of the displacement from hand to racket and the displacement from racket to service court (Fig. 175). The displacement of the ball from the player's hand to contact with the racket is determined by the velocity and relative height at which the ball is released and by the air resistance it encounters during its flight (Fig. 175). The relative height of release is, by definition, equal to the difference between the height of the ball at release and at contact and is, therefore, completely determined by these two factors (Fig. 175).

The greater the height of contact, the lower the angle and the greater the speed of release at which the player can direct the ball toward the service court. The height of contact should thus be as high as possible consistent with the other requirements for an effective serve.

Although of some practical importance when playing under windy conditions, the air resistance encountered by the ball during its flight is almost completely beyond the control of the player.

With the height of contact and the air resistance thus dictated by other considerations or beyond the player's control, the only remaining factors that influence the displacement of the ball from hand to racket are the velocity and

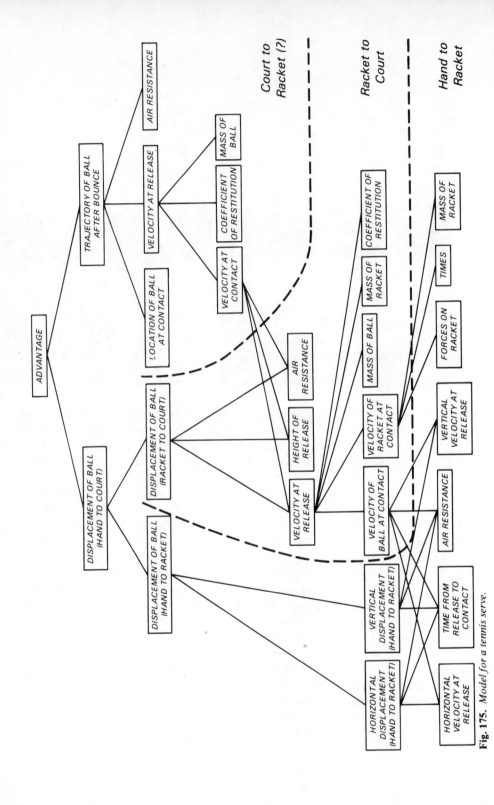

**Fig. 175.** *Model for a tennis serve.*

height of the ball at the instant of release (Fig. 175).

An infinite number of combinations of velocity of release and height of release will result in the ball reaching the desired point of contact with the racket. The ideal combination has certain distinguishing features. First, because the effects of small errors in the angle of release are magnified as the ball moves along its path—the ball simply deviates by a progressively greater distance from its intended trajectory—this path, and the resulting displacement from release to contact, should be as short as possible. This means that the player should have the ball as high as is feasible at the moment of release from the hand. It also means that, unless additional time is needed to execute the various movements that precede contact, the player should release the ball with just enough vertical velocity to carry it upward to the point of contact. Finally, the player should impart to the ball just sufficient horizontal velocity to carry it from its position at release to the ideal position for contact—a position somewhat forward of the baseline and on a line between the player and the service court.

With these general comments as background, an analysis of the initial part of the performance shown in Fig. 174 can be undertaken. An examination of the positions of the ball at release (b), at the peak of its flight (h), and at contact (n) reveal several ways in which this phase of the serve might be improved. Although rather better than is often seen, the height from which the ball was released (b) can be substantially improved (Fig. 176). The peak height to which the ball was tossed (h) was approximately 80–100 cm above its height at contact with the racket. This difference between the peak height and the height of contact is much greater than that recorded for top-class players. For example, of the thirty-five

**Fig. 176.** *The ball should be released from a point above head level.*

world-class professionals analyzed by Plagenhoef,[8] twenty tossed the ball to heights of less than 30 cm and fourteen to heights of 30–60 cm above the point of contact. Only one, Nancy Chaffee, tossed the ball higher and even she did not exceed 75 cm. Tossing the ball so that it attains a peak height greater than the height of contact can be justified if the player requires more time to execute the movements preceding contact than would otherwise be available. Most of the differences in height reported by Plagenhoef can almost certainly be justified on this basis. The same is most unlikely to be true, however, in the case of the player in Fig. 174. For, rather than needing the additional time provided by his very high toss, he appears to be waiting for the ball to return within range before completing his serving motion. While it is difficult to make such a judgment with complete confidence when the only available evidence is a photosequence, the positions shown in (h) through (k) appear to warrant such a conclusion. In summary, it appears that the accuracy with which the player is able to place the ball in the desired position for contact is likely to be improved if he reduces the height of the toss. Finally, from the position of the right arm in (n), it appears that the ball has been thrown upward and to the player's right and is, thus, not on a straight line between him and the service court.

The displacement of the ball from the moment it leaves the racket until the end of its flight is governed by its velocity and relative height at release and by the air resistance encountered in flight (Fig. 175). Although spin services may be used to manipulate the effects of air resistance to advantage, the influence of air resistance is effectively beyond the player's control when a flat serve is used. The player in Fig. 174 thus has only two factors that he can modify to bring about the required changes in the displacement of the ball from racket to court—he can modify the velocity and/or the height at which the ball leaves his racket.

The velocity with which the ball leaves the racket is governed by five factors—the velocity of the ball and the velocity of the racket at impact, the mass of the ball, the mass of the racket, and their mutual coefficient of restitution (Fig. 175).

The velocity of the ball at contact is determined primarily by how high the ball is tossed. If the ball is tossed so that its peak height and the height of contact are the same, its vertical velocity at contact is zero. This, coupled with a small horizontal velocity, means that the resultant velocity of the ball is nearly zero and of little influence in determining the outcome of the impact between racket and ball. If the ball is tossed well above the point of contact, as in the present case, the vertical velocity it has at contact will tend to cause the ball to leave the racket with slightly more downward velocity than would otherwise be the case. The effect, however, is very small and would not, in itself, warrant reducing the height to which the ball was tossed.

The velocity of the racket at contact—unquestionably the most important of

[8] Stanley Plagenhoef, *Fundamentals of Tennis* (Englewood Cliffs, N.J.: Prentice-Hall, 1970), p. 72.

the five factors—is determined by the forces that act on the racket prior to contact, by the times for which these forces act, and by the mass of the racket (Fig. 175).

The forces that act on the racket prior to contact are those exerted by the player via his right hand, the weight of the racket acting downward through its center of gravity, and the air resistance that opposed its motion (Fig. 177).

There are numerous ways in which the player in Fig. 174 might modify the forces that he exerts on the racket and thus obtain a more favorable velocity of the racket at contact.

He might place his feet farther apart and thus facilitate the shifting of his weight from right to left during the serving motion. The player's feet are approximately 25 cm apart in his initial stance, and this is somewhat less than that recommended by most leading authorities. A review conducted by Baker[9] revealed that, of thirty-eight authors who made specific recommendations concerning the distance between the feet, twenty-two said the feet should be 12–18 in. (30–46 cm) apart, seven said they should be shoulder width or approximately 18–24 in. (46–61 cm) apart, seven said they should be "comfortably spread," and two said that the stance should be "wide." The player might also place his feet so that they are turned a little more in the general direction of the service court and thus facilitate still further the shifting of weight from right to left. In this respect, it is of some interest to note how the position of the left foot, parallel with the baseline, appears to inhibit the forward motion of the player's body immediately prior to contact (m)–(n).

The player keeps his legs straight as the ball travels upward to the peak of its flight (a–g), flexes them a little at the knees as it begins to descend (h–k), and keeps them flexed as he brings the racket up to make contact with the ball (l–n). In this process, the contribution that extension of the legs might make to the velocity of the racket is largely lost. The player might thus increase the velocity of the racket at impact by increasing the extent to which he flexes his knees during the serve and by vigorously extending his legs to initiate the final forward and upward motion of the racket.

The direction in which the racket is moving at impact—and thus, of course, the direction in which the ball moves after impact—is influenced by the player's body position at that instant. This position, in turn, is influenced by the forces to which the body has been subjected prior to impact. In the case of the player in Fig. 174 a premature flexion of the hips and simultaneous lowering of the head placed the body in a poor position at impact—a position that almost certainly had an adverse effect on the angle at which the ball left the racket. These inappropriate movements are very likely due to the player taking his eyes off the

[9] John A. W. Baker, "Suggested Method of Executing the Tennis Serve," (M. A. non-thesis paper, University of Iowa, 1976), p. 16.

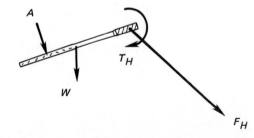

**Fig. 177.** *Forces acting on the racket during the final part of the serving motion.*

ball in an attempt to generate a maximum velocity of the racket at impact. They might, therefore, be corrected by having the player concentrate his attention on the ball right up to the moment of impact.

No discussion of the forces exerted on the racket would be complete without some reference to the unusual position of the racket in (a) and the strange manner in which the first part of the backswing was performed (a–g). Given that these characteristics of the player's service differ greatly from the orthodox, it would be quite understandable if a teacher or coach suggested they be changed. Such action does not seem warranted, however, for these deviations from the orthodox do not appear to have had any adverse effect on the overall purpose of the backswing—that is, to place the player's body and the racket in the ideal position from which to initiate the final hitting action. The excellent position of body and racket in (j) suggests that this purpose has been well served. (No pun intended!)

There is one other aspect of the player's backswing that deserves consideration. Because the initial motion of his racket and hand are clearly not in line with the intended direction of the serve, there is a distinct possibility that he may have some difficulty in getting the racket aligned correctly for contact with the ball. Indeed, it is possible that the out-to-the-side position of the right arm at impact could be due to an error in alignment during the backswing. Such an error would be very difficult to detect from the position taken to obtain the photosequence of Fig. 174. The teacher or coach should, therefore, take a position behind the player and in line with the intended direction of the serve to determine whether something in the alignment of arm, hand, and racket during the backswing was responsible for the fault observed later.

The weight of the racket is constant during any given performance. Thus, the only way in which a player can modify the performance by altering the weight of the racket is to use a different racket—one that is either lighter or heavier than the one used at present. There is some evidence to suggest that such a change might be appropriate in the case of the player in Fig. 174. During the final hitting action, the weight of the racket acts to oppose the player's efforts to swing the racket forward and upward to the point of contact. If the player has sufficient strength and skill to overcome this opposition, he is able to bring the racket around to contact the ball at the appropriate time and place. If he does not, he is likely to make contact with the ball before the racket is correctly positioned or, in extreme cases, to miss it entirely. Although the photograph was taken a split second before the racket made contact with the ball, the position of the player in Fig. 174 (n) suggests that he had difficulty in getting the racket around to make contact with the ball. As already implied, this might well be due to the racket being too heavy for him. There are at least two other indications that the player's racket may be too heavy for him. First, the palm-up position of the hand in (a) and (b) and the apparently close-to-the-side position of the right elbow in (b) through (k) are logical adjustments to make when a racket is too heavy to handle in more orthodox fashion. They may have been made for that reason. Second, the very long follow-through of the racket after impact (t) is exactly what might be expected if the racket were too heavy for the player to control properly. There is, in short, ample evidence to warrant checking to see whether the weight of the racket is consistent with the strength and skill levels of the player.

Of the remaining factors that influence the velocity of the racket at release (Fig. 175), the times for which the applied forces act are determined largely by the magnitudes of these factors—the larger the forces, the shorter the times for which they act—and have thus already been accounted for indirectly; the mass of the racket has been considered indirectly in the discussion of the weight of the racket; and the mass of the ball and the coefficient of restitution are subject to only minor variations that are of no practical significance in the present case.

As already mentioned, the height of release is much lower than it should be for best results. The player might improve his height of release in a number of ways. He might more completely extend the hip, knee, and ankle joints of his left leg. He might keep his trunk more nearly erect and his head up—the latter, incidentally, will also permit him to keep his eyes focused on the ball. Finally, he might extend his right arm and the racket vertically upward to meet the ball.

In summary, there are several ways in which the player of Fig. 174 might improve his performance:

Stance: Increase distance between feet
        Turn feet in direction of service court

Toss: Increase height of release
      Decrease height of toss
      Toss ball to contact point on line between player and service court

Swing: Increase knee flexion and extension
       Focus eyes on ball up to moment of impact
       Extend hip, knee, and ankle joints of left leg prior to contact
       Keep trunk erect and head up at contact
       Extend right arm and racket vertically upward at contact

### Evaluation of Faults

Of the various faults identified in the preceding analysis, three appear to be effects produced by other faults. These are the forward position of the head and trunk at contact (apparently the result of the player's failure to keep his eyes focused on the ball), the out-to-the-side position of the player's right arm and racket at contact (apparently due to an inappropriate toss of the ball), and the lack of extension of the left leg at contact (apparently due to the lack of knee extension prior to contact). These faults can thus be eliminated from further consideration at this point.

Before one can establish an order of priority among the remaining seven faults, it is necessary to know how much time is available in which to bring about the required corrections. For this purpose, assume that the sequence of Fig. 174 was taken during competition in the middle of the tennis season and that the player's goal is to perform well enough to retain his place on his school team for the remainder of the season. Under these circumstances, the faults identified must be placed in an order of priority based on the amount of improvement that might be obtained by working to correct them over a relatively short period. Fortunately, almost all the faults identified can usually be corrected with appropriate concentration and a limited amount of practice.

The greatest improvement in the player's performance is almost certain to come from correcting the fault (or faults) responsible for his failure to execute a legal service—that is, one that passes over the net and lands in the service court. Of the seven faults remaining, the player's failure to keep his eyes focused on the ball right up to the moment of contact appeals as the one most likely to be responsible for the ball's landing outside the service court. Correction of this fault should, therefore, be given top priority.

There are at least two possible explanations for the player taking his eyes off the ball immediately prior to contact. First, he may have been trying to hit the ball much harder than he usually did and, in an effort to bring more forces to bear, flexed his hips, trunk, and neck prior to contact. These actions would lead inevitably to his taking his eyes off the ball. Second, if the racket was too heavy for him, he may have used this same flexion of the hips, trunk, and neck in an attempt to overcome his inability to handle the racket by more orthodox means. Once again, these actions would inevitably lead to his taking his eyes off the ball. Given these two possibilities, a teacher or coach working with this player might instruct him to keep his eyes focused on the ball throughout the serve and not strive for maximum speed. If the problems evident in Fig. 174—and, in particular, the player's failure to get the racket around to contact the ball at the appropriate time and place—were resolved in this manner, the teacher or coach could conclude that they were not due to the racket being too heavy and proceed to the correction of the other faults identified. If the problems persisted despite the player keeping his eyes on the ball, the teacher or coach might conclude that the player's choice of racket was probably at fault and take steps to verify this. These might include, for example, observing the effects produced when the player serves with rackets of different weights.

It is very difficult to estimate how the other faults compare in terms of the improvement that might be expected if they were corrected. Increasing the flexion–extension action of the knees appeals as the source of a significant increase in the velocity of the racket at impact. However, it also seems likely to be one of the more difficult faults to correct. Conversely, although the suggested changes in the initial position of the feet can probably be made with ease, they seem less likely to lead to dramatic improvements in the overall results. Under

(a)                                    (b)

circumstances like these, the remaining faults should be placed in an order of priority consistent with the order in which they occur in the performance of the skill in question—that is, in the present case, in the order in which they were listed previously. With these several decisions made, the order is as follows:

1. Focus eyes on ball up to moment of impact
2. Increase distance between feet
3. Turn feet in direction of service court
4. Increase height of release
5. Decrease height of toss
6. Toss ball to contact point on line between player and service court
7. Increase knee flexion and extension

## CONCLUSION

Proficiency in the use of any effective method of qualitative analysis requires extensive practice. The method described in Chapter 15 and demonstrated in the preceding examples is no exception. To gain the level of proficiency necessary to use the method with confidence in the practical situation, one must repeatedly and thoughtfully practice each and every one of the basic steps involved. One

|        |        |
|--------|--------|
| (c)    | (d)    |

**Fig. 178.** *Sequences for analysis of Fosbury-flop-style high jump. A. The final strides of the approach.*

must become adept at identifying the result, at developing an appropriate model, at perceiving movements accurately, at establishing an order of priority and, finally, at giving appropriate instruction. It is only by such extended, intelligent practice that the usefulness of the method as a practical tool can be realized.

## EXERCISES

1.  Modify the model for a tennis serve shown in Fig. 175 to include the effects produced by spin imparted to the ball by the racket. (The effect that spin has on the ball due to the Magnus effect is already accounted for by the inclusion of air resistance in the model. What is not included is the effect that spin has on the velocity of the ball after impact—pp. 149–150.)
2.  List the faults evident in the performance of the woman high jumper in Fig. 178 (A)–(C) (p. 351–358) and, assuming there are two months left before the championships for which she is training, establish an order of priority among these faults.

(e)

(f)

(i)

(j)

(g)

(h)

(k)

(l)

*(m)*

**Fig. 178B.** *The position at takeoff.*

*(b)*

**Fig. 178C.** *The actions over the bar.*

(a)

(c)

*(d)*

*(f)*

356

(e)

(g)

357

(h)

3. Identify the faults in the performance of the tennis player in Fig. 179 and, assuming that he is practicing for an intramural tournament due to start in ten days, establish an order of priority among these faults.

*(i)*

*(a)*          *(b)*

**Fig. 179.** *Sequence for analysis of a tennis serve.*

*(c)*

*(d)*

*(g)*

*(h)*

*(e)*

*(f)*

*(i)*

*(j)*

(k)

(l)

(o)

(p)

362

(m)

(n)

(q)

(r)

(s)

(t)

(a)

(b)

**Fig. 180.** *Sequence for analysis of a hang-style long jump.*

4. Using the model developed previously (Exercise 3, Chapter 15), list the faults evident in the performance of the long jumper in Fig. 180. Assuming that she has two months left before the championships for which she is training—as indeed she had at the time these photographs were taken—establish an order of priority among these faults.

(c)

(d)

(e)

(f)

(i)

(j)

(g)                                    (h)

(k)                                    (l)

(m)

(n)

(q)

(r)

(o)

(p)

(s)

(t)

(u)

(v)

(y)

(z)

(w).

(x)

(a')

(b')

(a)                                                                    (b)

**Fig. 181.** *Sequence for analysis of a discus throw.*

5. Conduct a qualitative analysis of the performance of the discus thrower shown in Fig. 181—omitting, of course, the final step of giving instruction to the performer. (Assume that the photographs of Fig. 181 were taken during the athlete's final meet of the season, that the throw in question was a good one for him, and that he is now about to start training for the next season.)

(c)

(d)

(e)

(f)

(g)

(h)

(k)

(l)

*(i)*

*(j)*

*(m)*

*(n)*

(o)            (p)

6. The trampolinist in Fig. 182 set out to perform a $1\frac{1}{2}$ turntable—that is, a front drop followed by $1\frac{1}{2}$ turns about his frontal axis to a front drop—but aborted the attempt shortly before landing. Conduct a qualitative analysis of his performance concluding with what you would tell him to do on his next attempt to complete the stunt successfully. (Assume that you have worked with him over an extended period and have no difficulty getting him to understand what you would like him to do. Assume, too, that he is a gifted athlete who can make corrections and learn new movements very quickly.)

(q)

(r)

(a)

(b)

**Fig. 182.** *Sequence for analysis of a 1½ turntable.*

(c)

(d)

(g)

(h)

(e)

(f)

(i)

(j)

(k)

(l)

(o)

(p)

(m)

(n)

(q)

(r)

# Appendices

# Appendix A
# Elementary Anatomy

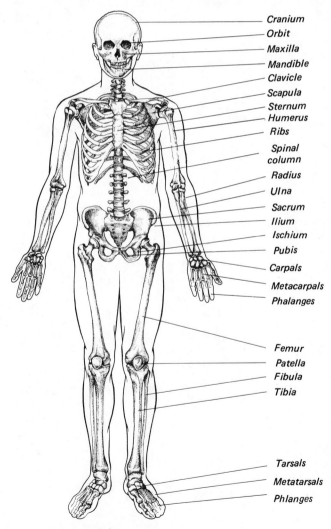

Cranium
Orbit
Maxilla
Mandible
Clavicle
Scapula
Sternum
Humerus
Ribs
Spinal column
Radius
Ulna
Sacrum
Ilium
Ischium
Pubis
Carpals
Metacarpals
Phalanges
Femur
Patella
Fibula
Tibia
Tarsals
Metatarsals
Phlanges

**Fig. 183.** *The human skeleton.*

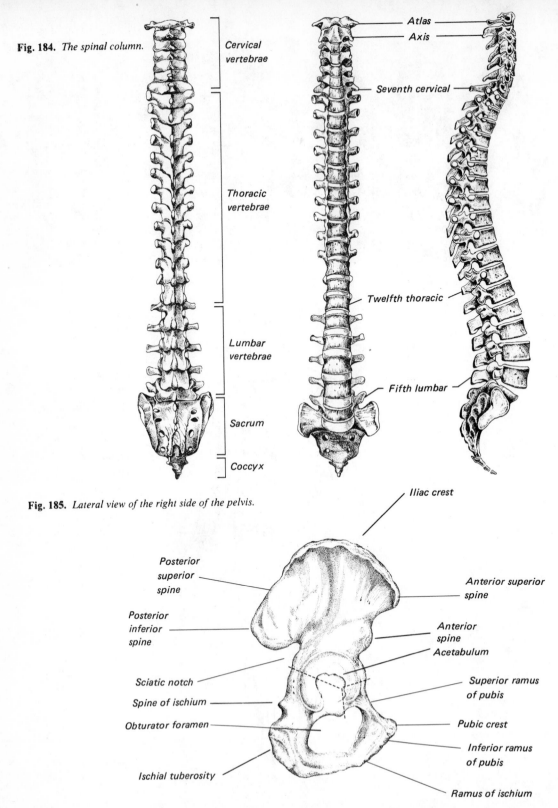

**Fig. 184.** *The spinal column.*

Cervical vertebrae

Thoracic vertebrae

Lumbar vertebrae

Sacrum

Coccyx

Atlas

Axis

Seventh cervical

Twelfth thoracic

Fifth lumbar

**Fig. 185.** *Lateral view of the right side of the pelvis.*

Iliac crest

Posterior superior spine

Posterior inferior spine

Sciatic notch

Spine of ischium

Obturator foramen

Ischial tuberosity

Anterior superior spine

Anterior spine

Acetabulum

Superior ramus of pubis

Pubic crest

Inferior ramus of pubis

Ramus of ischium

386

**Fig. 186.** *The shoulder girdle, shoulder joint, and humerus.*

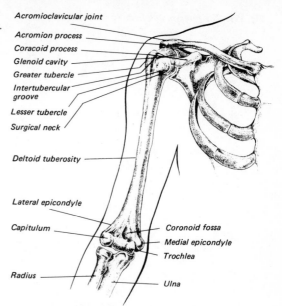

Acromioclavicular joint
Acromion process
Coracoid process
Glenoid cavity
Greater tubercle
Intertubercular groove
Lesser tubercle
Surgical neck

Deltoid tuberosity

Lateral epicondyle

Capitulum

Radius

Coronoid fossa
Medial epicondyle
Trochlea
Ulna

**Fig. 187.** *The scapula.*

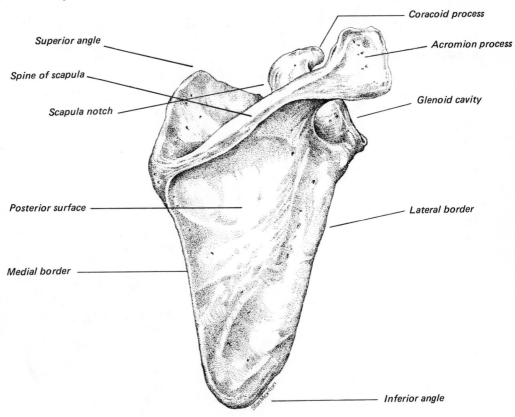

Coracoid process
Acromion process
Superior angle
Spine of scapula
Scapula notch
Glenoid cavity
Posterior surface
Lateral border
Medial border
Inferior angle

**Fig. 188.** *The forearm and hand (anterior view).*

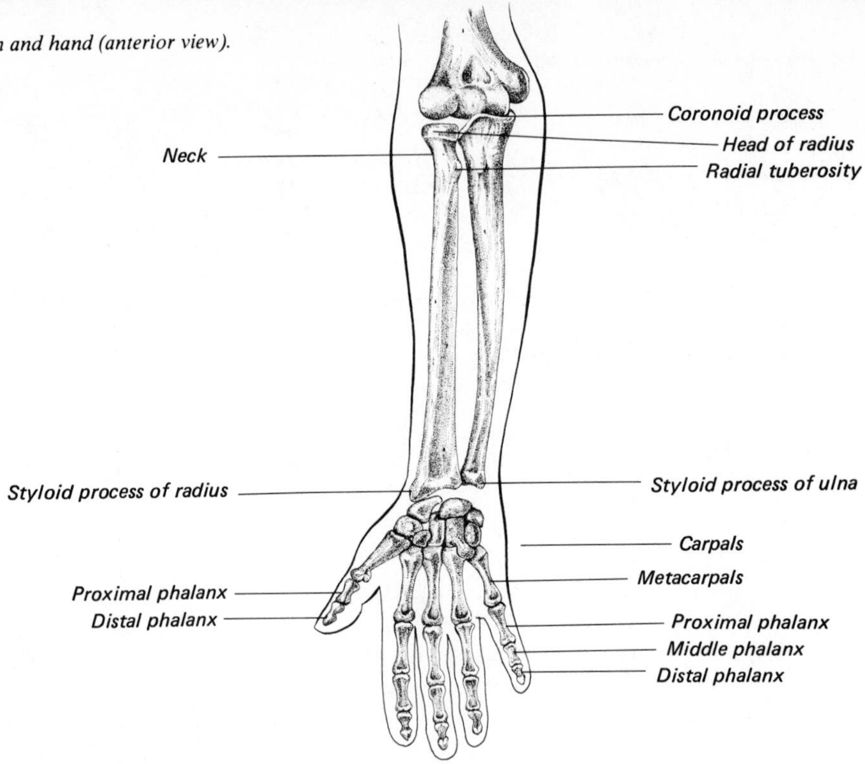

Neck

Coronoid process
Head of radius
Radial tuberosity

Styloid process of radius

Styloid process of ulna

Carpals
Metacarpals

Proximal phalanx
Distal phalanx

Proximal phalanx
Middle phalanx
Distal phalanx

**Fig. 189.** *The forearm and hand (posterior view).*

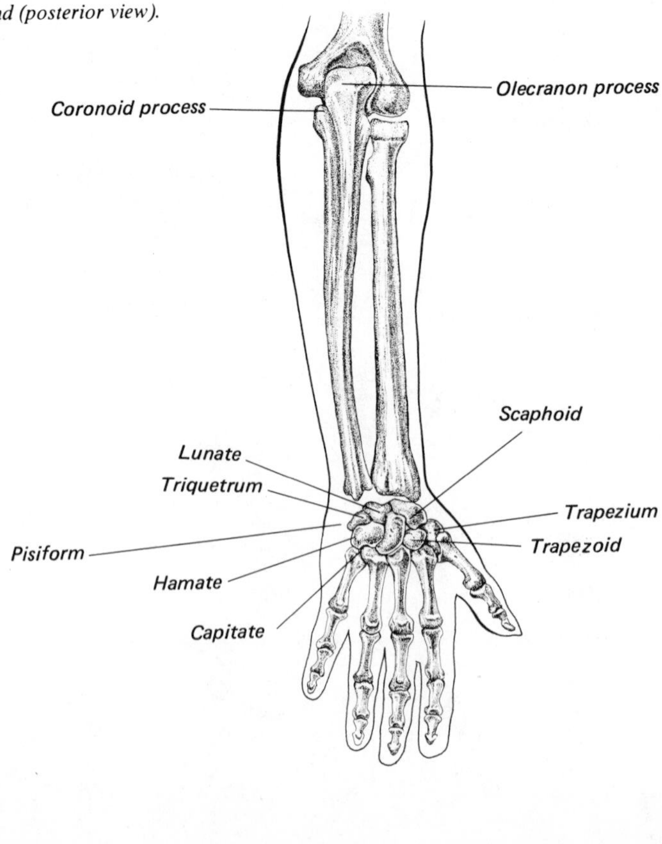

Coronoid process

Olecranon process

Scaphoid

Lunate
Triquetrum

Pisiform

Trapezium
Trapezoid

Hamate

Capitate

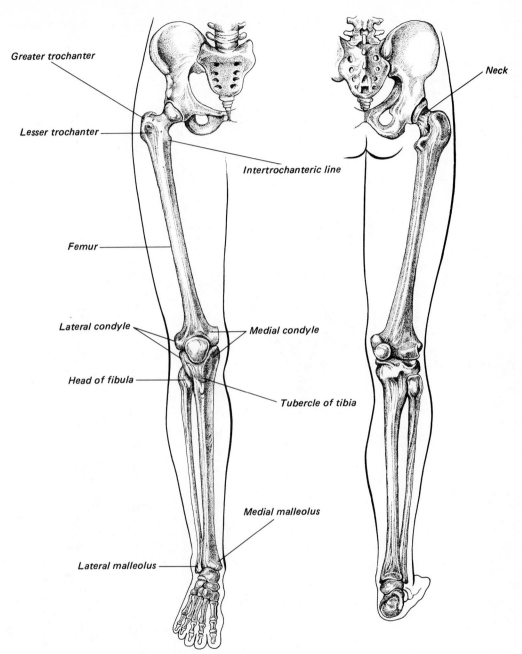

Greater trochanter

Lesser trochanter

Intertrochanteric line

Femur

Lateral condyle

Medial condyle

Head of fibula

Tubercle of tibia

Medial malleolus

Lateral malleolus

Neck

**Fig. 190.** *The lower extremity.*

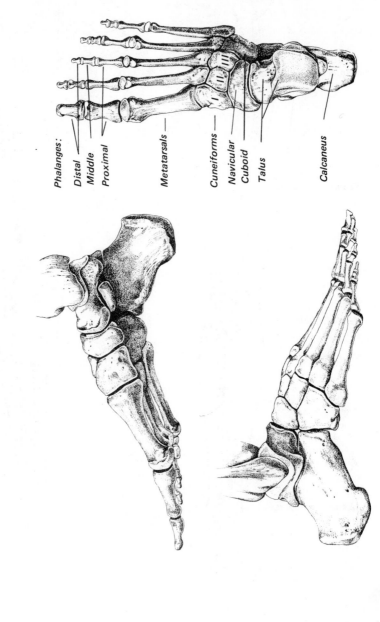

Phalanges :
Distal
Middle
Proximal

Metatarsals

Cuneiforms
Navicular
Cuboid
Talus

Calcaneus

**Fig. 191.** *The ankle joint and foot.*

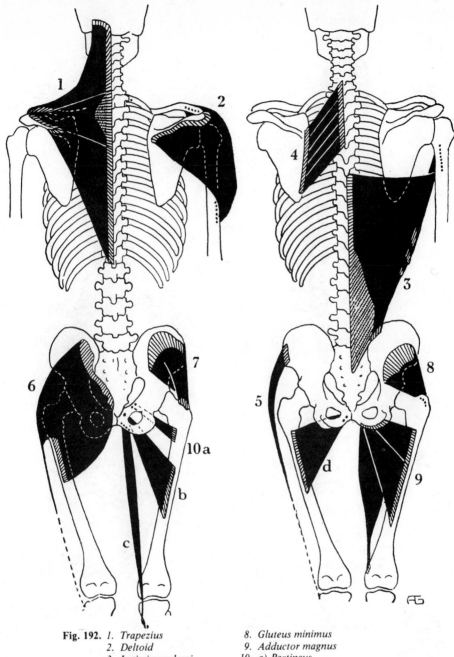

**Fig. 192.**
1. *Trapezius*
2. *Deltoid*
3. *Latissimus dorsi*
4. *Rhomboids*
5. *Tensor fasciae latae*
6. *Gluteus maximus*
7. *Gluteus medius*
8. *Gluteus minimus*
9. *Adductor magnus*
10. a) *Pectineus*
    b) *Adductor longus*
    c) *Adductor gracilis*
    d) *Adductor brevis*

Adapted from Aage Gotved and Richardt Pedersen: *Skematisk Huskeloversigt.* Danske Boghandleres Kommissionsanstalt, København.

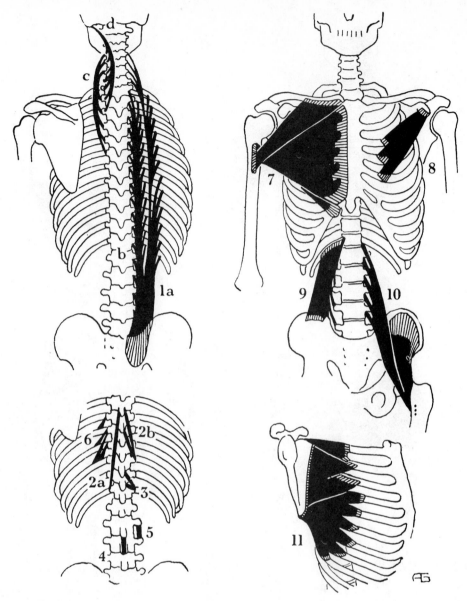

**Fig. 193.** *1. Sacrospinalis*
    *a) Iliocostalis*
    *b, c, d) Longissimus dorsi*
  *2. a) Semispinalis*
    *b) Multifidus*
  *3. Rotators*
  *4. Interspinalis*
  *5. Intertransversarii*

  *6. Levator costarum*
  *7. Pectoralis major*
  *8. Pectoralis minor*
  *9. Quadratus lumborum*
 *10. Iliopsoas*
 *11. Serratus anterior*

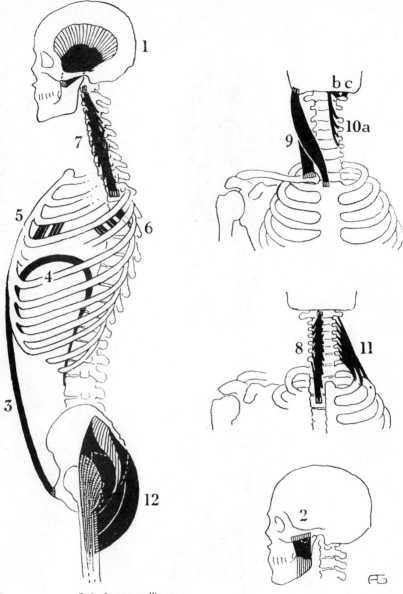

**Fig. 194.**
1. *Temporalis*
2. *Masseter*
3. *Rectus abdominis*
4. *Diaphram*
5. *External intercostals*
6. *Internal intercostals*
7, 8. *Longus colli*
9. *Sternocleidomastoid*
10. *a) Longus capitis*
   *b, c) Rectus capitus*
11. *Scalenius*
12. *Gluteus maximus*

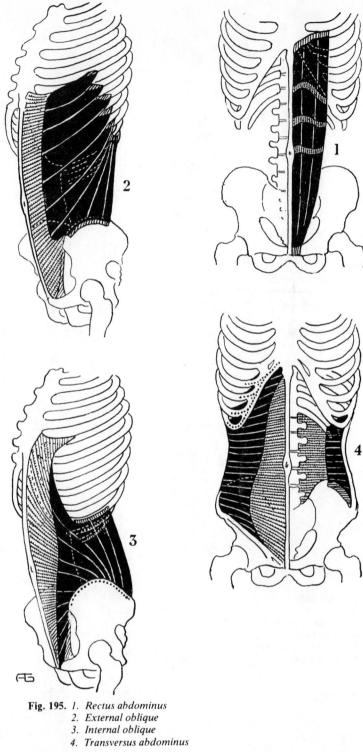

**Fig. 195.** *1. Rectus abdominus*
*2. External oblique*
*3. Internal oblique*
*4. Transversus abdominus*

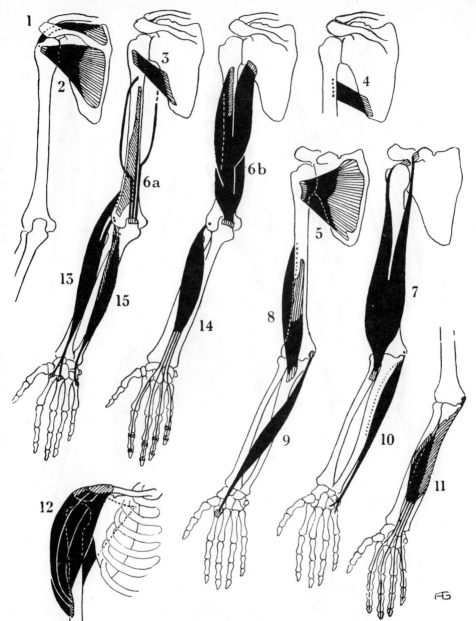

**Fig. 196.** 1. *Supraspinatus*
2. *Infraspinatus*
3. *Teres minor*
4. *Teres major*
5. *Subscapularis*
6. *Triceps brachii*
7. *Biceps brachii*
8. *Brachialis*
9. *Flexor carpi radialis*
10. *Flexor carpi ulnaris*
11. *Flexor digitorum sublimis and profundus*
12. *Deltoid*
13. *Extensor carpi radialis longus and brevis*
14. *Extensor digitorum communis*
15. *Extensor carpi ulnaris*

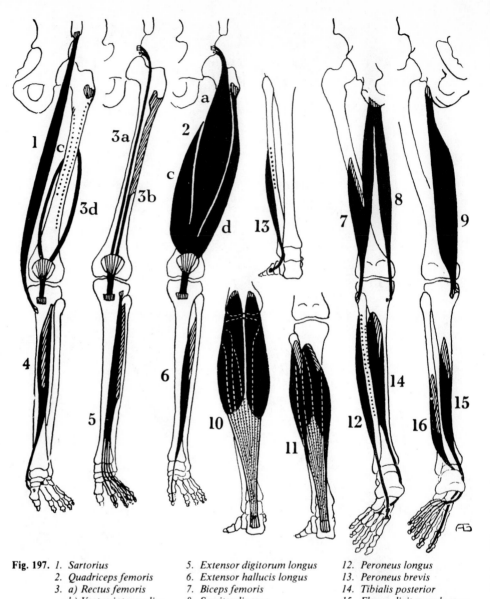

**Fig. 197.** *1. Sartorius*
*2. Quadriceps femoris*
*3. a) Rectus femoris*
*b) Vastus intermedius*
*c) Vastus medialis*
*d) Vastus lateralis*
*4. Tibialis anterior*

*5. Extensor digitorum longus*
*6. Extensor hallucis longus*
*7. Biceps femoris*
*8. Semitendinosus*
*9. Semimembranosus*
*10. Gastrocnemius*
*11. Soleus*

*12. Peroneus longus*
*13. Peroneus brevis*
*14. Tibialis posterior*
*15. Flexor digitorum longus*
*16. Flexor hallucis longus*

**TABLE 15  Muscles Acting on the Neck**

| Muscle | Location | Origin | Insertion | Action |
|---|---|---|---|---|
| Longissimus capitis | Posterior–Lateral neck | Articular processes of lower 3 or 4 cervical vertebrae and transverse processes of upper 4 or 5 thoracic vertebrae | Temporal bone (mastoid process) | Extends neck. Singly: Bends neck to side and rotates head toward contracting side. |
| Semispinalis capitis | Posterior neck | Transverse processes of seventh cervical and superior 6 or 7 thoracic vertebrae | Occipital bone | Extends neck. Singly: Rotates head to same side. |
| Splenius capitis | Posterior neck | Inferior half ligamentum nuchae, spine of seventh cervical and superior 3 or 4 thoracic vertebrae | Occipital bone, temporal bone (mastoid process) | Extends neck. Singly: Rotates head to same side. |
| Sternocleidomastoid | Lateral neck | Sternum (manubrium), clavicle | Temporal bone (mastoid process) | Flexes neck. Singly: Rotates head toward opposite side of contracting muscle. |

**TABLE 16   Muscles of the Abdominal Wall (Anterior Trunk)**

| Muscle | Location | Origin | Insertion | Action |
|---|---|---|---|---|
| External oblique | Anterior abdominal (superficial) | Anterior inferior aspect of lower 8 ribs | Iliac crest, linea alba, pubic tubercle | Compresses, supports abdominal viscera, flexion, and rotation. Singly: Lateral bending of spine. |
| Internal oblique | Anterior and lateral abdominal deep to external oblique | Lateral portion of inguinal ligament, iliac crest, and thoracolumbar fascia | Cartilage of lower three ribs, linea alba | Compresses abdomen; assists in forced breathing (expiration). Singly: Assists in lateral trunk bending; assists in flexion of spine. |
| Rectus abdominis | Anterior midline of abdomen (superficial) | Pubic symphysis and crest of pubis | Sternum (xiphoid process) and cartilage of fifth, sixth, and seventh ribs | Flexes vertebral column; compresses abdominal viscera; assists in breathing. |
| Transversus abdominis | Abdominal deep layer | Inguinal ligament, iliac crest, cartilages of lower 6 ribs, and lumbar fascia | Xiphoid process, linea alba, and pubis | Compresses abdomen; assists in forced expiration. |

## TABLE 17  Muscles of the Trunk (Posterior)

| Muscle | Location | Origin | Insertion | Action |
|---|---|---|---|---|
| Erector spinae (sacrospinalis); includes iliocostalis, longissimus, spinalis | Length of back (sacrum to skull) | Iliac crest, sacrum, ribs, transverse and spinous processes of most lumbar and thoracic vertebrae | Ribs; transverse processes of second to sixth cervical, all thoracic, and the upper lumbar vertebrae; spines of upper thoracic vertebrae; temporal and occipital bones | Extends, bends laterally, and rotates vertebral column. |
| Quadratus lumborum | Between iliac crest and lower rib | Iliac crest (posterior), iliolumbar ligament | Lower rib, transverse processes of upper four lumbar vertebrae | Extends lumbar spine. Singly: Bends spinal column laterally. |

## TABLE 18  Muscles Used in Respiration*

| Muscle | Location | Origin | Insertion | Action |
|---|---|---|---|---|
| Diaphragm | Between abdominal and thoracic cavities | Xiphoid process, costal cartilages of lower six ribs, and lumbocostal arches | Central tendon | Contraction; increases thoracic cavity; aids in respiration. |
| External intercostals (eleven pairs) | Between ribs (superficial) | Inferior border of rib above | Superior border of rib below | Elevates ribs in inspiration. |
| Internal intercostals | Between ribs (deep) | Superior border of rib below | Inferior border of rib above | Draws adjacent ribs together during forced expiration. |

*Note Table 16, muscles of the abdominal wall.

399

**TABLE 19  Muscles Acting on the Shoulder Girdle**

| Muscle | Location | Origin | Insertion | Action |
|---|---|---|---|---|
| Levator scapulae | Neck (posterior) | First 4 cervical vertebrae | Medial border of scapula from the spine to the superior angle | Elevates scapula. |
| Pectoralis minor | Chest (deep to pectoralis major) | Ribs (third to fifth) | Scapula (coracoid process) | Depresses scapula; pulls shoulder forward. |
| Rhomboid major | Deep upper back | Spinous processes (second to fifth thoracic vertebrae) | Medial border of scapula (spine to inferior angle) | Adducts and rotates scapula. |
| Rhomboid minor | Deep upper back, superior and superficial to major | Ligamentum nuchae (lower part), seventh cervical and first thoracic vertebrae | Scapula spine (root) | Adducts scapula. |
| Serratus anterior | Lateral thorax | Upper 8 or 9 ribs | Medial border of scapula | Abducts scapula. |
| Trapezius | Upper back and neck (superficial) | Occipital protuberance, ligamentum nuchae, spine of seventh cervical and all thoracic vertebrae | Clavicle, spine of scapula, and acromion process | Adducts and rotates scapula; elevates and depresses scapula; extends neck. Singly: Bends neck laterally. |

**TABLE 20   Muscles Acting on the Upper Arm (Shoulder Joint)**

| Muscle | Location | Origin | Insertion | Action |
|---|---|---|---|---|
| Coracobrachialis | Upper arm (medial) | Scapula (coracoid process) | Humerus (middle of medial surface) | Flexes and adducts. |
| Deltoid | Anterior, lateral, and posterior upper surface of humerus | Clavicle, scapula (acromion and spine) | Deltoid tuberosity of humerus | Abducts arm. Parts: flexes, extends, and rotates. |
| Infraspinatus | Posterior surface of scapula below spine | Scapula (infraspinous fossa) | Greater tuberosity of humerus | Rotates humerus laterally. |
| Latissimus dorsi | Lower back (superficial) | Vertebrae spines (thoracic sixth through twelfth lumbar and sacral), lumbosacral fascia, crest of ilium, muscular slips from lower 3 or 4 ribs | Humerus (bicipital groove) | Adducts, extends, and medially rotates humerus. |
| Pectoralis major | Chest | | | |
| Clavicular pectoralis | | Clavicle (medial half) | Humerus (lateral lip of bicipital groove) | Flexes and medially rotates humerus. |
| Sternocostal pectoralis | | Sternum and costal cartilages of true ribs | Humerus (lateral lip of bicipital groove) | Extends, adducts, and medially rotates humerus. |
| Supraspinatus | Posterior surface of scapula above spine | Scapula (supraspinous fossa) | Humerus (greater tuberosity) | Adducts humerus (assists). |
| Teres major | Inferior angle of scapula to humerus | Scapula (dorsal surface, inferior angle) | Humerus (bicipital groove) | Adducts, extends, and medially rotates humerus. |
| Teres minor | Immediately superior to teres major | Scapula (dorsal surface of lateral border) | Humerus (greater tuberosity) | Adducts and rotates humerus laterally. |

**TABLE 21   Muscles Acting on the Forearm**

| Muscle | Location | Origin | Insertion | Action |
|---|---|---|---|---|
| Biceps brachii | Upper arm (anterior and superficial) | Short head: scapula (Coracoid process); long head: scapula (supraglenoid tuberosity) | Radial tuberosity, deep fascia of forearm | Flexes and supinates forearm. |
| Brachialis | Upper arm (lower two thirds, anterior and deep) | Humerus (lower two thirds of anterior surface) | Ulna (coronoid process and tuberosity) | Flexes forearm. |
| Brachioradialis | Lateral forearm | Humerus (superior two thirds of lateral supracondylar ridge) | Radius (styloid process) | Flexes forearm. |
| Pronator quadratus | Anterior forearm (distal end and deep) | Ulna (distal portion on shaft, anterior surface) | Radius (distal fourth, lateral border) | Pronates forearm. |
| Pronator teres | Anterior upper forearm | Humerus (medial epicondyle), ulna (coronoid process) | Radius (middle portion, lateral side) | Pronates and flexes forearm. |
| Supinator | Anterior upper forearm (deep to pronator teres and brachioradialis) | Humerus (lateral epicondyle), radius (ridge below radial notch) | Radius (proximal third lateral and anterior surface) | Supinates forearm. |
| Triceps brachii | Posterior humerus | Longhead: scapula (infraglenoid tuberosity); medial head: humerus (lower surface); lateral head (above radial groove) | Ulna (olecranon process) | Extends forearm. |
| Anconeus | Posterior elbow | Lateral epicondyle of humerus | Olecranon process and ulna | Extends elbow; pronates forearm. |

**TABLE 22  Muscles that Move the Wrist and Hand**

| Muscle | Location | Origin | Insertion | Action |
|---|---|---|---|---|
| Extensor carpi radialis longus | Posterior forearm | Humerus (lateral supracondylar ridge) | Second metacarpal (base) | Extends wrist; abducts hand. |
| Extensor carpi radialis brevis | Posterior forearm | Humerus (lateral epicondyle) | Third metacarpal (base) | Extends wrist; abducts hand. |
| Extensor carpi ulnaris | Posterior forearm | Humerus (lateral epicondyle), ulna (proximal half of shaft) | Fifth metacarpal (base) | Extends wrist; adducts hand. |
| Extensor digitorum | Posterior forearm | Humerus (lateral epicondyle) | Fingers (middle and distal phalanges) | Extends phalanges. |
| Extensor indicis | Posterior forearm (distal end) | Ulna (distal surface) | Index finger (proximal phalanx) | Extends proximal phalanx of index finger. |
| Flexor carpi radialis | Anterior forearm (superficial) | Humerus (medial epicondyle) | Second and third metacarpals (base) | Flexes wrist; abducts hands; assists elbow flexion. |
| Flexor carpi ulnaris | Anterior forearm (superficial) | Humerus (medial epicondyle), ulna (proximal two thirds) | Pisiform, hamate, base of fifth metacarpal | Flexes wrist; adducts hands; assists in elbow flexion. |
| Flexor digitorum profundus | Anterior forearm (deep) | Ulna (medial and anterior surface), deep fascia | Fingers (base of distal phalanges) | Flexes distal phalanges. |
| Flexor digitorum superficialis | Anterior forearm (medium) | Humerus (medial epicondyle), ulna (coronoid process), radius (oblique line) | Fingers (base of second phalanx) | Flexes second phalanx of fingers; flexes hand; assists in elbow flexion. |
| Palmaris longus | Anterior forearm | Humerus (medial epicondyle) | Palmar aponeurosis | Flexes wrist. |

**TABLE 23   Muscles Acting on the Thigh (Hip Joint)**

| Muscle | Location | Origin | Insertion | Action |
|---|---|---|---|---|
| **Adductor group** | | | | |
| Magnus | Medial thigh | Rami of ischium and pubis, ischial tuberosity | Femur (greater trochanter to linea aspera), linea aspera, supracondylar line, adductor tubercle | Adducts, flexes, extends, and rotates thigh. |
| Longus | Medial thigh | Pubis (crest and symphysis) | Femur (linea aspera, middle third) | Adducts, flexes, and rotates thigh. |
| Brevis | Medial thigh | Pubis (inferior ramus) | Femur (linea aspera) | Adducts, flexes, and rotates thigh. |
| **Gluteal group** | | | | |
| Maximus | Buttocks (superficial) | Ilium (posterior gluteal line), sacrum and coccyx (posterior surface), sacrotuberous ligament | Femur (greater tuberosity), iliotibial tract | Extends, abducts, and laterally rotates thigh; extends lower trunk. |
| Medius | Buttocks (intermediate) | Ilium (outer surface and crest) | Femur (greater trochanter) | Abducts and medially rotates thigh. |
| Minimus | Buttocks (deep) | Ilium (inferior to medius) | Femur (anterior surface of greater trochanter) | Abducts and medially rotates thigh. |
| Gracilis | Medial thigh | Symphysis pubis (inferior aspect) | Tibia (proximal end medial surface) | Adducts thigh; flexes and medially rotates leg. |
| Iliopsoas | Posterior wall of pelvic cavity | Vertebrae (transverse processes and bodies of twelfth thoracic and all lumbar) iliac fossa | Femur (lesser trochanter) | Flexes thigh; flexes lumbar region on thigh. |
| Pectineus | Medial thigh | Pubis (pectineal line and fascia) | Femur (pectineal line) | Flexes and adducts thigh. |

| Muscle | Location | Origin | Insertion | Action |
|---|---|---|---|---|
| Piriformis | Posterior pelvis (deep) | Sacrum (anterior surface), sacrotuberous ligament | Femur (greater trochanter) | Abducts and laterally rotates thigh. |
| Rectus femoris | Anterior thigh (superficial) | Ilium (anterior inferior iliac spine), groove on acetabulum | Tibia (patella tendon) | Flexes thigh; extends leg. |
| Sartorius | Anterior and medial thigh (superficial) | Ilium (anterior superior iliac spine) | Tibia (proximal end, medial surface) | Flexes thigh and rotates it laterally; flexes leg. |
| Tensor fasciae latae | Lateral hip and proximal thigh | Ilium (anterior part of crest) | Iliotibial tract of fascia lata | Flexes, abducts, and medially rotates thigh; tenses fascia lata. |
| Hamstring group | Posterior thigh | (See Table 24, muscles acting on the leg) | | |

**TABLE 24   Muscles Acting on the Leg (Knee Joint)**

| Muscle | Location | Origin | Insertion | Action |
|---|---|---|---|---|
| Hamstring group Biceps femoris | Posterior thigh (superficial) | Ischium (tuberosity and sacrotuberous ligament), femur (linea aspera) | Fibula (head), tibia (lateral condyle) | Flexes lag and extends thigh (rotates leg laterally when knee is flexed). |
| Semitendinosus | Posterior thigh (medial) | Ischium (tuberosity) | Tibia (proximal end, medial surface) | Flexes leg and extends thigh, (rotates leg medially when knee is flexed). |
| Semimembranosus | Posterior thigh | Ischium (tuberosity) | Tibia (medial condyle) | Flexes leg and extends thigh (rotates leg medially when knee is flexed). |

**TABLE 24** (cont.)

| | | | |
|---|---|---|---|
| Quadriceps femoris group<br>Rectus femoris | Anterior thigh (superficial) | Ilium (anterior inferior iliac spine), groove on acetabulum | Patella (patellar ligament to tibial tubercle) | Extends leg and flexes thigh. |
| Vastus lateralis | Lateral thigh | Femur (linea aspera and greater trochanter) | Lateral border of patellar ligament (to tibial tubercle) | Extends leg. |
| Vastus medialis | Medial thigh | Femur (upper portion of medial aspect of shaft) | Medial border of patellar ligament to tibial tubercle | Extends leg. |
| Vastus intermedius | Anterior thigh (deep) | Femur (anterior surface) | Patellar ligament to tibial tubercle | Extends leg. |
| Gracilis | Medial thigh (superficial) | Symphysis pubis | Tibia (proximal end, medial surface) | Flexes and medially rotates leg; adducts thigh. |
| Sartorius | Anterior and medial thigh (superficial) | Ilium (anterior superior iliac spine) | Tibia (proximal end, medial surface) | Flexes leg; flexes thigh and rotates it laterally. |
| Popliteus | Posterior knee | Femur (lateral condyle and popliteal ligament) | Tibia (superior to popliteal line, posterior aspect) | Rotates femur laterally on tibia. |

**TABLE 25  Muscles that Move the Foot**

| Muscle | Location | Origin | Insertion | Action |
|---|---|---|---|---|
| Extensor digitorum longus | Anteriolateral leg | Tibia (lateral condyle), fibula (anterior surface), interosseous membrane | Four lateral toes (second and third phalanges) | Extends phalanges; dorsiflexes foot. |
| Extensor hallucis longus | Anterior fibula | Fibula (middle half of shaft), interosseous membrane | Great toe (base of distal phalanx) | Extends great toe; aids dorsiflexion of foot. |
| Flexor digitorum longus | Posterior tibia | Tibia (posterior shaft) | Distal phalanges of lateral four toes | Flexes distal phalanges; aids in plantar flexion. |
| Flexor hallucis longus | Posterior fibula | Fibula (distal two thirds of shaft), interosseous membrane | Distal phalanx of great toe | Flexes great toe; aids in plantar flexion. |
| Gastrocnemius | Calf of leg (superficial) | Lateral head: femur (lateral condyle); Medial head: femur (medial condyle) | Calcaneus (by calcaneal tendon) | Plantar flexes the foot; flexes the leg. |
| Peroneus brevis | Lateral leg | Fibula (distal two thirds of lateral surface of shaft) | Fifth metatarsal (base on lateral side) | Plantar flexes; everts foot. |
| Peroneus longus | Lateral leg | Tibia (lateral condyle) fibula (head and proximal two thirds of shaft) | Medial cuneiform and base of first metatarsal | Plantar flexes; everts foot. |
| Peroneus tertius | Anterior fibula (distal end) | Fibula (distal anterior surface) | Fifth metatarsal (base) | Dorsiflexes and everts foot. |
| Soleus | Calf of leg (deep to gastocnemius) | Tibia (middle third), Fibula (head and proximal third of shaft) | Calcaneus (by calcaneal tendon) | Plantar flexes foot. |
| Tibialis anterior | Anterior fibula | Tibia (lateral condyle, proximal two-thirds of shaft) | Medial cuneiform and metatarsal (base of first) | Dorsiflexes and inverts foot. |
| Tibialis posterior | Posterior leg (deep to soleus) | Tibia and fibula (shaft) interosseous membrane | Navicular, cuneiforms, second, third, and fourth metatarsals (base) and cubloid | Plantar flexes and inverts foot. |

# Appendix B
# Elementary Mathematics

A knowledge of elementary concepts in mathematics is essential to the study of kinesiology. This appendix contains a concise summary of the most important of these concepts.

## SIGNED NUMBERS

Numbers—or letters representing numbers—are generally preceded by a positive or negative sign. Positive signs are sometimes omitted for the sake of brevity. Where no sign is included, the number involved is understood to be positive. Operations involving signed numbers proceed according to the following rules:

1. The sum of two numbers is preceded by the sign they have in common, if they have *like* signs (that is, if they are both positive or both negative), and by the sign of the larger number, if they have *unlike* signs (that is, if one is positive and the other negative). Examples:

$$(+5) + (+8) = +13$$
$$(-5) + (-8) = -13$$
$$(+5) + (-8) = -3$$
$$(-5) + (+8) = +3$$
$$5 + 8 = 13$$
$$5 + (-8) = -3$$
$$(-5) + 8 = +3$$

2. The difference between two numbers is preceded by the sign of the larger number. A minus sign followed immediately by a negative number is effectively the same as a plus sign. Thus, for example, $-(-3) = +3$. A minus sign followed by a positive number—or vice versa—is effectively the same as a minus sign. Thus, for example, $-(+3) = -3$ and $+(-3) = -3$. Examples:

$$(+7) - (+3) = +4$$
$$(-7) - (-3) = -4$$
$$(+7) - (-3) = +10$$
$$(-7) - (+3) = -10$$

3. The product of two numbers is preceded by a positive sign if the two numbers have like signs and by a negative sign if they have unlike signs. Examples:

$$(+5)(+3) = +15$$
$$(-5)(-3) = +15$$
$$(+5)(-3) = -15$$
$$(-5)(+3) = -15$$

4. The quotient of two numbers is preceded by a positive sign if the two numbers have like signs and by a negative sign if they have unlike signs. Examples:

$$\frac{+12}{+6} = +2$$

$$\frac{-12}{-6} = +2$$

$$\frac{+12}{-6} = -2$$

$$\frac{-12}{+6} = -2$$

## SQUARES AND SQUARE ROOTS

A number is said to be *squared* when it is multiplied by itself. Thus, for example, $3 \times 3$ is "three squared" or $3^2$.

The square root of a number is another number that, when multiplied by itself, yields the first number. Each positive number has two square roots, one positive and one negative. For example, the positive square root of 9 is 3:

$$3 \times 3 = 9$$

and the negative square root is $-3$:

$$(-3) \times (-3) = 9$$

These results can be summarized in the form

$$\sqrt{9} = \pm 3$$

which reads "the square root of nine is equal to plus or minus three." Examples:

$$2 \times 2 = 2^2 = 4 \qquad \sqrt{4} = \pm 2$$

$$4 \times 4 = 4^2 = 16 \qquad \sqrt{16} = \pm 4$$

$$(-5) \times (-5) = (-5)^2 = 25 \qquad \sqrt{25} = \pm 5$$

$$8 \times 8 = 8^2 = 64 \qquad \sqrt{64} = \pm 8$$

$$(-11) \times (-11) = (-11)^2 = 121 \qquad \sqrt{121} = \pm 11$$

## ORDER OF PRECEDENCE

The order in which the operations involved in a computation are performed is important in determining whether the correct result is obtained.

1. When a computation involves only addition and subtraction, the computation must proceed in order from left to right. Examples:

$$12 - 3 + 7 = 9 + 7$$
$$= 16$$
$$5 - 10 - 3 + 4 = -5 - 3 + 4$$
$$= -8 + 4$$
$$= -4$$
$$9 + 2 - 8 - 5 + 3 = 11 - 8 - 5 + 3$$
$$= 3 - 5 + 3$$
$$= -2 + 3$$
$$= 1$$

2. When a computation involves only multiplication and division, the computation must proceed from left to right. Examples:

$$6 \div 3 \times 4 = 2 \times 4$$
$$= 8$$
$$9 \div 3 \times 8 \div 6 = 3 \times 8 \div 6$$
$$= 24 \div 6$$
$$= 4$$
$$4 \times 9 \div 6 \div 2 \times 3 = 36 \div 6 \div 2 \times 3$$
$$= 6 \div 2 \times 3$$
$$= 3 \times 3$$
$$= 9$$

3. When a computation involves some combination of addition and subtraction, on the one hand, and multiplication and division, on the other, the latter take precedence—that is, those steps involving multiplication and division are performed before those involving addition and subtraction. Where there are two or more steps involving either addition and subtraction or multiplication and division, those steps are performed in order from left to right. Examples:

$$3 \times 7 + 5 = 21 + 5$$
$$= 26$$
$$15 - 3 \times 4 = 15 - 12$$
$$= 3$$
$$10 \div 5 + 5 = 2 + 5$$
$$= 7$$
$$16 + 8 \div 2 = 16 + 4$$
$$= 20$$
$$4 \times 3 \times 2 \div 8 + 2 = 12 \times 2 \div 8 + 2$$
$$= 24 \div 8 + 2$$
$$= 3 + 2$$
$$= 5$$

## GROUPING OF TERMS

Parentheses ( ), brackets [ ], and braces { } are frequently used to indicate the order in which algebraic operations are to be performed. When parentheses, or either of the other grouping symbols, are removed, each of the terms within is multiplied by the term that stands immediately in front of the parentheses. Where there is only a positive or negative sign in front of the parentheses, each of the terms within is multiplied by $+1$ or $-1$, respectively. Examples:

$$3x + (2x - 10) = 3x + 2x - 10$$
$$3x - (2x - 10) = 3x - 2x + 10$$
$$5y + 2(y + 3) = 5y + 2y + 6$$
$$5y - 2(y + 3) = 5y - 2y - 6$$

## FRACTIONS

Fractions have a *numerator* (above the line) and a *denominator* (below the line). If the numerator and the denominator of a fraction are both multiplied or divided by the same number, the value of the fraction remains unaltered. Examples:

$$\frac{1\ (\times\ 7)}{2\ (\times\ 7)} = \frac{7}{14}$$

$$\frac{3\ (\times\ 10)}{5\ (\times\ 10)} = \frac{30}{50}$$

$$\frac{16\ (\div\ 16)}{64\ (\div\ 16)} = \frac{1}{4}$$

$$\frac{75\ (\div 25)}{100\ (\div\ 25)} = \frac{3}{4}$$

The smallest number into which the denominators of several fractions can be divided without a remainder is called the *lowest common multiple* (LCM),—also known as the least, or lowest, common denominator. Example:

| | |
|---|---|
| Denominators: | 2, 4, |
| LCM: | 4 |
| Denominators: | 2, 4, 6 |
| LCM: | 12 |

## Addition

The sum of two or more fractions is obtained by converting each fraction to that form in which it has the LCM as denominator and then adding the numerators. Examples:

$$\frac{1}{2} + \frac{1}{4} = \frac{2}{4} + \frac{1}{4} \qquad \frac{1}{2} + \frac{3}{4} + \frac{5}{6} = \frac{6}{12} + \frac{9}{12} + \frac{10}{12}$$

$$= \frac{2+1}{4} \qquad\qquad = \frac{6+9+10}{12}$$

$$= \frac{3}{4} \qquad\qquad = \frac{25}{12}$$

$$\qquad\qquad\qquad = 2\frac{1}{12}$$

## Subtraction

The difference between two fractions is obtained in similar fashion. Example:

$$\frac{3}{8} - \frac{5}{16} = \frac{6}{16} - \frac{5}{16}$$

$$= \frac{6-5}{16}$$

$$= \frac{1}{16}$$

## Multiplication

The product of two or more fractions is obtained by expressing the product of their numerators as a fraction of the product of their denominators. To simplify this procedure, any numerator and denominator may be divided by a factor they have in common. Examples:

$$\frac{1}{3} \times \frac{1}{5} = \frac{1 \times 1}{3 \times 5} \qquad\qquad \frac{\overset{1}{\cancel{6}}}{7} \times \frac{2}{\underset{3}{\cancel{15}}} \times \frac{1}{3} = \frac{1 \times 2 \times 1}{7 \times 3 \times 3}$$

$$= \frac{1}{15} \qquad\qquad\qquad\qquad = \frac{2}{63}$$

## Division

The quotient of two fractions is obtained by inverting the divisor and multiplying. Example:

$$\frac{5}{8} \div \frac{1}{8} = \frac{5}{\cancel{8}} \times \frac{\cancel{8}}{1}$$

$$= \frac{5 \times 1}{1 \times 1}$$

$$= 5$$

$$\frac{2}{3} \div \frac{5}{12} = \frac{2}{\cancel{3}} \times \frac{\cancel{12}}{5}$$

$$= \frac{2 \times 4}{1 \times 5}$$

$$= \frac{8}{5}$$

$$= 1\tfrac{3}{5}$$

## Square Root

The square root of a fraction is equal to the square root of the numerator expressed as a fraction of the square root of the denominator. Example:

$$\frac{9}{16} = \frac{\sqrt{9}}{\sqrt{16}}$$

$$= \frac{\pm 3}{\pm 4}$$

$$= \pm \frac{3}{4}$$

## DECIMALS

Fractions may be expressed in several alternative forms of which the decimal form is probably the most convenient and widely used. In the decimal system, fractions are expressed in tenths (.1, .2, .3, ..., .9), hundredths (.01, .02, .03, ..., .09), thousandths (.001, .002, .003, ...., .009), and so on.

Fractions are converted to decimal fractions by dividing the denominator into the numerator. Examples:

$$\frac{1}{4} \qquad \begin{array}{r} .25 \\ 4\overline{)1.00} \\ \underline{8} \\ 20 \\ \underline{20} \end{array}$$

$$\frac{3}{8} \qquad \begin{array}{r} .375 \\ 8\overline{)3.000} \\ \underline{2\,4} \\ 60 \\ \underline{56} \\ 40 \\ \underline{40} \end{array}$$

Decimal fractions are converted to fractions by placing the numbers after the decimal point in the numerator and a 1 followed by the same number of zeros as there are numbers in the numerator in the denominator. Examples:

$$.25 = \frac{25}{100}$$

$$= \frac{1}{4}$$

$$.375 = \frac{375}{1000}$$

$$= \frac{3}{3}$$

## Addition and Subtraction

The sum of (or difference between) two numbers expressed in decimal form is obtained by aligning the decimal points of the numbers and adding (or subtracting) in the usual manner. Examples:

$$\begin{array}{r} 6.3 \\ +2.4 \\ \hline 8.7 \end{array} \qquad \begin{array}{r} 5.246 \\ +0.132 \\ \hline 5.378 \end{array} \qquad \begin{array}{r} 4.94 \\ -1.02 \\ \hline 3.92 \end{array} \qquad \begin{array}{r} 8.875 \\ -7.986 \\ \hline 0.889 \end{array}$$

## Multiplication

The product of two numbers expressed in decimal form is obtained by multiplying the numbers as if the decimal points had been omitted and then inserting a decimal point in the answer. This point is placed so that the number of figures after the point is equal to the *sum* of the number of figures after the decimal points in the numbers being multiplied. If there are insufficient figures in the answer for this purpose, zeros are added to the left-hand end of the answer until there are sufficient. Examples:

$$\begin{array}{r} 0.3 \\ \times 0.4 \\ \hline .12 \end{array} \qquad \begin{array}{r} 11.25 \\ \times .2 \\ \hline 2.250 \end{array} \qquad \begin{array}{r} 0.025 \\ \times .003 \\ \hline .000075 \end{array}$$

## Division

The quotient of two numbers expressed in decimal form is obtained by first multiplying both numbers by whatever is necessary (10, 100, 1000, etc.) to change the divisor into a whole number and then proceeding in the manner shown in the examples on p. 414. Example:

$$\frac{4.35 \ (\times \ 10)}{1.5 \ (\times \ 10)} = \frac{43.5}{15}$$

$$\begin{array}{r} 2.9 \\ 15)\overline{43.5} \\ \underline{30} \\ 135 \\ \underline{135} \end{array}$$

## Rounding

Decimal fractions often have a large number of figures following the decimal point. Because there is usually little to be gained by writing out more than the first few of these figures, such fractions are "rounded off" to a given number of decimal places. To do this, the number occupying the next place is examined. If this number is less than 5, it and all the numbers that follow it are disregarded. Example:

| | | |
|---|---|---|
| 1.8341 | correct to one decimal place is | 1.8 |
| 1.8341 | correct to two decimal places is | 1.83 |
| 1.8341 | correct to three decimal places is | 1.834 |

If this number is 5 or more than 5, the preceding number is increased by 1 and all those that follow it are disregarded. Example:

| | | |
|---|---|---|
| 3.4659 | correct to one decimal place is | 3.5 |
| 3.4695 | correct to two decimal places is | 3.47 |
| 3.4695 | correct to three decimal places is | 3.470 |

## Percentages

The term percent (%) is often used instead of a direct statement that the denominator of a fraction is 100. Thus, for example, $10/100$ is referred to as 10%, $20/100$ as 20%, and so on. A fraction may be converted to a percentage by multiplying it by 100%. Examples:

$$\frac{1}{2} \times 100\% = \frac{100\%}{2} \qquad\qquad \frac{5}{8} \times 100\% = \frac{500\%}{8}$$

$$= 50\% \qquad\qquad\qquad = 62.5\%$$

A percentage may be converted to a fraction by dividing it by 100 and omitting the % sign. The fraction formed in this manner may then be reduced to its lowest form by dividing numerator and denominator by factors they have in common. Examples:

$$50\% = \frac{\overset{1}{\cancel{50}}}{\underset{2}{\cancel{100}}} \qquad\qquad 16\% = \frac{\overset{4}{\cancel{16}}}{\underset{25}{\cancel{100}}}$$

$$= \frac{1}{2} \qquad\qquad\qquad = \frac{4}{25}$$

416

## EQUATIONS

An equation is a mathematical statement of equality that usually includes some terms of known value and some of unknown value.

An equation that contains only one unknown (say, $n$), and in which this unknown is of the same order ($n$ or $n^2$ or $n^3$, ..., etc.) throughout, can be solved by rearranging the terms of the equation. This rearranging process—designed to place the unknown on the left-hand side of the equation and all the other terms on the right-hand side—is based on the fact that the equality that exists between the two sides of an equation is preserved if both sides are operated on in the same manner. Thus, if the same number is added to (or subtracted from) both sides, the equality between the two is preserved. Similarly, if both sides are multiplied (or divided) by the same number, the equality is again preserved. Finally, if both sides are squared, or if the square root of each side is taken, the equality is maintained. Examples:

$n - 2 = 8$    Add 2 to both sides
$$n - 2 + 2 = 8 + 2$$
$$n = 10$$

$n + 5 = 7$    Subtract 5 from both sides
$$n + 5 - 5 = 7 - 5$$
$$n = 2$$

$\dfrac{n}{4} = 20$    Multiply both sides by 4
$$\frac{n}{4} \times 4 = 20 \times 4$$
$$n = 80$$

$3n = 15$    Divide both side by 3
$$\frac{3n}{3} = \frac{15}{3}$$
$$n = 5$$

$n^2 = 16$    Take square root of both sides
$$\sqrt{n^2} = \sqrt{16}$$
$$n = \pm 4$$

In many cases, a series of these steps is necessary before the unknown can be isolated on the left-hand side of the equation. Examples:

$5n - 9 = 16$    Add 9 to both sides
$$5n - 9 + 9 = 16 + 9$$
$$5n = 25$$

Divide both side by 5
$$\frac{5n}{5} = \frac{25}{5}$$
$$n = 5$$

$$\frac{n}{4} + 8 = 11 \qquad \text{Subtract 8 from both sides} \qquad \frac{n}{4} + 8 - 8 = 11 - 8$$

$$\frac{n}{4} = 3$$

$$\text{Multiply both sides by 4} \qquad \frac{n}{4} \times 4 = 3 \times 4$$

$$n = 12$$

$$3n^2 - 4 = 23 \qquad \text{Add 4 to both sides} \qquad 3n^2 - 4 + 4 = 23 + 4$$

$$3n^2 = 27$$

$$\text{Divide both sides by 3} \qquad \frac{3n^2}{3} = \frac{27}{3}$$

$$n^2 = 9$$

$$\text{Take square root of both sides} \qquad \sqrt{n^2} = \sqrt{9}$$

$$n = \pm 3$$

## GRAPHS

A graph is a pictorial representation of the relationship between two or more quantities. Graphs vary greatly in form and complexity. There are line graphs, bar graphs, column graphs, and circle graphs; there are two-dimensional and three-dimensional graphs; and so on. Of these, the two-dimensional line graph is almost certainly the simplest, most versatile, and most widely used.

The drawing of a two-dimensional line graph proceeds in the following series of steps:

1. The data are examined to determine whether $\llcorner$ or $\vdash$ or $\lrcorner$ or $+$ axes should be used. The first of these shapes is used if there are only positive values for the quantities to be graphed; the second or third shape if there are only positive values for one of the quantities and positive and negative for the other; and the fourth if there are positive and negative values for both quantities.

2. A decision is made concerning which quantity should be associated with each axis. When one of the quantities depends on the other—that is, when it is determined to some extent by the other—this quantity (or dependent variable) is plotted on the vertical axis (or ordinate). When no such dependency exists, the selection of which quantity to plot on each axis is purely arbitrary. (When

there are positive and negative values for one quantity and only positive for the other, the decision concerning which quantity should be plotted on which axis determines the general shape of the axes to be used—see the preceding step.)

3. The data are examined to determine the maximum and minimum values for each quantity and thus the range of values that the graph must be designed to accommodate.

4. A scale is selected for each quantity. In a well-planned graph these scales are as large as they can be, given the space available and the ranges of the respective sets of data.

5. The axes are ruled, graduated, and labeled.

6. Each pair of values of the data is represented on the graph by a point plotted at the intersection of a horizontal line through the appropriate point on the vertical axis and a vertical line through the appropriate point on the horizontal axis. Thus, for example, if a force-time graph were to be drawn for an experiment in which a force of 500 N had been recorded 3 s after the start of the experiment, these data would be represented on the graph by a point plotted at the intersection of horizontal and vertical lines through these respective points (Fig. 198).

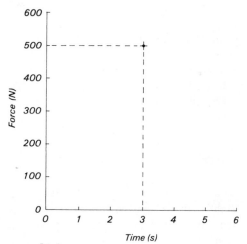

**Fig. 198.**

7. Straight lines are ruled between adjacent plotted points (Fig. 199). To eliminate the undesirable effects of chance errors in measurement (Fig. 200(a)), a smooth curve is sometimes drawn through the data to provide what is considered to be the best available estimate of the true relationship between the quantities concerned (Fig. 200(b)).

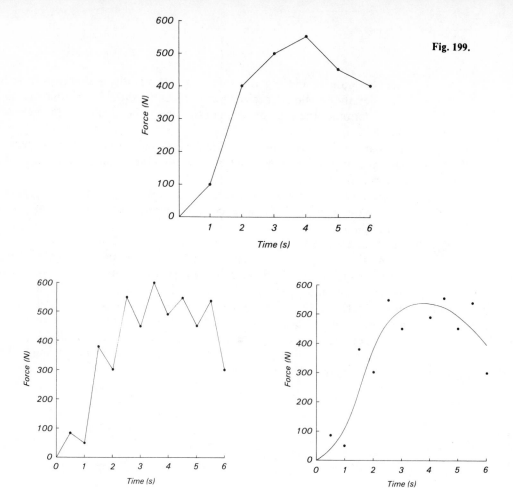

Fig. 199.

(a)

(b)

Fig. 200.

## THEOREM OF PYTHAGORAS

In a right-angled triangle, the square on the hypotenuse (the side opposite the right angle) is equal to the sum of the squares on the other two sides. Thus, if the lengths of the two sides enclosing the right angle are a and b and the length of the hypotenuse is c (Fig. 201):

$$c^2 = a^2 + b^2$$

This relationship can be used to determine the length of any one side of a right-angled triangle if the lengths of the other two sides are known. For example, if $a = 3$ m, $b = 4$ m and $c$ is unknown, $c$ can be found as follows:

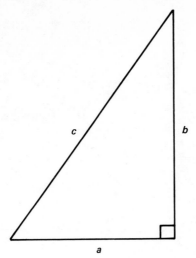

Fig. 201.

$$c^2 = a^2 + b^2$$
$$= 3^2 + 4^2$$
$$= 9 + 16$$
$$= 25$$
$$\therefore \ c = \sqrt{25}$$
$$= 5 \text{ m}$$

Similarly, if $a = 5$ m, $c = 13$ m, and $b$ is unknown,

$$c^2 = a^2 + b^2$$

rearranging

$$b^2 = c^2 - a^2$$
$$= 13^2 - 5^2$$
$$= 169 - 25$$
$$= 144$$
$$\therefore \ b = 12 \text{ m}$$

# Appendix C
# Metric Conversion Tables

Appendix C contains tables for the conversion of metric units to equivalent linear (or American) units and vice versa.

Values that fall between those given in the tables can be determined by addition. For example, to convert 5493 N to pounds, the pound equivalents of 5000 N, 400 N, 90 N, and 3 N are summed:

$$5000 \text{ N} = 1124.0 \text{ lb}$$
$$400 \text{ N} = \phantom{0}90.0 \text{ lb}$$
$$90 \text{ N} = \phantom{0}20.2 \text{ lb}$$
$$3 \text{ N} = \phantom{00}0.7 \text{ lb}$$
$$\overline{5493 \text{ N} \quad 1234.9 \text{ lb}}$$

The same procedure can be used to convert quantities that are greater than the maximum given in the appropriate table.

**TABLE 26   Conversion of Force Units: Newtons and Pounds**

| Newtons to Pounds (multiply by 0.2248) | | Pounds to Newtons (multiply by 4.448) | |
|---|---|---|---|
| Newtons | Pounds | Pounds | Newtons |
| 1 | 0.2 | 1 | 4.4 |
| 2 | 0.4 | 2 | 8.9 |
| 3 | 0.7 | 3 | 13.3 |
| 4 | 0.9 | 4 | 17.8 |
| 5 | 1.1 | 5 | 22.2 |
| 6 | 1.3 | 6 | 26.7 |
| 7 | 1.6 | 7 | 31.1 |
| 8 | 1.8 | 8 | 35.6 |
| 9 | 2.0 | 9 | 40.0 |
| 10 | 2.2 | 10 | 44.5 |
| 20 | 4.5 | 20 | 89.0 |
| 30 | 6.7 | 30 | 133.4 |
| 40 | 9.0 | 40 | 177.9 |
| 50 | 11.2 | 50 | 222.4 |
| 60 | 13.5 | 60 | 266.9 |
| 70 | 15.7 | 70 | 311.4 |
| 80 | 18.0 | 80 | 355.8 |
| 90 | 20.2 | 90 | 400.3 |
| 100 | 22 | 100 | 445 |
| 200 | 45 | 200 | 890 |
| 300 | 67 | 300 | 1 334 |
| 400 | 90 | 400 | 1 779 |
| 500 | 112 | 500 | 2 224 |
| 600 | 135 | 600 | 2 669 |
| 700 | 157 | 700 | 3 114 |
| 800 | 180 | 800 | 3 558 |
| 900 | 202 | 900 | 4 003 |
| 1 000 | 225 | 1,000 | 4 448 |
| 2 000 | 450 | 2,000 | 8 896 |
| 3 000 | 674 | 3,000 | 13 344 |
| 4 000 | 899 | 4,000 | 17 792 |
| 5 000 | 1,124 | 5,000 | 22 240 |
| 6 000 | 1,349 | 6,000 | 26 688 |
| 7 000 | 1,574 | 7,000 | 31 136 |
| 8 000 | 1,798 | 8,000 | 35 584 |
| 9 000 | 2,023 | 9,000 | 40 032 |
| 10 000 | 2,248 | 10,000 | 44 480 |

## TABLE 27  Conversion of Mass Units: Kilograms and Slugs

| Kilograms to Slugs (multiply by 0.0685) | | Slugs to Kilograms (multiply by 14.59) | |
|---|---|---|---|
| Kilograms | Slugs | Slugs | Kilograms |
| 1 | 0.07 | 1 | 14.6 |
| 2 | 0.14 | 2 | 29.2 |
| 3 | 0.21 | 3 | 43.8 |
| 4 | 0.27 | 4 | 58.4 |
| 5 | 0.34 | 5 | 73.0 |
| 6 | 0.41 | 6 | 87.5 |
| 7 | 0.48 | 7 | 102.1 |
| 8 | 0.55 | 8 | 116.7 |
| 9 | 0.62 | 9 | 131.3 |
| 10 | 0.69 | 10 | 146 |
| 20 | 1.37 | 20 | 292 |
| 30 | 2.06 | 30 | 438 |
| 40 | 2.74 | 40 | 584 |
| 50 | 3.43 | 50 | 730 |
| 60 | 4.11 | 60 | 875 |
| 70 | 4.80 | 70 | 1 021 |
| 80 | 5.48 | 80 | 1 167 |
| 90 | 6.17 | 90 | 1 313 |
| 100 | 6.9 | 100 | 1 459 |
| 200 | 13.7 | 200 | 2 918 |
| 300 | 20.6 | 300 | 4 377 |
| 400 | 27.4 | 400 | 5 836 |
| 500 | 34.3 | 500 | 7 295 |
| 600 | 41.1 | 600 | 8 754 |
| 700 | 48.0 | 700 | 10 213 |
| 800 | 54.8 | 800 | 11 672 |
| 900 | 61.7 | 900 | 13 131 |
| 1 000 | 68.5 | 1,000 | 14 590 |

## TABLE 28   Conversion of Length Units: Meters and Feet

| Meters to Feet (multiply by 3.281) | | Feet to Meters (multiply by 0.3048) | |
|---|---|---|---|
| Meters | Feet | Feet | Meters |
| 1 | 3.3 | 1 | 0.30 |
| 2 | 6.6 | 2 | 0.61 |
| 3 | 9.8 | 3 | 0.91 |
| 4 | 13.1 | 4 | 1.22 |
| 5 | 16.4 | 5 | 1.52 |
| 6 | 19.7 | 6 | 1.83 |
| 7 | 23.0 | 7 | 2.13 |
| 8 | 26.2 | 8 | 2.44 |
| 9 | 29.5 | 9 | 2.74 |
| 10 | 32.8 | 10 | 3.0 |
| 20 | 65.6 | 20 | 6.1 |
| 30 | 98.4 | 30 | 9.1 |
| 40 | 131.2 | 40 | 12.2 |
| 50 | 164.1 | 50 | 15.2 |
| 60 | 196.9 | 60 | 18.3 |
| 70 | 229.7 | 70 | 21.3 |
| 80 | 262.5 | 80 | 24.4 |
| 90 | 295.3 | 90 | 27.4 |
| 100 | 328 | 100 | 30.5 |
| 200 | 656 | 200 | 61.0 |
| 300 | 984 | 300 | 91.4 |
| 400 | 1,312 | 400 | 121.9 |
| 500 | 1,641 | 500 | 152.4 |
| 600 | 1,969 | 600 | 182.9 |
| 700 | 2,297 | 700 | 213.4 |
| 800 | 2,625 | 800 | 243.8 |
| 900 | 2,953 | 900 | 274.3 |
| 1 000 | 3,281 | 1,000 | 304.8 |

**TABLE 29    Conversion of Length Units: Kilometers and Miles**

| Kilometers to Miles (multiply by 0.6214) | | Miles to Kilometers (multiply by 1.609) | |
|---|---|---|---|
| Kilometers | Miles | Miles | Kilometers |
| 1 | 0.62 | 1 | 1.6 |
| 2 | 1.24 | 2 | 3.2 |
| 3 | 1.86 | 3 | 4.8 |
| 4 | 2.49 | 4 | 6.4 |
| 5 | 3.11 | 5 | 8.0 |
| 6 | 3.73 | 6 | 9.7 |
| 7 | 4.35 | 7 | 11.3 |
| 8 | 4.97 | 8 | 12.9 |
| 9 | 5.59 | 9 | 14.5 |
| 10 | 6.2 | 10 | 16 |
| 20 | 12.4 | 20 | 32 |
| 30 | 18.6 | 30 | 48 |
| 40 | 24.9 | 40 | 64 |
| 50 | 31.1 | 50 | 80 |
| 60 | 37.3 | 60 | 97 |
| 70 | 43.5 | 70 | 113 |
| 80 | 49.7 | 80 | 129 |
| 90 | 55.9 | 90 | 145 |
| 100 | 62.1 | 100 | 161 |

**TABLE 30    Conversion of Length Units: Centimeters and Inches**

| Centimeters to Inches (multiply by 0.3937) | | Inches to Centimeters (multiply by 2.54) | |
|---|---|---|---|
| Centimeters | Inches | Inches | Centimeters |
| 1 | 0.39 | 1 | 2.5 |
| 2 | 0.79 | 2 | 5.1 |
| 3 | 1.18 | 3 | 7.6 |
| 4 | 1.57 | 4 | 10.2 |
| 5 | 1.97 | 5 | 12.7 |
| 6 | 2.36 | 6 | 15.2 |
| 7 | 2.76 | 7 | 17.8 |
| 8 | 3.15 | 8 | 20.3 |
| 9 | 3.54 | 9 | 22.9 |
| 10 | 3.94 | 10 | 25.4 |
| 20 | 7.87 | 20 | 50.8 |
| 30 | 11.81 | 30 | 76.2 |
| 40 | 15.75 | 40 | 101.6 |
| 50 | 19.69 | 50 | 127.0 |
| 60 | 23.62 | 60 | 152.4 |
| 70 | 27.56 | 70 | 177.8 |
| 80 | 31.50 | 80 | 203.2 |
| 90 | 35.43 | 90 | 228.6 |
| 100 | 39.37 | 100 | 254.0 |

**TABLE 31  Conversion of Speed Units: Meters/Second and Miles/Hour**

| Meters/Second to Miles/Hour (multiply by 2.237) | | Miles/Hour to Meters/Second (multiply by 0.447) | |
|---|---|---|---|
| Meters/Second | Miles/Hour | Miles/Hour | Meters/Second |
| 1 | 2.24 | 1 | 0.45 |
| 2 | 4.47 | 2 | 0.89 |
| 3 | 6.71 | 3 | 1.34 |
| 4 | 8.95 | 4 | 1.79 |
| 5 | 11.19 | 5 | 2.24 |
| 6 | 13.42 | 6 | 2.68 |
| 7 | 15.66 | 7 | 3.13 |
| 8 | 17.90 | 8 | 3.58 |
| 9 | 20.13 | 9 | 4.02 |
| 10 | 22.4 | 10 | 4.5 |
| 20 | 44.7 | 20 | 8.9 |
| 30 | 67.1 | 30 | 13.4 |
| 40 | 89.5 | 40 | 17.9 |
| 50 | 111.9 | 50 | 22.4 |
| 60 | 134.2 | 60 | 26.8 |
| 70 | 156.6 | 70 | 31.3 |
| 80 | 179.0 | 80 | 35.8 |
| 90 | 201.3 | 90 | 40.2 |
| 100 | 224 | 100 | 45 |
| 200 | 447 | 200 | 89 |
| 300 | 671 | 300 | 134 |
| 400 | 895 | 400 | 179 |
| 500 | 1,119 | 500 | 224 |

# Appendix D
# Center of Gravity Mannikin

The center of gravity mannikin shown in Figs. 120 and 121 can be constructed as follows:

## EQUIPMENT AND MATERIALS

Opaque projector, pantograph, or some other means to enlarge the segments of the mannikin shown in Fig. 202.

A large sheet of reasonably heavy cardboard. Avoid the cardboard with internal corrugations that is often used in packing and in cardboard boxes.

Heavy-duty scissors.

Some means of punching circular holes of approximately 3 mm diameter.

Twelve $\frac{3}{4}$ in. (1.9 cm) round-head, paper fasteners or their equivalent.

Thumbtacks or tape to fasten the cardboard to a wall or bulletin board.

## INSTRUCTIONS[1]

1. Fasten the large sheet of cardboard to a wall or bulletin board so that it can serve as a screen for the opaque projector.
2. Using the opaque projector, project Fig. 202 onto the cardboard and adjust the position of the projector to obtain an acceptable image size. (The mannikin shown in Fig. 120 is actually 38 cm long when posed in an erect, standing position.)

[1]The following instructions are presented with the assumption that an opaque projector will be used to enlarge Fig. 202. Steps 1–3 must be modified if some other device or procedure is used.

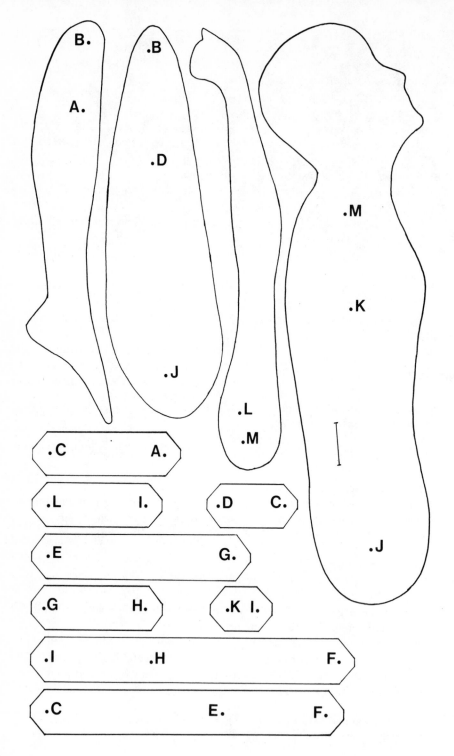

Fig. 202.

3. Trace the outline for each segment of the mannikin and mark and label each of the points indicated in Fig. 202.
4. Cut out each of the twelve segments of the mannikin and punch a hole at each of the labeled points.
5. Lay the thigh segment (the segment with the points labeled B, D, and J) on a flat surface and place the trunk segment (MKJ) on top of it so that the holes marked J on each segment are superimposed. Push a paper fastener downward through these J holes and fasten.
6. Place the upper arm segment (LM) on the trunk segment with M holes superimposed and fasten.
7. Place the lower leg segment (AB) on the thigh segment with B holes superimposed and fasten.
8. Place segment IL on the arm segment (L holes superimposed) and fasten. Place segment KI on the trunk segment (K holes superimposed) and fasten. Place segment CD on the thigh segment (D holes superimposed) and fasten. Place segment AC on the lower leg segment (A holes superimposed) and fasten.
9. Place segment CEF on a flat surface with C on the right and F on the left. Place segment FHI on top of it (F holes superimposed) and fasten. Place segment GE on top of segment CEF (E holes superimposed) and fasten. Place segment GH on top of segment FHI (H holes superimposed) and fasten. Fasten G holes of segments GH and GE with GH on top.
10. Place the linkage constructed in step 9 on the rest of the mannikin so that the C holes of segments CEF, AC, and CD are superimposed. Check that the order from top to bottom is CEF, AC, and CD and fasten. Similarly align the I holes of segments FHI, LI, and KI and fasten.
11. The mannikin, with the paper fastener at G indicating the center of gravity, is now ready for use.

[Notes: (a) One arm (or leg) of the mannikin is equivalent to two arms (or legs) of the human subject it represents. (b) When moving the arm segment from in front of to behind the trunk segment, or vice versa, the segment IL should always remain parallel with a line joining the M and K holes. It should never lie diagonally between these two holes. (c) There is a short line marked on the trunk segment. This line, which has a length equal to 1/25 of the "standing height" of the mannikin, can be used to make conversions from mannikin distance to real-life distances.]

# Indexes

## AUTHOR INDEX

**A**

Abbot, L. R. C., 78
American Academy of Orthopedic Surgeons, 32
Anderson, P., 35
Asmussen, E., 71
Astrand, P.-O., 35
Atwater, A. E., 314

**B**

Baker, J. A., 345
Barnett, C. H., 32
Barret, K. R., 301
Basmajian, J. V., 48, 71, 72, 80, 81, 85, 86, 91, 96, 97, 109
Batterman, C., 331
Bearn, J. G., 91
Behnke, A. R., 240
Berardi, A. C., 91
Bergan, J. J., 72
Berzin, F., 91
Bierman, W., 109
Bishop, R. D., 206, 237
Bojsen-Moller, F., 107
Bragin, S., 110
Brandell, B. R., 109
Bresler, B., 72
Brown, R. M., 264
Brumlik, J., 72
Burke, R. K., 29

**C**

Carlsöö, S., 56, 109
Cavanagh, P. R., 48, 193, 244, 237, 253

Chapman, A. E., 302
Cochran, A., 190, 238
Cooper, L., 249
Counsilman, J. E., 264

**D**

Daish, C. B., 172, 192, 238, 253
Dalzell, D., 249
Dapena, J., 135, 223
Davidovits, P., 22
Davies, D. V., 32
Davies, J. M., 242
De Luca, C. J., 91
Deshon, D. E., 315
de Sousa, O. M., 71, 91
de Vries, H. A., 31
Drowatzky, J. N., 302
Duvall, E. N., 90
Dyatchkov, V. M., 288, 292
Dyson, G. H. G., 144, 186, 192, 237, 265

**E**

Ebert, L. J., 142
Ecker, T., 193, 233, 237, 265

**F**

Faria, I. E., 193, 237, 244, 253
Feinstein, B., 41, 42, 48
Fetz, F., 135
Fielding, J. W., 54
Fischer, F. J., 109
Flint, M. M., 71
Floyd, W. R., 72

Forrest, W. J., 91
Francis, P. R., 160, 162
Fujiwara, M., 72, 109
Furlani, J., 71, 91

**G**

Ganslen, R. V., 241
Gardner, E. B., 49
Gillespie, J. A., 22
Godfrey, K. E., 72
Gratz, C. M., 40
Gray, H., 68
Greenlaw, R. K., 71
Gudgell, J., 71

**H**

Hall, K. G., 241
Hall, M. C., 102
Hamilton, S. G. I., 32
Hamilton, W. J., 32
Hanson, B. D., 163
Harden, J. P., 110
Harris, M. L., 32
Hay, J. G., 127, 151, 193, 206, 207, 216, 223, 237, 253, 265
Heger, W., 135
Heiskanen, W. A., 158
Henriksson, J., 35
Herman, R., 110
Heusner, W. W., 135
Hill, A. V., 45, 48
Hippocrates, 28
Hochmuth, G., 220
Hoerler, E., 207
Holland, G. J., 29, 32
Holt, L. E., 31
Houtz, S. J ., 109
Hromas, L. A., 241
Hubbard, A. W., 48
Hygaard, E., 35

**I**

Iida, M., 72
Inman, V. T., 78
Ishimura, K., 78
Ito, N., 78
Iuchi, M., 242

**J**

Johns, R. J., 28
Johnson, B. L., 317
Jonsson, B., 72
Joseph, J., 36, 72

**K**

Kadefors, R., 49

Karlsson, J., 36
Keagy, R. D., 72
Kelly, D. L., 117, 193
Kentzer, C. P., 241
Kindig, L. E., 72
Kirby, R. F., 136
Klausen, K., 72
Knight, S., 135
Komi, P., 135
Kottke, F. J., 29
Kraus, H., 31
Kuhlow, A., 135
Kuwahara, H., 78
Kuzneyetsov, W., 135

**L**

Lamoreux, L., 107
Larson, L., 36
Larson, R. L., 21, 32
Latif, A., 85, 91
Lee, H. Y., 135
Lieb, F. J., 110
Lindegard, B., 41, 42, 48
Lovejoy, F., 110
Lucas, D. B., 72
Luttgens, K., 127, 186, 237
Lyttleton, R. A., 241

**M**

MacConaill, M. A., 32
Magill, R. A., 302
Malina, R. M., 32
Marteniuk, R. G., 299
McElroy, G. K., 135
Miller, D. I., 223
Mood, D. P., 328
Morehouse, C. A., 223
Morris, J. M., 72

**N**

Nelson, G., 302
Nelson, J. K., 317
Nelson, R. C., 135, 223, 315
Nett, T. 135, 265
Newton, I., 154
Nigg, B. M., 135
Nightingale, A., 36
Nordstrand, A., 109
Nyman, E., 41, 42, 48

**O**

O'Connell, A. L., 49
Ohtsuki, T., 242
Okita, T., 31
Opavsky, P., 135
Owens, M. S., 135

**P**

Paré, E. B., 98, 110
Pauley, D. L., 29
Perey, O., 56
Perry, J., 110
Petak, R. A., 29
Plagenhoef, S., 344
Pulli, M., 135

**R**

Rackham, G., 144, 237
Raine, A. E., 242, 244
Ralston, H. J., 109
Ramey, M. R., 223
Rasch, P. J., 29
Rau, G., 49
Reid, J. G., 31
Rischer, O., 207
Rodahl, K., 35
Roethlin, K., 135
Rogers, E. M., 144, 193, 253
Rusk, H. A., 30

**S**

Saltin, B., 35
Saunder, J. B. DeC. M., 78
Schreck, R. M., 186
Schwartz, J. M., 110
Silver, P. H. S., 72
Silverman, E., 249
Simon, G., 32
Simons, D. G., 91
Singer, R. N., 299
Sloan, A. W., 240
States, J. D., 186
Steindler, A., 15, 55, 56, 65
Stern, J. T., Jr., 110
Stobbs, J., 190, 238
Suzuki, R., 78

**T**

Tani, I., 242
Terauds, J., 135, 241
Tesch, P., 36
Thorstensson, A., 36
Travill, A., 86, 91
Travis, T. M., 31
Tricker, B. J. K., 144, 193, 206, 237, 253
Tricker, R. A. R., 144, 193, 206, 237, 253

**V**

van Huss, W., 30
Viano, D. C., 186
Vitti, M, 72

**W**

Walmsley, R. P., 110
Wartenweiler, J., 135
Watanabe, K., 242
Weber, S., 31
Webster, F. A. M., 283
Wells, K. F., 127, 186, 237
Wessel, J. A., 30
Whiting, H. T. A., 240, 241, 254
Williams, P. L., 36
Wilmore, J. H., 240
Wilson, B. D., 135, 223
Wilt, F., 265
Windell, E. J., 72
Winter, D. A., 49
Withington, E. T., 28
Wohlfart, G., 41, 42, 48
Wright, V., 28

**Z**

Zuniga, E. N., 91

# SUBJECT INDEX

Abduction, 13
Acceleration, 121–24
  angular, 147
  average, 121
  gravitational, 124
  instantaneous, 121
  negative, 123–24
  positive, 123–24
Achilles tendon, 104
Action potential, 33, 35
Adduction, 13
Air resistance, 242
All-or-nothing law, 41
Anatomical axes, 11
  anteroposterior, 11
  longitudinal, 11
  transverse, 11
Anatomical position, 8, 9, 12
  definition, 8
Anatomy, 1, 5–6
  definition, 5
  functional, 5
  gross, 5
  surface, 5, 6
Angle of attack, 247–48
Aponeurosis, 40–41
Archery, 132–33, 184–85
Arches, of the foot, 107–108
  longitudinal, 107
  pes cavus, 108
  pes planus, 108
  transverse, 107
Arthrology, 16
Articulations (see Joints)
Attitude angle, 246
Autonomic nervous system, 34

**B**

Ballet, 227
Ballistic pendulum, 184–85
Baseball, 186, 212, 244–45, 249–50, 284
Basketball, 157–58, 171–74, 176–77,
  206–207, 266, 285
Bone, 16–24
  growth and development, 20–24
    center of ossification, 20–21, 22
    endochondral bone, 20
    epiphyses, pressure, 21
    epiphyses, traction, 21
    epiphysial disc, 20–21
    indexing age, 22
    intramembranous bone, 20
    mesenchyme, 20
    ossification, 20–21, 22
    osteoblasts, 20
  structure, 18–19

  cancellous tissue, 18
  compact tissue, 18
  diaphysis, 18, 19, 20–21, 25
  epiphysis, 18, 19, 20–21, 25
  marrow, 18, 19
  medullary cavity, 18, 19
  periosteum, 18, 19
  types, 16–18
    flat, 17
    irregular, 17–18
    long, 16
    short, 16–17
Bones:
  of the arm, 73–76
  atlas, 52–53
  axis, 52–53
  carpals, 76
    capitate, 76
    hamate, 76
    lunate, 76
    pisiform, 76
    scaphoid, 76
    trapezium, 76
    trapezoid, 76
    triquetrum, 76
  clavicle, 57, 68
  coccyx, 65, 392
  femur, 92–93, 395
  fibula, 93, 395
  humerus, 73–74
  ilium, 64–65
  ischium, 64, 65
  of the leg, 92–94, 395
  metacarpals, 76
  metatarsals, 94
  patella, 93
  of the pelvic girdle, 64–65
  phalanges:
    of the foot, 94
    of the hand, 76
  pubis, 64, 65
  radius, 75–76
  ribs, 63
    false, 63
    floating, 63
    true, 63
  sacrum, 65, 392
  scapula, 67
  of the shoulder girdle, 67–68
  sternum, 62–63
  tarsals, 94
    calcaneus, 94
    cuboid, 94
    cuneiform, 94
    navicular, 94
    talus, 94
  of the thorax, 62–63
  tibia, 93
  ulna, 74–75

vertebrae, 51–54
 atlas, 52–53, 392
 axis, 52–53, 392
 cervical, 52–53, 392
 lumbar, 53–54, 392
 thoracic, 53, 392
 of the vertebral column, 51–54
Boundary layer, 249
Bowling, 177–79
"Broken hip," 92
Buoyancy, center of, 241

**C**

Calcaneal tendon, 109
Center of gravity, 9, 202–16
 human body, 206–16
 mannikin for determination of, 207–208
 segmentation method, 216
 women, 212
Circumduction, 13, 14
Couple, 195
Cricket, 186, 244–45
Curling, 155
Cycling, 132, 244

**D**

Density, 239–40
Depression, 13, 14, 69
Directional terms, 8–9
 anterior, 8
 deep, 9
 distal, 8
 inferior, 8
 lateral, 8
 medial, 8
 posterior, 8–9
 proximal, 8
 superficial, 9
 superior, 8
Displacement, 118
 angular, 145–46
Distance, 118
 angular, 145–46
Diving, 2, 148–49, 185–87, 318–34
Dorsiflexion, 12, 13, 103, 105
Drag, 242
 form, 244–46
 surface, 243–44

**E**

Elasticity, 167
Elevation, 13, 69
Energy, 183–87
 kinetic, 183
 mechanical, 183
  conservation law, 184
 potential, 85

strain, 186–87
 work-energy relationship, 185–86
Equilibrium, 198–202
 dynamic, 200
 neutral, 216–18
 stable, 216–18
 static, 199
 unstable, 216–18
Eversion, 13, 14
Exercise:
 for abdominals, 60–61
 bench press, 82
 joint mobility, 30–31
  active, 30, 31
  concentric, 30
  eccentric, 30
  forced, 30
  isometric, 30
  isotonic, 30
  kinetic, 30
  non-forced, 30
  passive, 30–31
  static, 30
 for lordosis, 66
 prescription, 6
 pull-ups, 84
 sit-ups, 60–61, 97

**F**

Fascia, 40
 classification, 40
  deep, 40
  subserous, 40
  superficial, 40
 definition, 40
 intramuscular septa, 40
Field hockey, 204–205
Figure skating, 284
Flat feet, 108
Flexion, 12, 13
Flotation, 238–41
Fluid mechanics, 238–54
Fluid resistance, 242–51
Football, 125–27, 154, 155, 183
Force, 153
 bouyant, 238–39, 241
 contact, 158
 eccentric, 194–95
 external, 153
 internal, 153
Force platform, 164
Free body diagram, 255–58
 space diagram, 256
Friction, 160–64
 law of, 162–63
 limiting, 160–63
  coefficient of, 163–64
 sliding, 164

coefficient of, 164
Friction:
  skin, 244
Frontal plane (*see* Planes)

**G**

Glycolytic enzymes, 35
Golf, 172, 219, 227, 244–45, 249, 284
Goniometer, 31
Gravity, 42–43, 124, 158
Gravity line, 202
Gymnastics, 55, 81, 145–47, 179–80, 186, 199, 212,
  224–26, 231, 285

**H**

Horizontal abduction, 13
Horizontal adduction, 13
Hurdling, 265, 284
Hyperextension, 12

**I**

Impact, 166–79
  angle of incidence, 176–77
  angle of reflection, 176–77
  between moving bodies, 177–79
  coefficient of restitution, 168–72
  elastic, 167
  inelastic, 166
  velocity of impact, 168
  velocity of separation, 168
  with a fixed surface, 172–77
Impulse, 164
Impulse-momentum relationship, 165
Inertia, 152–53, 219
Interosseous membrane, of the forearm, 83
Intramuscular septa, 40
Inversion, 13, 14, 106

**J**

Joint, 24–31
  classification, 24, 25–28
    amphiarthrodial, 24, 25–26
    arthrodial, 28
    diarthrodial, 24, 25, 26–28
    ellipsoidal, 28
    enarthrodial, 27–28
    fibrocartilaginous, 26
    ginglymus, 28
    sellar, 28
    sutures, 24, 25
    symphysis, 24, 25, 26
    synarthrodial, 24, 25
    synchrondroses, 24, 25
    syndesmoses, 24, 25
    trochoid, 28
  components, 26–30
    capsular ligament, 26

connective tissue, 29–30
  fascia, 26, 27
  fat pads, 26
  fibrocartilaginous discs, 26, 27, 29
  joint cavity, 26, 27
  ligaments, 26, 29
  synovial fluid, 26
  synovial membrane, 26
  tendons, 26, 27
disease, 28
dislocation, 26
flexibility, 29, 30–31
  exercise, 30–31 (*see also* Exercise)
mobility, 6
range of motion, 28–31
Joints:
  acromioclavicular, 68
  ankle, 103–105
  atlanto-axial, 54
  carpometacarpal, 89–90
  elbow, 82–86
  of the foot, 106
  hip, 95–100
  interphalangeal, 90, 106
    foot, 106
    hand, 90
  intertarsal, 106
  knee, 100–103
  metacarpophalangeal, 90
  metatarsalphalangeal, 106
  occipito-atlantal, 54
  of the pelvic girdle, 65
  pubic symphysis, 64, 65
  radioulnar, 83
    distal, 83
    intermediate, 83
    proximal, 83
  sacroiliac, 65
  shoulder, 77
  of the shoulder girdle, 68
  of the spine, 54–55
    cervical spine, 54–55
    lumbar spine, 55
    thoracic spine, 55
  sternoclavicular, 68
  tarsometatarsal, 106
  of the thorax, 63
  wrist, 87–89
Judo, 180
Jumping, 285
  high jump, 157–58, 159, 271–77, 287–89, 290–93,
    295–98
  long jump, 138, 157–58, 159, 231, 268–69
  pole vault, 159, 187, 200–202, 227–29, 269, 294
  triple jump, 285

**K**

Karate, 194

Kinematics:
    angular, 145–51
    linear, 118–44
Kinesiology, 1
Kinetics:
    angular, 194–237
    linear, 152–93

**L**

Laminar flow, 245
Lateral bending, 14
Lift, 242, 246–49
Lift-drag ratio, 248–49
Ligaments:
    acromioclavicular, 68
    of the ankle, 103
    anterior, 82
    anular, 75, 83
    calcaneofibular, 103
    coracoclavicular, 68
    coracohumeral, 77
    costoclavicular, 68
    cruciate, 100, 101
        anterior, 100, 101
        posterior, 100, 101
    deltoid, 103
    of the elbow, 82–83
    of the foot, 108
    glenohumeral, 77
        inferior, 77
        middle, 77
        superior, 77
    of the hip, 95–96
    iliofemoral, 95–96
    inguinal, 62
    interclavicular, 68
    ischiofemoral, 95, 96
    of the knee, 100, 101
    lateral collateral, 100
    medial collateral, 100
    nape, 57
    patellar, 100
    plantar, 108
    plantar calcaneocuboid, 108
    plantar calcaneonavicular, 108
    posterior, 82
    pubofemoral, 95, 96
    radial collateral, 82
    of the shoulder, 77
    of the shoulder girdle, 68
    sternoclavicular, 68
    talofibular, 103
        anterior, 103
        posterior, 103
    transverse humeral, 77
    ulnar collateral, 82
Linea alba, 60
Linea semi lunaris, 60

Line of gravity, 202
Little Leaguer's elbow, 21
Lordosis, 55, 56

**M**

Magnus effect, 249
Mass, 153, 159
Mechanics, 1
Medicine, practices, 5
Meniscus, 100, 101
    lateral, 100, 101
    medial, 100, 101
Midfrontal plane (*see* Planes)
Midsagitall plane (*see* Planes)
Midtransverse plane (*see* Planes)
Moment, 196–98
    of applied force, 196–97
    arm, 196
    resultant, 198
Moment of inertia, 219–21
    human body, 221
Momentum, 153–54
    angular, 222–23
        conservation principle, 224
        transfer of, 231
    conservation principle, 178–79
Motion:
    angular, 115–16
    general, 116
    linear, 113–14, 116
    projectile, 129–39
Motor behavior, 5
Motor skill, 261
Motor unit, 41–42
Movement:
    abduction, 13
    adduction, 13
    of the ankle, 103–105
    of the carpometacarpal joints, 89–90
    circumduction, 13, 14
    depression, 13, 14, 69
    dorsiflexion, 12, 13, 103, 105
    of the elbow, 83–86
    elevation, 13, 69
    eversion, 13, 14, 106
    extension, 12, 13
    flexion, 12, 13
    of the foot, 106–107
    of the hip, 96–100
    horizontal abduction, 13
    horizontal adduction, 13
    hyperextension, 12
    of the interphalangeal joints, 90
    inversion, 13, 14, 106
    of the knee joint, 102–103
    lateral bending, 14, 57
    of the metacarpophalangeal joints, 90
    of the pelvic girdle, 65–66

plantar flexion, 12, 13, 103, 104–105
pronation, 13, 14, 84, 86
  of the radioulnar joints, 84, 86
  of respiration, 63–64
    expiration, 64
    inspiration, 63
  rotation, 13, 14
    lateral or outward, 14
    medial or inward, 14
  of the shoulder girdle, 68–69
  of the shoulder joint, 78–82
  supination, 13, 14
  of the thumb, 90
    abduction, 90
    adduction, 90
    circumduction, 90
    extension, 90
    flexion, 90
    opposition, 90
  tilt, 66
  of the vertebral column, 54–57
    cervical spine, 54–55
    lumbar spine, 55
    thoracic spine, 55
  of the wrist, 87–89
Muscle:
  actions, 46–47
    antagonists, 46
    assistant movers, 46
    prime movers, 46
    stabilizers, 46–47
    synergists, 47
  attachments, 40–41
  contraction, 34, 41–45
    all-or-nothing law, 41, 43
    concentric, 42–43, 45
    definition, 42
    eccentric, 42–43
    energetics, 34
    isometric, 42, 43, 45
    isotonic, 42, 45
    static, 42
  cross-sectional area, 38–39
  fiber characteristics, 33
    contractility, 33
    elasticity, 33
    extensibility, 33
    irritability, 33
  fiber composition, 35–36
  fiber types, 35–36
    fast-twitch, 35–36
    red, 35
    slow-twitch, 35–36
    white, 35
  force, 38–39
  force-velocity relationship, 45
  gross structure, 36–39
    bipennate, 37, 38, 39

    fusiform, 36–37, 39
    longitudinal, 36, 37, 38–39
    multipennate, 37, 38, 39
    penniform, 37–38, 39
    radiate, 37, 39
    unipennate, 37–38, 39
  insertion, 41
  length-tension relationship, 44–45
  microstructure, 34–35
    endomysium, 35
    epimysium, 35
    fasciculi, 35, 36
    nuclei, 35
    perimysium, 35
    sarcolemma, 35
  origin, 41
  oxidative capacity, 35
  stimulus response, 43–44
    all-or-nothing law, 43
    latent period, 43
    summation, 43
    tetanus, 43
    threshold, 43
    twitch, 43
  study methods, 47–48
    electrical stimulation, 47
    electromyography, 47–48
    observation, 47
    palpation, 47
  types, 33–34
    cardiac, 34
    involuntary, 33–34
    skeletal, 34
    smooth, 33–34
    striated, 34
    unstriated, 33–34
    voluntary, 34
Muscles:
  of the abdominal wall, 404
  adductor brevis, 100, 397, 410
  adductor longus, 100, 397, 410
  adductor, magnus, 100, 397, 410
  anconeus, 84, 408
  acting on the ankle, 104–105
  biceps brachii, 46, 79, 80, 81, 82, 84, 401, 408
    long head, 79, 81
    short head, 80, 81, 82
  biceps femoris, 99, 100, 102–103, 402, 411
  brachialis, 84, 85, 401, 408
  brachioradialis, 84, 85, 408
  clavicular pectoralis, 79, 80, 407
  coracobrachialis, 79, 80, 81, 82, 407
  deltoid, 77, 80, 81, 82, 397, 401, 407
  diaphragm, 63, 64, 399, 405
  acting on the elbow, 84–86
  erector spinae, 59, 66, 398, 405
  extensor carpi radialis, 88, 401, 409
  extensor carpi ulnaris, 88, 401, 409
  extensor digitorum, 409

extensor digitorum communis, 401
extensor digitorum longus, 105, 107, 402, 413
extensor hallucis longus, 107, 402, 413
extensor, indicis, 409
extensor peroneus tertius, 105
external oblique, 60, 62, 400, 404
flexor carpi radialis, 88, 401, 409
flexor carpi ulnaris, 88, 401, 409
flexor digitorum longus, 107, 402, 413
flexor digitorum profundus, 401, 409
flexor digitorum superficialis, 401, 409
flexor hallucis longus, 107, 402, 413
of the foot, 106–107, 413
of the forearm, 408
gastrocnemius, 103, 104, 402, 413
gluteus maximus, 66, 98, 397, 399, 410
gluteus medius, 100, 397, 410
gluteus minimus, 100, 397, 410
gracilis, 100, 103, 397, 410, 412
hamstrings, 99, 102–103, 411
  biceps femoris, 99, 100, 102–103, 402, 411
  semimembranosus, 99, 102–103, 402, 411
  semitendonosus, 99, 102–103, 402, 411
of the hand, 90, 409
acting on the hip, 96–100, 410–11
iliacus, 97
iliocostalis, 398, 405
iliopsoas, 97, 398, 410
infraspinatus, 77, 79, 82, 401, 407
intercostal, 63, 64, 399, 405
internal oblique, 62, 400, 404
interspinalis, 398
intertransversarii, 398
acting on the knee, 102–103, 411–12
latissimus dorsi, 79, 80–81, 82, 397, 407
levator costarum, 398
levator scapulae, 69, 406
longissumus capitis, 403, 405
longissimus dorsi, 397, 405
longus capitis, 399
longus, colli, 399
masseter, 399
multifidus, 398
of the neck, 57–58, 403
palmaris, longus, 409
pectineus, 100, 397, 410
pectoralis major, 71, 79, 80, 81, 82, 398, 407
  clavicular pectoralis, 79, 80, 407
  sternocostal pectoralis, 79, 80, 81, 407
pectoralis minor, 70, 71, 398, 406
acting on the pelvic girdle, 66
peronius brevis, 104, 106, 402, 413
peronius longus, 104, 106, 402, 413
peronius tertius, 107, 413
piriformis, 411
plantaris, 104
platysma, 58
popliteus, 103, 412
pronator quadratus, 84, 86, 408

pronator teres, 84, 86, 408
psoas major, 97
quadratus lumborum, 397, 405
quadriceps femoris, 103, 402, 412
  rectus femoris, 97, 103, 402, 411, 412
  vastus intermedius, 103, 402, 412
  vastus lateralis, 103, 402, 412
  vastus medialis, 103, 402, 412
acting on the radioulnar joints, 84–86
rectus abdominus, 60, 399, 400, 404
rectus capitus, 399
rectus femoris, 97, 103, 402, 411, 412
of respiration, 405
rhomboid major, 69, 397, 406
rhomboid minor, 69, 397, 406
"rotator cuff," 77, 398
sacrospinalis, 397, 405
sartorius, 103, 402, 411, 412
scalenius, 399
semimembranosus, 99, 102–103, 402, 411
semispinalis capitus, 57, 398, 403
semitendonosus, 99, 102–103, 402, 411
serratus, anterior, 70, 398, 406
acting on the shoulder girdle, 69–71, 406
acting on the shoulder joint, 79, 407
soleus, 104, 105, 402, 413
splenius capitis, 403
sternocleidomastoid, 57, 399, 403
sternocostal pectoralis, 79, 80, 81, 407
subscapularis, 77, 79, 82
supinator, 84, 86, 408
supraspinatus, 79, 81, 401, 407
temporalis, 399
tensor facia lata, 98, 397, 411
teres major, 79, 80, 81, 82, 401, 407
teres minor, 77, 79, 82, 401, 407
tibialis anterior, 105, 107, 402, 413
tibialis posterior, 104, 107, 402, 413
tranversus abdominis, 64, 400, 404
trapezius, 57, 69, 397, 406
triceps brachii, 79, 84, 85, 401, 408
  long head, 79, 81
of the trunk, 59–62
vastus intermedius, 103, 402, 412
vastus lateralis, 103, 402, 412
vastus medialis, 103, 402, 412
acting on the wrist, 88–89, 409
Myofibrillar ATPase, 35

**N**

Neuromusculoskeletal system, 34
Newton, 156
Newton's laws:
  angular motion, 223–30
    first, 224–26
    second, 226–27
    third, 227–30
  gravitation, 158–59

impact, 168
motion, 154–58
    first, 154–55
    second, 155–56
    third, 156–58

**O**

Osteology, 16

**P**

Palpation, 6, 47
Parabola, 131–32
Performance, 261
Pes cavus, 108
Pes planus, 108
Planes, 10–14
    frontal, 10
        association with anteroposterior axis, 11
        movements in, 13
    midfrontal, 10
    midsagittal, 10
    midtransverse, 10
    sagittal, 10
        association with transverse axis, 11
        movements in, 12
    transverse, 10
        association with longitudinal axis, 11
        movements in, 14
Plantar flexion, 12, 13, 103, 104–105
Posture, 6, 30
    abnormalities, 6
Power, 182
Pressure, 179–80
Principal axes, 221–22
Projectile, 129
    flight time, 132–33, 139
    range, 132–33, 139
    release
        angle of, 133–37, 139
        relative height of, 133
        speed of, 133
        velocity at, 139
    takeoff
        angle of, 135
Pronation, 13, 14, 84, 86

**Q**

Qualitative analysis, 2, 3, 261–387
    basic steps, 265–66
Quantitative analysis, 2, 3, 262–63

**R**

Racing:
    Grand Prix, 246–47

Radian, 150
Radiography, 19–20, 22
    use for indexing age, 22
Reaction board, 209–14
Relative motion, 241–42
Respiration, 6, 63–64
    expiration, 64
    inspiration, 63
    muscles of, 405
    respiratory patterns, 6
Result, 261
Right-hand thumb rule, 148–49
Rotation, 13, 14, 115
    axis of, 115
    inward, 14
    lateral, 14
    medial, 14
    outward, 14
Running, 121, 164–66, 264–65, 303–18

**S**

Sagittal plane (*see* Planes)
Scalars, 124–28
Scoliosis, 57
Semilunar cartilages, 100, 101
Sheep-shearing, 263
Sit-ups, 60–61, 97
Skiing, 154, 244, 284
    cross-country, 118
Ski-jumping, 113, 248
Sky diving, 200
Soccer, 249, 251
Softball, 227, 244–45
Speed, 119–21
    angular, 146–47
    average, 120
    instantaneous, 120
Spinal column (*see* Vertebral column)
Squash, 172
Stability, 216–18
Supination, 13, 14
Swimming, 121, 123–24, 256–57, 238–41, 247–48,
    269–70

**T**

Table tennis, 175–76
Tendon, 40–41
    definition, 40
Tennis, 172, 174–75, 202–204, 249, 334–350
Throwing:
    discus throw, 229, 248–49
    hammer throw, 159
    javelin throw, 159, 205–206, 242–43, 246, 248, 284
    shot put, 127–31, 136, 269
Tightrope walking, 195–96
Torque, 196

Track and field, 119
Trajectory, 129–32
Translation, 113
Transverse plane (*see* Planes)
Traumatic epiphysitis, 21
Turbulent flow, 244
Turntable demonstration, 226

**V**

Vectors, 124–28
  angular motion, 147–49
  components, 127–29
  parallelogram, 125–27
  resultant, 124–27
Velocity, 119–21, 149–50
  angular, 146–47
  and angular velocity, 149–50
  average, 120
  instantaneous, 120
  terminal, 200
Vertebral column, 50–62, 392
  abnormalities, 55, 57

  lordosis, 55
  scoliosis, 57
composition, 50–51, 392
curves, 50–51
  primary, 51
  secondary, 51
intervertebral discs, 56
  annulus fibrosus, 56
  nucleus pulposus, 56
  importance for movement, 56
Volleyball, 149–50, 157–58, 285

**W**

Weight, 159
Weight training, 152–53, 181–82, 197–98
Winged scapula, 71
Work, 181–82
  mechanical, 181
Wrestling, 217–18

**XYZ**

X-rays (*see* Radiography)